Lecture Notes in Computer Science 11407

Commenced Publication in 1973
Founding and Former Series Editors:
Gerhard Goos, Juris Hartmanis, and Jan van Leeuwen

More information about this series at http://www.springer.com/series/7409

Éric Renault · Paul Mühlethaler ·
Selma Boumerdassi (Eds.)

Machine Learning for Networking

First International Conference, MLN 2018
Paris, France, November 27–29, 2018
Revised Selected Papers

Springer

Editors
Éric Renault
Télécom SudParis
Évry, France

Paul Mühlethaler
Inria
Paris, France

Selma Boumerdassi
CNAM/CEDRIC
Paris, France

ISSN 0302-9743 ISSN 1611-3349 (electronic)
Lecture Notes in Computer Science
ISBN 978-3-030-19944-9 ISBN 978-3-030-19945-6 (eBook)
https://doi.org/10.1007/978-3-030-19945-6

LNCS Sublibrary: SL3 – Information Systems and Applications, incl. Internet/Web, and HCI

This Springer imprint is published by the registered company Springer Nature Switzerland AG
The registered company address is: Gewerbestrasse 11, 6330 Cham, Switzerland

Preface

The rapid development of new network infrastructures and services has led to the generation of huge amounts of data, and machine learning now appears to be the best solution to process these data and make the right decisions for network management. The International Conference on Machine Learning for Networking (MLN) aimed at providing a top forum for researchers and practitioners to present and discuss new trends in deep and reinforcement learning, pattern recognition and classification for networks, machine learning for network slicing optimization, 5G system, user behavior prediction, multimedia, IoT, security and protection, optimization and new innovative machine learning methods, performance analysis of machine learning algorithms, experimental evaluations of machine learning, data mining in heterogeneous networks, distributed and decentralized machine learning algorithms, intelligent cloud-support communications, ressource allocation, energy-aware/green communications, software defined networks, cooperative networks, positioning and navigation systems, wireless communications, wireless sensor networks, underwater sensor networks. In 2018, MLN was hosted by Inria Paris, France, one of the top-level research center in Paris.

The call for papers resulted in a total of 48 submissions from all around the world: Algeria, Australia, Canada, Chile, China, Colombia, France, India, Italy, Morocco, Pakistan, Saoudi Arabia, Tunisia, Turkey, UAE, UK, and USA. All submissions were assigned to at least three members of the Program Committee for review. The Program Committee decided to accept 22 papers. Two intriguing keynotes from Pierre Gaillard, Inria, France, and Franck Gaillard, Microsoft, France, complete the technical program.

We would like to thank all who contributed to the success of this conference, in particular the members of the Program Committee and the reviewers for carefully reviewing the contributions and selecting a high-quality program. Our special thanks go to the members of the Organizing Committee for their great help.

Thursday morning was dedicated to the First International Workshop on Networking for Smart Living. The technical program of NSL included five presentations and keynotes by Mérouane Debbah, Huawei, France and Kevin Curran, Ulster University, UK.

We hope that all participants enjoyed this successful conference, made many new contacts, engaged in fruitful discussions, and had a pleasant stay in Paris, France.

November 2018

Selma Boumerdassi
Paul Mühlethaler
Éric Renault

Organization

MLN 2018 was organized by the EVA Project of Inria Paris and the Wireless Networks and Multimedia Services (RS2M) Department of Télécom SudParis (TSP), a member of Institut Mines-Télécom (IMT) and University of Paris-Saclay.

General Chairs

Selma Boumerdassi	CNAM, France
Paul Mühlethaler	Inria, France
Éric Renault	IMT-TSP, France

Steering Committee

Selma Boumerdassi	CNAM, France
Éric Renault	IMT-TSP, France

Tutorial and Workshop Chair

Nardjes Bouchemal	University Center of Mila, Algeria

Publicity Chair

Abdallah Sobehy	IMT-TSP, France

Organization Committee

Lamia Essalhi	ADDA, France

Technical Program Committee

Nadjib Ait Saadi	ESIEE, France
Claudio A. Ardagna	Università degli Studi di Milano, Italy
Mohamad Badra	Zayed University, UAE
Maxim Bakaev	NSTU, Russia
Aissa Belmeguenai	University of Skikda, Algeria
Indayara Bertoldi Martins	PUC Campinas, Brazil
Luiz Bittencourt	University of Campinas, Brazil
Naïla Bouchemal	ALTRAN, France
Nardjes Bouchemal	University Center of Mila, Algeria
Tahar Z. Boulmezaoud	UVSQ, France
Selma Boumerdassi	CNAM, France

Ahcene Bounceur	University of Brest, France
Rajkumar Buyya	University of Melbourne, Australia
Luca Caviglione	National Research Council, Italy
Hervé Chabanne	Télécom ParisTech, France
De-Jiu Chen	KTH, Sweden
Yuh-Shyan Chen	National Taipei University, Taiwan
Yun-Maw Kevin Cheng	Tatung University, Taiwan
Yeh-Ching Chung	National Tsing Hua University, Taiwan
Domenico Ciuonzo	Network Measurement and Monitoring (NM2), Italy
Dabin Ding	University of Central Missouri, MO, USA
Soufiene Djahel	Manchester Metropolitan University, UK
Paul Farrow	BT, UK
Hacène Fouchal	Université de Reims Champagne-Ardenne, France
Kaori Fujinami	Tokyo University of Agriculture and Technology, Japan
Nidhi Gautam	Panjab University Chandigarh, India
Aravinthan Gopalasingham	NOKIA Bell Labs, France
Viet Hai Ha	Hue University, Vietnam
Sun-Yuan Hsieh	National Cheng Kung University, Taiwan
Alberto Huertas Celdran	University of Murcia, Spain
Youssef Iraqi	Khalifa University, UAE
Euee S. Jang	Hanyang University, South Korea
Ahmed Khattab	Cairo University, Egypt
Donghyun Kim	Kennesaw State University, GA, USA
Adam Krzyzak	Concordia University, QC, Canada
Cherkaoui Leghris	Hassan II University, Morocco
Pang-Feng Liu	National Taiwan University, Taiwan
Zoltán Mann	University of Duisburg-Essen, Germany
Natarajan Meghanathan	Jackson State University, MS, USA
Saurabh Mehta	Vidyalankar Institute of Technology, Mumbai, India
Ruben Milocco	Universidad Nacional des Comahue, Argentina
Pascale Minet	Inria, France
Paul Mühlethaler	Inria, France
Frank Phillipson	TNO, The Netherlands
Sabine Randriamasy	NOKIA Bell Labs, France
Éric Renault	IMT-TSP, France
Leonardo Rey Vega	Universidad de Buenos Aires, Argentina
Jihene Rezgui	LRIMA Lab, Maisonneuve, QC, Canada
Hamida Seba	University of Lyon, France
Cristiano M. Silva	Universidade Federal de São João del-Rei, Brazil
Vinod Kumar Verma	SLIET, India
Corrado Aaron Visaggio	Università degli Studi del Sannio, Italy
Chuan Yue	Colorado School of Mines, USA
Xuyun Zhang	The University of Auckland, New Zealand
Djemel Ziou	Sherbrooke University, Canada

Sponsoring Institutions

CNAM, Paris, France
Inria, Paris, France
IMT-TSP, Évry, France

Contents

Learning Concave-Convex Profiles of Data Transport over Dedicated Connections

Nageswara S. V. Rao[1(✉)], Satyabrata Sen[1], Zhengchun Liu[2], Rajkumar Kettimuthu[2], and Ian Foster[2]

[1] Oak Ridge National Laboratory, Oak Ridge, TN, USA
{raons,sens}@ornl.gov

[2] Argonne National Laboratory, Argonne, IL, USA
{zhengchun.liu,kettimut,foster}@anl.gov

Abstract. Dedicated data transport infrastructures are increasingly being deployed to support distributed big-data and high-performance computing scenarios. These infrastructures employ data transfer nodes that use sophisticated software stacks to support network transport among sites, which often house distributed file and storage systems. Throughput measurements collected over such infrastructures for a range of round trip times (RTTs) reflect the underlying complex end-to-end connections, and have revealed dichotomous throughput profiles as functions of RTT. In particular, concave regions of throughput profiles at lower RTTs indicate near-optimal performance, and convex regions at higher RTTs indicate bottlenecks due to factors such as buffer or credit limits. We present a machine learning method that explicitly infers these concave and convex regions and transitions between them using sigmoid functions. We also provide distribution-free confidence estimates for the generalization error of these concave-convex profile estimates. Throughput profiles for data transfers over 10 Gbps connections with 0–366 ms RTT provide important performance insights, including the near optimality of transfers performed with the XDD tool between XFS filesystems, and the performance limits of wide-area Lustre extensions using LNet routers. A direct application of generic machine learning packages does not adequately highlight these critical performance regions or provide as precise confidence estimates.

Keywords: Data transport · Throughput profile · Concavity-convexity · Generalization bounds

This work is funded by RAMSES project and Applied Mathematics program, Office of Advanced Computing Research, U.S. Department of Energy, and by Extreme Scale Systems Center, sponsored by U.S. Department of Defense, and performed at Oak Ridge National Laboratory managed by UT-Battelle, LLC for U.S. Department of Energy under Contract No. DE-AC05-00OR22725.

E. Renault et al. (Eds.): MLN 2018, LNCS 11407, pp. 1–22, 2019.
https://doi.org/10.1007/978-3-030-19945-6_1

1 Introduction

There have been unprecedented increases in the volume and types of data transfers over long-distance network connections in a number of scenarios, such as transfers of partial computations over geographically dispersed cloud-server sites for big data computations. Extensive throughput measurements collected over a range of testbed and production infrastructures (for example, [9,19]) show that performance can vary greatly as a result of both transfer characteristics and choices made along four dimensions:

(i) Data Transfer Node (DTN) host systems, which can vary significantly in terms of the number of cores, Network Interface Card (NIC) capability, and connectivity;
(ii) file and disk systems, such as Lustre [10], GPFS [5], and XFS [26], installed on Solid State Disk (SSD) or hard disk arrays;
(iii) network protocols, for example, CUBIC [20], H-TCP [23], and BBR [4] versions of Transmission Control Protocol (TCP); and
(iv) file transfer software, such as Globus [2], GridFTP [1], XDD [22,25], UDT [6], MDTM [11], Aspera [3], and LNet extensions of Lustre [16].

Hence, the throughput measurements need to be analyzed systematically to reveal performance trends of the components and infrastructures.

We denote the throughput at time t over a connection of RTT τ as $\theta(\tau, t)$, for a given configuration of end-to-end connection. We call its expectation over an observation period T_O the *throughput profile* (a function of τ), given by

$$\Theta_A(\tau) = \frac{1}{T_O} \int_0^{T_O} \theta(\tau, t)\, dt,$$

where modality $A = T$ and $A = E$ correspond to memory transfers using TCP and end-to-end disk file transfers, respectively. Each modality embodies the combined effects of corresponding components and their configurations; for file transfers, it reflects the composition of filesystems, network connections, and their couplings through host buffers. In general, $\Theta_A(\tau)$ is a random quantity and its distribution $\mathbf{P}_{\Theta_A(\tau)}$ is complex, since it depends on the properties and instantaneous states of network connections, filesystems, and transfer hosts. Throughput measurements are used to estimate its approximation $\hat{\Theta}_A(\tau)$, in order to characterize transport performance as RTT is varied.

To illustrate the importance of throughput profile properties, we show in Figs. 1(a)–(c) "measured" throughput profiles for three scenarios: (a) XDD transfers between sites with XFS filesystems mounted on SSD storage, (b) file copy operations between Lustre filesystems extended over wide-area connections using LNet routers, and (c) memory transfers between transfer hosts with identical configurations. Data transfers in these scenarios are over dedicated 10GigE connections with RTT $\tau \in [0, 366]$ ms. The profiles in Figs. 1(a) and (b) are quite different not only in their peak throughput values (10 and 4.5 Gbps respectively)

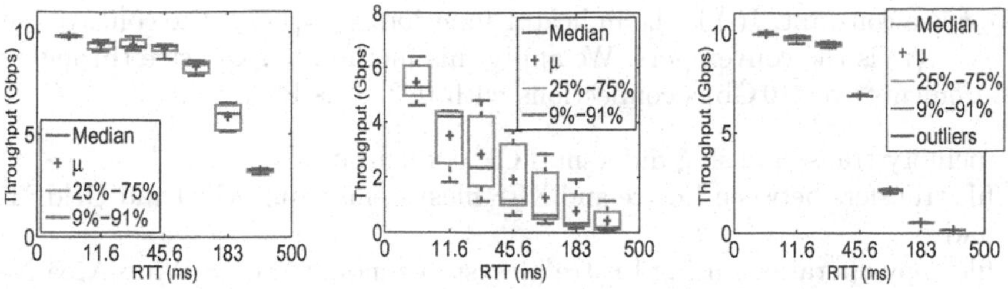

(a) XDD: XFS on SSD (b) LNet-routed Lustre (c) memory: Scalable TCP

Fig. 1. Throughput measurements over dedicated 10GigE connections for RTT $\in [0, 366]$ ms: (a) concave profile of near-optimal XDD transfers, (b) convex profile due to LNet router limits, and (c) intermediate concave-convex profile of memory transfer.

but also in their *concave* and *convex* shapes, respectively. The first represents a near-optimal performance achieved by balancing and tuning XFS and TCP parameters in XDD [18], whereas the second represents a performance bottleneck due to LNet credit limits [16]. On the other hand, the third profile, in Fig. 1(c), shows a combination of both concave and convex regions [17], wherein TCP buffers limit throughput values beyond a certain RTT at which the profile switches from concave to convex in shape.

From the perspective of network performance, a concave region is highly desirable since intermediate-RTT throughput values are, by definition, higher than the linear interpolation; this is often an indicator of near-optimal throughput performance. Now, obtaining a concave throughout profile requires (i) selection of TCP version and transport method parameters, (ii) selection of file, I/O, and storage system parameters, and (iii) joint optimization of these parameters by taking into account the interactions between the complex subsystems. On the other hand, in a convex region, intermediate-RTT throughput values can only be guaranteed to be higher than the minimum observed; this is often an indicator of system/component limits, for example, the LNet credits and TCP buffer sizes in Figs. 1(b) and (c), respectively [16]. Furthermore, the shape of $\hat{\Theta}_A(\tau)$ has a deeper connection to the time dynamics of data transfers [8,17]. For example, a concave profile requires a fast ramp-up followed by stable throughput rates, which in turn requires that file I/O and network connections be matched. Thus, profile estimates that accurately capture these concave and convex regions provide critical insights for performance optimization, prediction and diagnosis.

To estimate the concave-convex profile regions, we present a machine learning method based on the flipped sigmoid function $g_{a_1,\tau_1}(\tau) = 1 - \frac{1}{1+e^{-a_1(\tau-\tau_1)}}$. We use this method to estimate, given a set of performance measurements, a throughput profile using the function $f_{\Theta_A}(\tau)$, given by

$$f_{\Theta_A}(\tau) = b\left[g_{a_1,\tau_1}(\tau) I\left(\tau \leq \tau_T\right) + g_{a_2,\tau_2}(\tau) I\left(\tau \geq \tau_T\right)\right]$$

where b is a constant, $I(\cdot)$ is the indicator function, $g_{a_1,\tau_1}(\tau)$ is the concave part, and $g_{a_2,\tau_2}(\tau)$ is the convex part. We apply this method to extensive throughput measurements over 10 Gbps connections with 0–366 ms RTT for:

(i) memory transfers using different TCP configurations,
(ii) file transfers between Lustre and XFS filesystems using XDD and GridFTP, and
(iii) file copy operations using Lustre filesystems mounted over Wide-Area Network (WAN) connections using LNet routers.

In addition, we show analytically that the estimation of this concave-convex profile function is statistically sound [24] in that its expected error is close to optimal, with a probability that improves with the number of measurements. This guarantee is independent of the composite distribution $\mathbf{P}_{\Theta_A(\tau)}$ which is complex and unknown, since it depends on various hardware and software components and complex interactions between them.

To further evaluate our machine learning method, we also apply some generic machine learning modules from Python and MATLAB libraries to estimate the throughput profiles. We find that these methods do not, in most cases, adequately capture the critical concave-convex regions of the throughput profiles. The confidence probability bounds for their generalization errors are not readily available, but can be derived by using properties such as bounded variance or smoothness [15,21]. The generalization bounds for our sigmoid-based method (derived in Sect. 4) use the Lipschitz property and are sharper, as they require only a two-dimensional metric cover, rather than the higher-dimensional covers needed in generic learning techniques. For example, the dimensionality for estimates based on feedforward sigmoid neural networks is proportional to the total number of weights [15].

The rest of this paper is as follows. A brief description of our testbed used for measurements is provided in Sect. 2. The concave and convex regions of memory and file transfer throughput profiles are discussed in Sect. 3, along with the proposed profile-estimate $f_{\Theta_A}(\tau)$ computed based on measurements. A bound for confidence probability of the generalization error of $f_{\Theta_A}(\tau)$ is derived in Sect. 4. Application of some generic machine learning modules is described in Sect. 5. Conclusions and open areas are briefly described in Sect. 6.

2 Testbed and Measurements

We collected the measurements used in this paper on a testbed consisting of multiple data transfer servers, 10 Gbps wide-area hardware connection emulators, and a distributed Lustre filesystem with LNet routers. The testbed consists of 32-core (`feynman1`, `feynman2`, `tait1`, and `tait2`) and 48-core (`bohr05` and `bohr06`) Linux workstations, QDR Infiniband (IB) switches, and 10 Gbps Ethernet switches. For various network connections, hosts with identical configurations are connected in pairs over a back-to-back fiber connection with negligible 0.01 ms RTT and a physical 10GigE connection with 11.6 ms RTT via

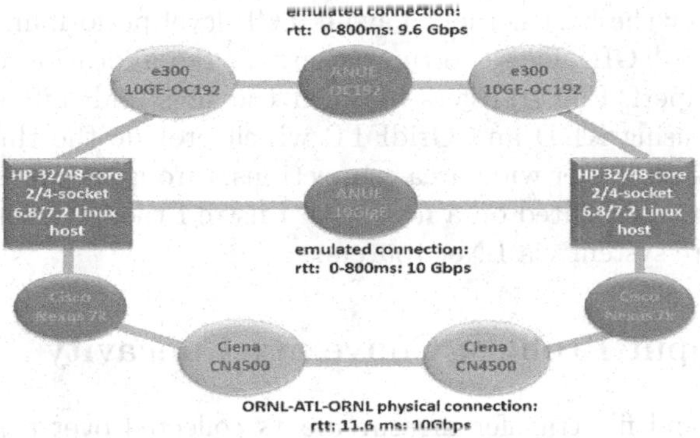

(a) physical and emulated connections between hosts

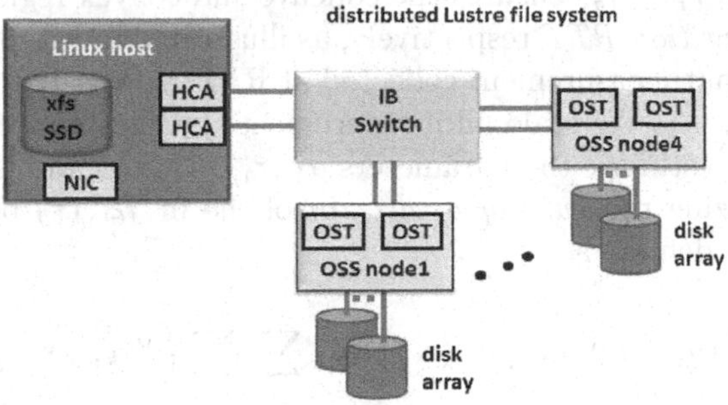

(b) Lustre and XFS filesystems

Fig. 2. Testbed network connections and filesystems.

Cisco and Ciena devices, as shown in Fig. 2(a). We use ANUE devices to emulate (in hardware) 10GigE connections with RTTs $\tau \in [0,366]$ ms. These RTT values are strategically chosen to represent a global infrastructure with three scenarios of interest: (a) smaller values represent cross-country connections, for example, facilities distributed across the US; (b) 93.6 and 183 ms represent inter-continental connections, for example, among US, Europe, and Asia; and (c) 366 ms represents a connection spanning the globe, which is mainly used as a limiting case.

The Lustre filesystem is supported by eight OSTs connected over IB QDR switch, as shown in Fig. 2(b). Host systems (`bohrs` and `taits`) are connected to IB switch via HCA and to Ethernet via 10 Gbps Ethernet NICs. In addition, our SSD drives are connected over PCI buses on the hosts `bohr05` and `bohr06`, which mount local XFS filesystems. We also consider configurations wherein Lustre is mounted over long-haul connections using LNet routers on `tait1` and `bohr06`.

Memory-to-memory throughput measurements for TCP are collected using *iperf*. Typically, 1–10 parallel streams are used for each configuration, and throughput measurements are repeated 10 times. TCP buffer sizes are set at largest allowed by the host kernel to avoid TCP-level performance bottlenecks, which for iperf is 2 GB. These settings result in the allocation of 1 GB socket buffer sizes for iperf. File transfers between the sites and different filesystems are carried out using XDD and GridFTP which provide the throughput measurements. For Lustre over wide-area connections, throughput measurements are made using IOzone executed on a host with Lustre Ethernet clients that access remote IB Lustre system via LNet routers.

3 Throughput Profiles: Convexity-Concavity

Using memory and file transfer measurements collected over connections with RTT τ_k, $k = 1, 2, \ldots, n$, we derive the estimate $\hat{f}_{\Theta_A}(\tau)$ of $\Theta_A(\tau)$. The flipped sigmoid function g_{a,τ_a} is concave for $\tau \leq \tau_a$ and switches to convex for $\tau \geq \tau_a$. The condition $\tau_{a_2} \leq \tau_T \leq \tau_{a_1}$ ensures the concave and convex regions to the left and right of the *transition-RTT* respectively, as illustrated in Fig. 5. Let $\theta(\tau_k, t_i^k)$ denote ith throughput measurement collected at RTT τ_k, $k = 1, 2, \ldots, n$, and at time t_i^k, $i = 1, 2, \ldots, n_k$. We scale all measurements by the connection capacity such that $b = 1$. We estimate the parameters a_1, τ_{a_1}, a_2, τ_{a_2}, and the transition-RTT τ_T by minimizing the *sum-squared error* of the fit $f_{\Theta_A}(\tau)$ based on measurements, which is defined as

$$\hat{S}\left(f_{\Theta_A}\right) = \sum_{\tau_k \leq \tau_T} \sum_{i=1}^{n_k} \left(\theta(\tau_k, t_i^k) - g_{a_1,\tau_{a_1}}(\tau_k)\right)^2 + \sum_{\tau_k \geq \tau_T} \sum_{i=1}^{n_k} \left(\theta(\tau_k, t_i^k) - g_{a_2,\tau_{a_2}}(\tau_k)\right)^2, \tag{3.1}$$

where the parameters are bounded such that $a_1, a_2 \in [-A, A]$ and $\tau_{a_1}, \tau_{a_2} \in [-T, T]$, and $\tau_T \in \{\tau_k : k = 1, 2, \ldots, n\}$.

3.1 Various Throughput Profiles

The three different throughput profile shapes illustrated in Fig. 1 correspond to scenarios with disparate end-subsystems and configurations. However, profiles may vary significantly even when end-subsystems are identical, as a result of different transport parameters. For example, Fig. 3, which presents throughput measurements for memory transfers using CUBIC with large buffers, shows that more streams not only increase the aggregate throughput, but also extend the concave region; here, the convex region with a single stream mostly disappears with 10 streams. TCP/IP buffer sizes have a similar effect, with larger buffers both increasing mean throughput and extending the concave region. As seen from Fig. 4, for CUBIC with 10 streams, the default buffer size results in an entirely convex profile; with the normal buffer size recommended for 100 ms RTT, a concave region (leading up to 91.6 ms) is followed by a convex region; finally, a large buffer extends the concave region beyond 183 ms. We will show next that $\hat{f}_{\Theta_A}(\tau)$ estimated using measurements provides us a systematic way to identify the concave and convex regions, and in particular, all convex regions that indicate performance bottlenecks and potential improvements.

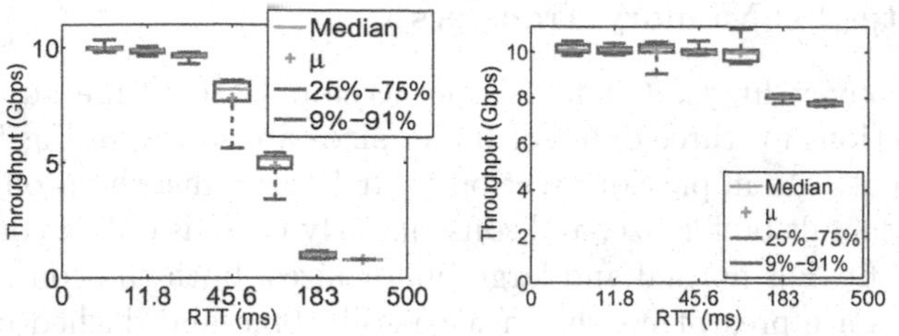

(a) $f1\_10gige\_f2$, 1 stream (b) $f1\_10gige\_f2$, 10 streams

Fig. 3. Throughput profiles improve with number of parallel streams.

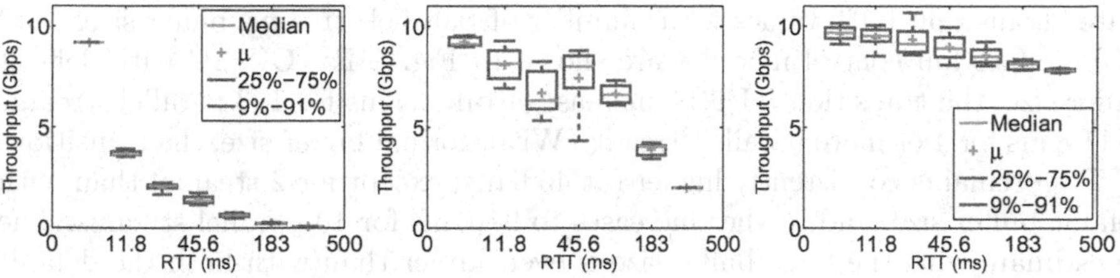

(a) default (b) normal (c) large

Fig. 4. Throughput profiles improve with larger buffer sizes.

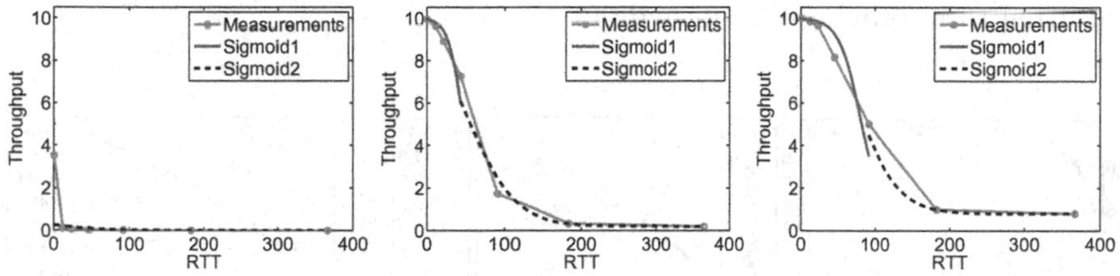

(a) default (b) normal (c) large

Fig. 5. Sigmoid-regression fits of throughput profiles with various buffer sizes for single CUBIC stream.

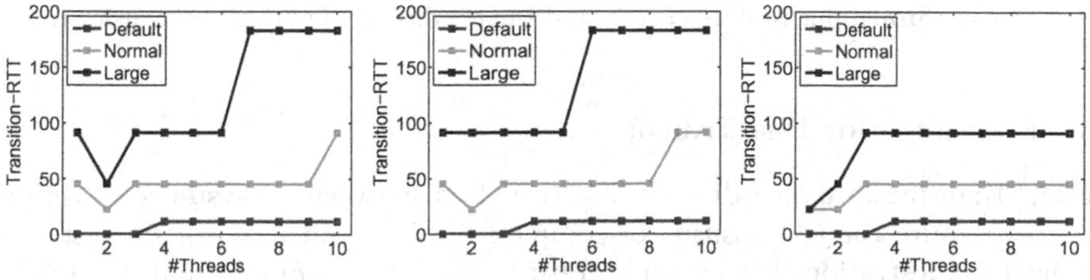

(a) CUBIC (b) HTCP (c) STCP

Fig. 6. Transition-RTT estimates with 1–10 streams and various buffer sizes for CUBIC, HTCP, and STCP.

3.2 Estimates for Memory Transfers

The concave-convex fit $\hat{f}_{\Theta_T}(\tau)$ for single stream CUBIC measurements over 10GigE connections for three different buffer sizes are shown in Fig. 5 along with the measurements. As in previous section with 10 streams, the profile is entirely convex at the default buffer size, and consequently there is only a convex portion to the sigmoid fit. For normal and large buffer sizes, both the concave and convex sigmoid fits are present, as shown with solid-blue and dashed-black curves, respectively. It is clear that τ_T increases, hence the concave region expands, as the buffer size is increased.

We estimate $\hat{f}_{\Theta_T}(\tau)$ for 1–10 parallel streams and three congestion control modules, namely, CUBIC, HTCP, and STCP. The overall variations of the estimated transition-RTT values w.r.t. number of parallel streams, buffer sizes, and TCP congestion control modules are shown in Fig. 6. For CUBIC with default buffer size, the transition-RTT τ_T increases from 0.4 ms for 1–3 parallel streams to 11.8 ms for 4 or more parallel streams. With normal buffer size, the transition-RTT τ_T remains consistently higher (at 45.6 ms, except for 2 steams) than with default buffer size, and further increases to 91.6 ms for 10 parallel streams. The τ_T estimate with the large buffer size is even larger than with both the default and normal buffer sizes; for example, 91.6 ms for 1–6 parallel streams (except 2 streams) and 183 ms for 7 or more parallel streams. Similar increasing trends of the estimate τ_T are also noted for HTCP and STCP, and thereby corroborate our inference that more streams and larger buffer sizes extend the concave region in addition to improving the throughput.

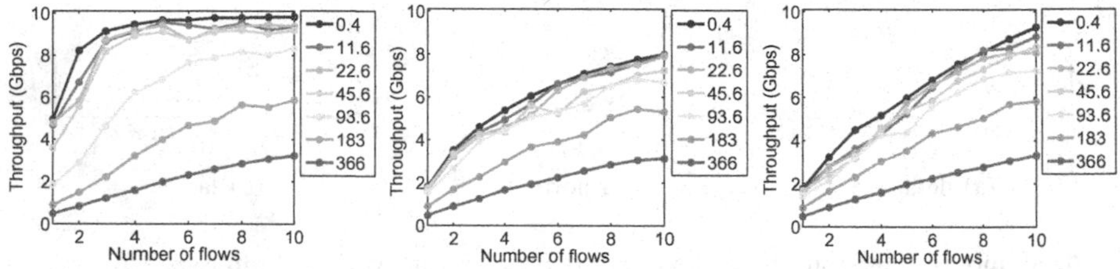

(a) XFS (b) direct I/O, 2 stripes (c) direct I/O, 8 stripes

Fig. 7. Mean throughput profiles of XFS and Lustre direct I/O file write transfer rates.

3.3 Estimates for File Transfers

XDD. High-performance disk-to-disk transfers between filesystems at different sites require the composition of complex file I/O and network subsystems, and host orchestration. For example, Lustre filesystem employs multiple OSTs to manage collections of disks, multiple OSSes to stripe file contents, and distributed MDSes to provide site-wide file naming and access. However, sustaining high file-transfer rates requires *joint* optimization of subsystem parameters to account for the impedance mismatches among them [22]. For Lustre filesystems, important parameters are the stripe size and number of stripes for the files, and these are typically specified at the creation time; the number of parallel I/O threads for read/write operations are specified at the transfer time. To sustain high throughput, I/O buffer size and the number of parallel threads are chosen to be sufficiently large, but this heuristic is not always optimal [18]. For instance, wide-area file transfers over 10 Gbps connections between two Lustre filesystems achieve transfer rates of only 1.5 Gbps, when striped across 8 storage servers, accessed with 8 MB buffers, and with 8 I/O and TCP threads [18], even though peak network memory-transfer rate and local file throughput are each close to 10 Gbps.

We measured file I/O and network throughput and file-transfer rates over Lustre and XFS filesystems for a suite of seven emulated connections in the 0–366 ms RTT range, which are used for memory transfer measurements in the previous section. We collected two sets of XDD disk-to-disk file transfer measurements, one from XFS to XFS and one from Lustre to Lustre, and each experiment is repeated 10 times. We considered both buffered I/O (the Linux default) and direct I/O options for Lustre. In the latter, XDD avoids the local copies of files on hosts by directly reading and writing into its buffers, which significantly improves the transfer rates. Results based on these measurements are summarized in [18]: (a) strategies of large buffers and higher parallelism do not always translate into higher transfer rates; (b) direct I/O methods that avoid file buffers at the hosts provide higher wide-area transfer rates, and (c) significant statistical variations in measurements, due to complex interactions of non-linear TCP dynamics with parallel file I/O streams, necessitate repeated measurements to ensure confidence in inferences based on them.

These file transfers are carried out using XDD, which spawns a set of *QThreads* to read a file from the source filesystem and subsequently transfer data over the network to the destination filesystem. We refer to each source-destination QThread connection as a *flow*. For XFS file transfers, the number of flows varies from 1 to 10, and the data transfer profile plot is shown in Fig. 7(a) for various RTTs. These profiles show monotonically increasing trends with respect to the number of flows. We use similar configuration for the Lustre experiments with the direct I/O option. In addition to varying the number of flows from 1 to 10, two different number of stripes were used: stripes 2 and 8, and the corresponding plots are respectively shown in Figs. 7(b) and (c). Similar to XFS profiles, the overall throughput exhibits predominantly increasing trends with respect to the number of flows; although compared to XFS, the Lustre throughput increases much slower with increasing flow counts in lower RTT cases. Comparing the performances of 2 vs. 8 striped Lustre, we notice that the use of 2 stripes yields somewhat higher transfer rates for lower flow counts. With more flows, overall throughput is higher, and 8 stripes is the better option.

To evaluate the transition-RTT values of file transfer profiles, we estimate the sigmoid fit $\hat{f}_{\Theta_E}(\tau)$ using measured XFS and Lustre throughput profiles. In Fig. 8, we demonstrate the fitted sigmoid plots overlayed on the mean profiles of measured throughput values with 6 parallel flows. We notice that all three

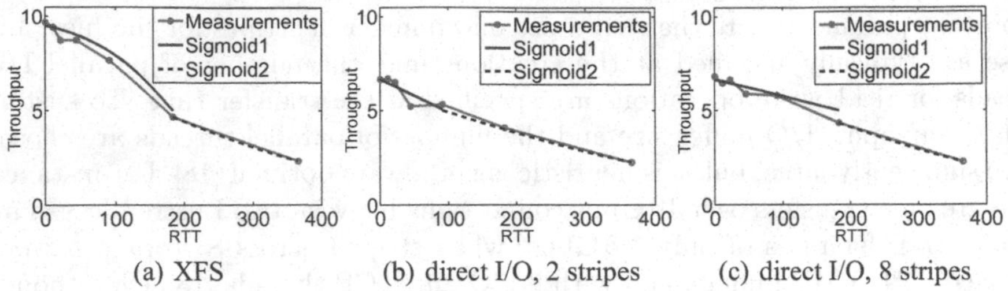

Fig. 8. Sigmoid regression fits of file transfer throughput profiles for XFS and Lustre.

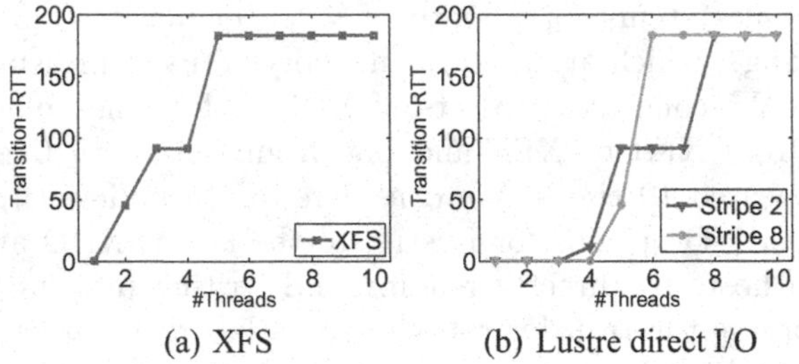

Fig. 9. Transition-RTT estimates with respect to the number of flows for XFS and Lustre.

variations of throughput profiles show concave and convex region to a different degree. For the XFS profile, the concave throughput profile at the smaller RTT values is more prominent than those in the Lustre profiles with both 2 and 8 stripes. In addition, we notice that transition-RTT values change depending on the file transfer throughput profiles; for example, both XFS and Lustre with 8 stripes show $\tau_T = 183$ ms with 6 flows, but Lustre with 2 stripes has $\tau_T = 91.6$ ms for this configuration.

The overall characteristic of the transition-RTT values for XFS and Lustre filesystems at different number of flows is summarized in Fig. 9. For XFS, the transition-RTT values steadily increases from 0.4 ms for 1 parallel flows to 45.6 ms for 2 flows, 91.6 ms for 3–4 flows, and 183 ms for 5 or more flows. Similar increasing trends of the estimated τ_T values are also noted for Lustre filesystem in Fig. 9(b), but comparing them with Fig. 9(a) it becomes quite evident that XFS profiles have higher τ_T values, and hence wider concave profile regions, at smaller number of flows. Among the Lustre profiles with 2 and 8 stripes in Fig. 9(b), both produce only convex profiles for 4 or less number of flows. Then, transition-RTT values of Lustre with 8 stripes rise sharply to 183 ms in contrast to that with 2 stripes, indicating wider concave regions for 8-striped Lustre profile. For 8 flows or more, both 2-striped and 8-striped Lustre throughput profiles show the same τ_T values, but the measured throughput values are higher for 8-striped Lustre (see Figs. 7(b) and (c)).

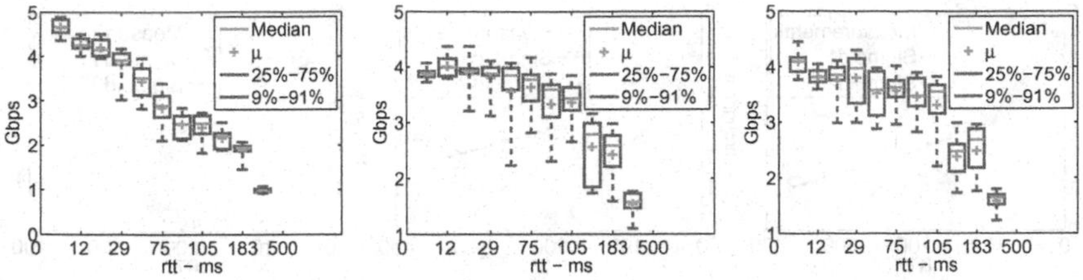

(a) cc=1, p=1 (b) cc=5, p=10 (c) cc=10, p=10

Fig. 10. GridFTP: Mean throughput profiles with CUBIC-Lustre-XFS configuration.

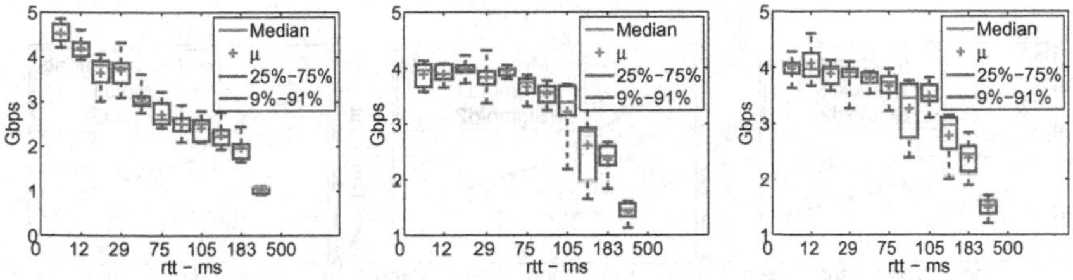

(a) cc=1, p=1 (b) cc=5, p=10 (c) cc=10, p=10

Fig. 11. GridFTP: Mean throughput profiles with HTCP-Lustre-XFS configuration.

GridFTP. GridFTP is an extension of the standard File Transfer Protocol (FTP) for high-speed, reliable, and secure data transfer [1]. It implements extensions to FTP, which provide support for striped transfers from multiple data sources. Data may be striped or interleaved across multiple servers, as in a parallel filesystem such as Lustre. GridFTP supports parallel TCP flows via FTP command extensions and data channel extensions. A GridFTP implementation can use long virtual round trip times to achieve fairness when using parallelism or striping. In general, GridFTP uses striping and parallelism in tandem, i.e., multiple TCP streams may be open between each of the multiple servers participating in a striped transfer. However, this process is somewhat different compared to XDD wherein each I/O stream is handled by a single TCP stream, whereas such association is less strict in GridFTP.

Figures 10 and 11 show GridFTP throughput performances for transfers between Lustre and XFS filesystems using CUBIC and HTCP congestion control modules. In each configuration, we vary the concurrency (*cc*) and parallelism (*p*) parameters, and collect throughput measurements across 11 different RTT values including the ones used in previous section. Here, XFS is mounted on SSD and provides throughput higher than 10 Gbps connection bandwidth. On the other hand, Lustre throughput is below 10Gbps and hence the transfer throughput is mainly limited by Lustre parameters. From these plots, it is quite evident that the overall throughput profiles of GridFTP are lower compared to XDD file transfer throughput values described in previous section.

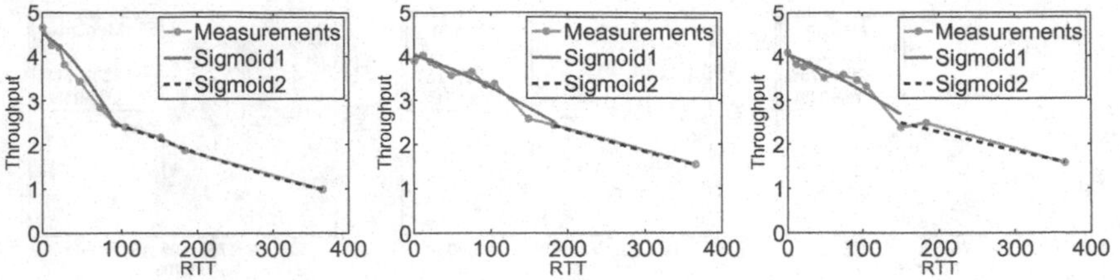

(a) cc=1, p=1 (b) cc=5, p=10 (c) cc=10, p=10

Fig. 12. Sigmoid fits of GridFTP throughput profiles with CUBIC-Lustre-XFS configuration.

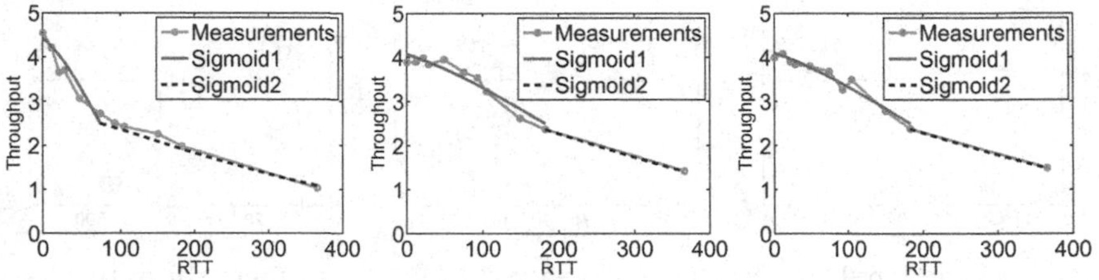

(a) cc=1, p=1 (b) cc=5, p=10 (c) cc=10, p=10

Fig. 13. Sigmoid fits of GridFTP throughput profiles with HTCP-Lustre-XFS configuration.

To evaluate the transition-RTT values for GridFTP profiles, we estimate the sigmoid-based fit $\hat{f}_{\Theta_E}(\tau)$ of the measured GridFTP throughput values, and the corresponding results are shown in Figs. 12 and 13. We notice from Fig. 12(a) that, when the values of *cc* and *p* parameters are small, the concave throughput profiles extend only up to 93 ms. However, Figs. 12(b) and (c) show that, with the increase in *cc* and *p* values, the concave throughput regions of the GridFTP profiles become wider resulting into $\tau_T = 183$ and 150 ms respectively at $cc = 5$ and 10 (along with $p = 10$). These variations of τ_T values with respect to *cc* and *p* parameters are plotted in Fig. 14(a) for the CUBIC-Lustre-XFS configuration of GridFTP profile. For the other GridFTP configuration, HTCP-Lustre-XFS, the sigmoid fitted models on the throughput profiles are shown in Fig. 13, and the resulting transition-RTT values are depicted in Fig. 14(b). Overall, they show similar trends as obtained with the CUBIC-Lustre-XFS configuration of GridFTP. The only notable difference is that, for the HTCP-Lustre-XFS configuration of GridFTP, the τ_T values stays at 183 ms for both $cc = 5$ and 10 (when $p = 10$), indicating monotonic trend with *cc* and *p* parameters.

Lustre over LNet. As mentioned earlier, Lustre filesystem employs multiple OSTs to manage collections of disks, and multiple OSSs to stripe file contents. Lustre clients and servers connect over the network, and are configured to match

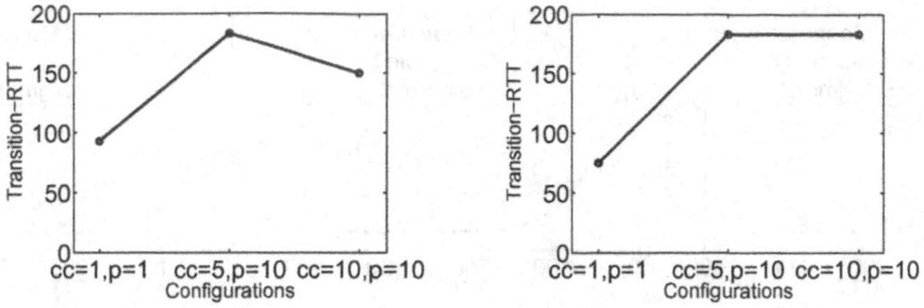

(a) CUBIC-Lustre-XFS (b) HTCP-Lustre-XFS

Fig. 14. Transition-RTT estimates with respect to the GridFTP configurations.

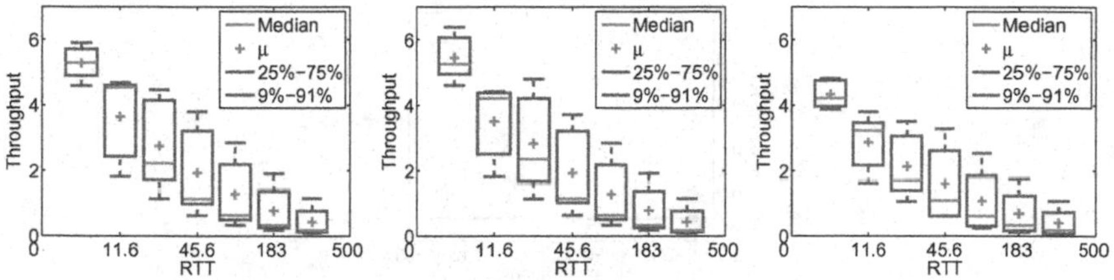

(a) b5-b6-2G-tuned (b) b5-b6-50M-tuned (c) t2-t1-2G-tuned

Fig. 15. LNet: Mean throughput profiles.

the underlying network modality, for example IB or Ethernet. Host systems are connected to IB switch via HCAs, and Lustre over IB clients is used to mount the filesystem on them over IB connections. Due to a latency limit of 2.5 ms, such deployments are limited to site-level access, and do not provide file access over wide-area networks. This IB-based Lustre filesystem is augmented with Ethernet Lustre clients, and LNet routers are utilized to make IB-based OSSs available over wide-area Ethernet connections [16]. Compared to GridFTP and XDD, which are software applications, the implementation of LNet routers requires more changes to the infrastructure.

The throughput profiles of Lustre over wide-area connections using LNet routers are shown in Fig. 15. For these measurements, we use two classes of hosts: (i) `bohr05` (b5) and `bohr06` (b6) are DTN servers with 48 cores, and (ii) `tait1` (t1) and `tait2` (t2) are compute nodes of a cluster with 32 cores. In each case, one of the node is used as a LNet router which extends IB network to wide-area TCP/IP connection. As shown in Sect. 3.1, TCP buffer size could have a significant impact on the concave-convex shape of the profile.

In Fig. 16, we demonstrate the sigmoid fit $\hat{f}_{\Theta_E}(\tau)$ along with the measured throughput profiles for LNet router configurations. As noted from Fig. 15, the profiles are entirely convex in all our LNet configurations, and consequently we get only the convex portions of the sigmoid fits. Therefore, the transition-RTT values for all LNet experiments are estimated to 0.1 ms, as shown in Fig. 17. We also used two LNet buffer sizes, 50M and 2G, which corresponded to convex

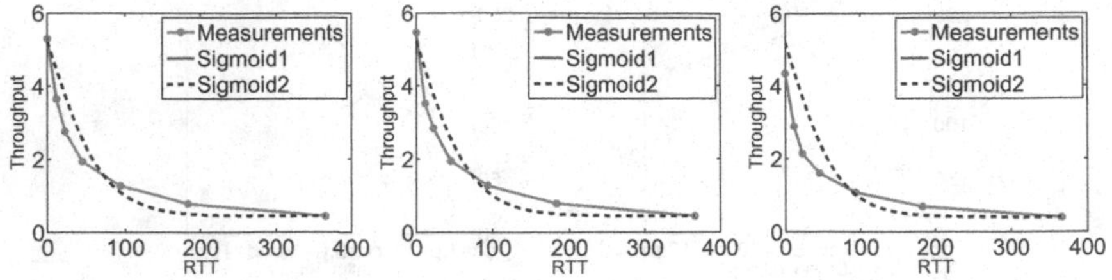

(a) b5-b6-2G-tuned (b) b5-b6-50M-tuned (c) t2-t1-2G-tuned

Fig. 16. Sigmoid regression fits of LNet throughput profiles.

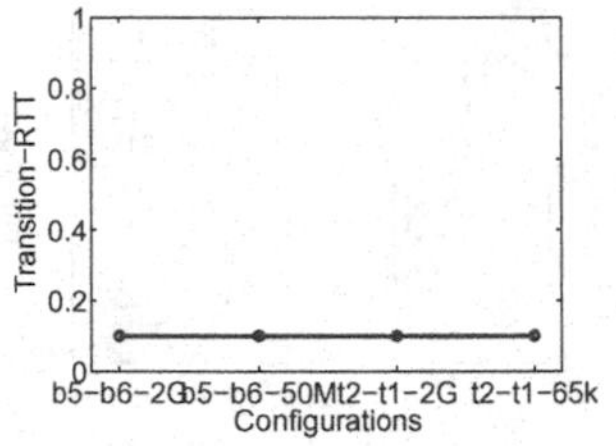

Fig. 17. Transition-RTT estimates with respect to various LNet setups.

and concave profiles for the underlying TCP throughput profiles, as illustrated in Fig. 4. However, for file transfers via LNet routers, in all the three configurations, the estimated throughput profile $\hat{f}_{\Theta_E}(\tau)$ as a function of RTT is entirely convex, which indicates that the source of convexity is elsewhere. Indeed, the LNet credits limit the number of packets in transit over the Lustre filesystem, which in turn limits the number of packets in transit over TCP connection. Such limit is equivalent to TCP buffer limit, which explains the convex profile indicated by our estimate $\hat{f}_{\Theta_E}(\tau)$.

4 Confidence Probability Estimates

In the previous section, the convex-concave estimate $\hat{f}_{\Theta_A}(\tau)$ based on measurements has been important in assessing the effectiveness of the transport methods and configurations. Now, we address its soundness from a finite sample statistics perspective [24], particularly in assessing properties of $\Theta_A(\tau)$ that depend on (potentially) infinite-dimensional $\mathbf{P}_{\Theta_A(\tau)}$. We consider the *profile regression* given by

$$\bar{\Theta}_A(\tau) = E\left[\Theta_A(\tau)\right] = \int \Theta_A(\tau) d\mathbf{P}_{\Theta_A(\tau)},$$

which is approximated by $\hat{f}_{\Theta_A}(\tau)$ using *finite* independent and identically distributed (i.i.d.) measurements $\theta(\tau_k, t_j^k)$ at τ_k, $k = 1, 2, \ldots, n$, collected at times t_j^k, $j = 1, 2, \ldots, n_k$. The critical concave-convex property provided by $\hat{f}_{\Theta_A}(\tau)$

is only an approximation of the true concave-convex property of $\bar{\Theta}_A(\tau)$, which depends on the complex distribution $\mathbf{P}_{\Theta_A(\tau)}$. In general, it is not practical to estimate $\mathbf{P}_{\Theta_A(\tau)}$ since it depends on hosts, filesystems, and network connections, as well as on software components, including TCP congestion control kernel modules, and file- and memory-transfer application modules. For the estimate $\hat{f}_{\Theta_A}(\tau)$ computed based solely on measurements, we derive probability that its expected error is close to the optimal error, independent of $\mathbf{P}_{\Theta_A(\tau)}$.

Consider an estimate $f(.)$ of $\bar{\Theta}_A(.)$ chosen from a function class $\mathcal{M}$. The *expected error* $I(f)$ of the estimator f is

$$I(f) = \int [f(\tau) - \theta(\tau,t)]^2 d\mathbf{P}_{\theta(\tau,t)}.$$

The *best estimator* f^* is given by $I(f^*) = \min_{f \in \mathcal{M}} I(f)$, which in general is not possible obtain since $\mathbf{P}_{\theta(\tau,t)}$ in unknown. The *empirical error* of estimator f based on measurements is given by

$$\hat{I}(f) = \frac{1}{n_T} \sum_{k=1}^{n} \sum_{j=1}^{n_k} [f(\tau_k) - \theta(\tau_k, t_j)]^2 ,$$

which is a scaled version of the sum-squared error $\hat{S}(f)$ in Eq. (3.1), that is $\hat{I}(f) = \frac{1}{n_T}\hat{S}(f)$, where $n_T = \sum_{k=1}^{n} n_k$. The *best empirical estimator* $\hat{f} \in \mathcal{M}$ minimizes the empirical error, that is, $\hat{I}(\hat{f}) = \min_{f \in \mathcal{M}} \hat{I}(f)$. Thus, $\hat{f}$ is estimated based entirely on measurements as an approximation to f^*.

Let us consider a class of flipped sigmoid functions with bounded weights $\mathcal{M}_{a,\tau} = \{g_{a,\tau} : a \in [-A, A], \tau \in [-T, T]\}$. The estimator $\hat{f}_{\Theta_A}$ is chosen from the set of estimators $\mathcal{M}_{\Theta_A} = \{f_{\Theta_A} : g_{a_1,\tau_1}, g_{a_2,\tau_2} \in \mathcal{M}_{a,\tau}\}$, which is not guaranteed to minimize the empirical error since τ_T is limited to τ_i. Then, the empirical error of our estimator $\hat{f}_{\Theta_A}$ is given by $\hat{I}\left(\hat{f}_{\Theta_A}\right) = \hat{I}(\hat{f}) + \hat{\epsilon}$ in terms of the empirical best estimator $\hat{f}$, where $\hat{\epsilon} \geq 0$. We now show that the expected error of $\hat{f}_{\Theta_A}$ is within $\epsilon + \hat{\epsilon}$ of optimal error $I(f^*)$, for any $\epsilon > 0$, with a probability that improves with the number of measurements and is independent of the complex, unknown $\mathbf{P}_{\theta(\tau,t)}$.

Theorem 1. *For $\hat{f}_{\Theta_A}$ estimated based on $n_T = \sum_{k=1}^{n} n_k$ i.i.d. measurements, the probability that its expected error is within $\epsilon + \hat{\epsilon}$ of the optimal error is upper bounded as*

$$\mathbf{P}\left\{I\left(\hat{f}_{\Theta_A}\right) - I(f^*) > \epsilon + \hat{\epsilon}\right\} \leq K\left(\frac{AT}{\epsilon^2}\right) e^{-\epsilon^2 n_T/512},$$

for any $\epsilon > 0$ and $K = 2048$.

Proof. We first note that

$$\mathbf{P}\left\{I\left(f_{\Theta_A}\right)-I(f^*)>\epsilon+\hat{\epsilon}\right\}\leq\mathbf{P}\left\{I\left(g_{a_1,\tau_{a_1}}\right)+I\left(g_{a_2,\tau_{a_2}}\right)-I(f^*)>\epsilon+\hat{\epsilon}\right\}$$

since $I\left(g_{a_1,\tau_{a_1}}\right)+I\left(g_{a_2,\tau_{a_2}}\right)\geq I\left(f_{\Theta_A}\right)$ wherein both $g_{a_1,\tau_{a_1}}$ and $g_{a_2,\tau_{a_2}}$ are expanded to the entire range of τ. Then, we can write

$$\mathbf{P}\left\{I\left(g_{a_1,\tau_{a_1}}\right)+I\left(g_{a_2,\tau_{a_2}}\right)-I(f^*)>\epsilon+\hat{\epsilon}\right\}\leq\mathbf{P}\left\{I\left(g_{a_1,\tau_{a_1}}\right)-I(g^*_{a,\tau_a})>\epsilon/2+\hat{\epsilon}/2\right\}$$
$$+\mathbf{P}\left\{I\left(g_{a_2,\tau_{a_2}}\right)-I(g^*_{a,\tau_a})>\epsilon/2+\hat{\epsilon}/2\right\},$$

where $I(g^*_{a,\tau_a})=\min_{g_{a,\tau_a}\in\mathcal{M}_{a,\tau}} I(g_{a,\tau_a})$. This inequality follows by noting that the condition

$$I\left(g_{a_i,\tau_{a_i}}\right)-I(g^*_{a,\tau_a})<\epsilon/2+\hat{\epsilon}/2$$

for both $i=1,2$ implies that $I\left(g_{a_1,\tau_{a_1}}\right)+I\left(g_{a_2,\tau_{a_2}}\right)-I(f^*)<\epsilon+\hat{\epsilon}$. Equivalently, the opposite of the latter condition implies that of the former, and thus the probability of the former is upper bounded by that of the latter, establishing the above inequality.

Next, by using Vapnik-Chervonenkis theory [24], we have

$$\mathbf{P}\left\{I\left(g_{a_1,\tau_{a_1}}\right)-I(g^*_{a,\tau_a})>\epsilon/2+\hat{\epsilon}/2\right\}$$
$$\leq\mathbf{P}\left\{\max_{h\in\mathcal{M}_{a,\tau}}\left|I\left(h\right)-\hat{I}(h)\right|>\epsilon/4\right\}\leq 8\mathcal{N}_\infty\left(\epsilon/32,\mathcal{M}_{a,\tau}\right)e^{-\epsilon^2 n_T/512}$$

where $\theta(\tau,t)\leq 1$, and $\mathcal{N}_\infty\left(\epsilon,\mathcal{M}_{a,\tau}\right)$ is the ϵ-cover of $\mathcal{M}_{a,\tau}$ under L_∞ norm (see [7,12] for the last bound).

We now show that $g_{a,\tau_a}\in\mathcal{M}_{a,\tau}$ is Lipschitz as follows: (i) $|g_{a,\tau}(x)-g_{a\prime,\tau}(x)|<AT/4\epsilon$ for all x under the condition $|a-a\prime|\leq\epsilon$, and (ii) $|g_{a,\tau}(x)-g_{a,\tau\prime}(x)|<AT/4\epsilon$ for all x under the condition $|\tau-\tau\prime|\leq\epsilon$. The Lipschitz constant in (i) is estimated by the maximum derivative with respect to a (part (ii) is similar). Let $z=a(\tau-\tau_a)$ and $\sigma(z)=1/(1+e^{-z})$, such that $g_{a,\tau_a}(\tau)=1-\sigma(z)$. Then, we have
$\frac{d\sigma(z)}{dz}=a\sigma(z)[1-\sigma(z)]\leq a/4$. Thus, we get [13]

$$\frac{dg_{a,\tau_a}}{da}=\frac{d\sigma(z)}{dz}\tau_a\leq AT/4.$$

We consider a two-dimensional grid $[-A,A]\times[-T,T]$ with $4AT/\epsilon^2$ points equally spaced in each dimension. Then, for any (a,τ_a), there exists a grid point (b,τ_b) such that $|a-b|\leq\epsilon$ and $|\tau_a-\tau_b|\leq\epsilon$, which in turn assures that $\| g_{a,\tau_a}-g_{b,\tau_b}\|_\infty\leq AT/4\epsilon$. Consequently, we have

$$\mathcal{N}_\infty\left(\epsilon/32,\mathcal{M}_{a,\tau}\right)\leq 128\left(\frac{AT}{\epsilon^2}\right).$$

By applying this bound for both $g_{a_1,\tau_{a_1}}$ and $g_{a_2,\tau_{a_2}}$, we obtain

$$\mathbf{P}\left\{I\left(\hat{f}_{\Theta_A}\right) - I(f^*) > \epsilon + \hat{\epsilon}\right\} < 16\mathcal{N}_\infty\left(\epsilon/32, \mathcal{M}_{a,\tau}\right) e^{-\epsilon^2 n_T/512},$$

which establishes the probability bound. □

This result ensures that $I\left(\hat{f}_{\Theta_A}\right) - I(f^*) < \epsilon + \hat{\epsilon}$ with a probability at least $\alpha = \left[1 - K\left(\frac{AT}{\epsilon^2}\right) e^{-\epsilon^2 n_T/512}\right]$, which approaches to 1 as $n_T \to \infty$. In particular, the exponential term decays faster in n_T than other terms, and hence for a sufficiently large n_T a given confidence probability α can be assured. This performance guarantee is independent of how complex the underlying distribution $\mathbf{P}_{\Theta_A(\tau)}$ is, and thus provides confidence in the inferences based on the concavity-convexity properties of the estimate $\hat{f}_{\Theta_A}$ derived entirely from measurements. Similar performance guarantees have been provided for a somewhat different problem of network throughput profile estimation in [14,17]. It is interesting to note that the exponent of ϵ in the bound for the cover size $\mathcal{N}_\infty\left(\epsilon/32, \mathcal{M}_{a,\tau}\right)$ is 2 in the above proof, although 5 parameters ($a_1, a_2, \tau_{a_1}, \tau_{a_2}$ and τ_T) are needed to estimate $\hat{f}_{\Theta_A}$. In this sense, the underlying structure of $\hat{f}_{\Theta_A}$ reduces the estimation dimensionality from 5 to 2. For a similar cover size estimate for feedforward sigmoidal neural networks, for example, the exponent of ϵ would be the total number of neural network parameters (much larger than 2).

5 Generic Machine Learning Methods

To gain further insights into throughput performance, we apply generic machine learning modules, available in Python and MATLAB, for obtaining regression-based estimates of throughput profiles. The regression-fits in general not only indicate the overall throughput trends as functions of various parameters, but also provide throughput predictions for configurations at which measurements have not been collected.

From the Python-based machine learning tools, we use three well-known learning approaches: (i) artificial neural network (ANN), (ii) random forest (bagging), and (iii) gradient boosting. From MATALB, we use (i) support vector machines (SVM) and (ii) Gaussian kernel estimates, for GridFTP measurements. We implement ANN in the `tensorflow` framework with the following specifications: number of hidden layers = 2, number of hidden units = 8 and 5, learning rate = 0.001, ReLU activation units, and Adam optimizer. Both the weights and bias terms of the ANN are initialized with normal random variables. The random forest model is also developed in the `tensorflow` framework, specifically with the *tensor_forest* module. In this setup, we use number of trees = 2 and number of maximum nodes = 10, and execute the tensor_forest graph in the regression mode. The gradient boosting method is implemented using the `scikit-learn` library, along with the following parameters: number of trees = 1500, maximum

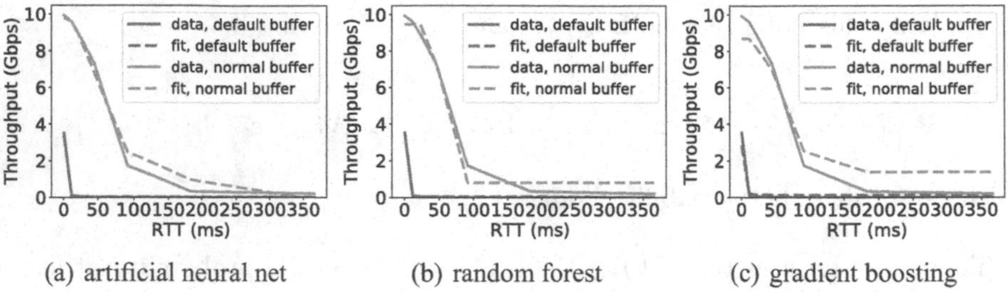

Fig. 18. CUBIC: Generic regression fits to default and normal buffer measurements with 1 flow.

depth = 2, learning rate = 0.001, and minimum samples per split = 10. Furthermore, in the all the three learning methods, we use mean-squared error as the cost metric for optimization purposes.

Memory. Figure 18 shows the fitted regression curves obtained via ANN, random forest, and gradient boosting methods, when they are applied to the CUBIC TCP throughput measurements. Specifically, in Fig. 18, we show only the cases with single flow using the default and normal buffer sizes. The mean throughput profiles of the measured data are shown with bold lines, whereas the corresponding fitted models are shown via dotted lines. The corresponding sigmoid-function based regression fits are shown in Figs. 5(a) and (b). At the chosen parameter setting for the generic learning methods, we notice from Fig. 18 that the ANN and random forest based regression models show better fits to these CUBIC TCP datasets than the gradient boosting based model. Particularly, the learning models show better accuracy (i.e., smaller error) with the default-buffer configuration and at somewhat small RTT values. However, from these generic regression fits, the concave-convex portions of throughput profiles and the associated transition-RTT values are not quite as clear as with sigmoid regression fits.

XDD. For the disk-to-disk file transfer throughput measurements, we also apply the ANN, random forest, and gradient boosting learning methods to obtain the regression-fit estimates. In Figs. 19 and 20, we depict the performances of these learning methods respectively for the XFS and Lustre (8-striped with direct I/O option) filesystems. To compare with the corresponding sigmoid-function based regression method (see Figs. 8(a) and (c)), we only show the fitted lines at 6 parallel flows. The parameters of the learning methods are kept the same as those used for the TCP throughput measurements. In general, ANN model show the best accuracies, followed by the random forest and gradient boosting models, for both XFS and Lustre direct I/O throughput measurements. The fitted lines via random forest and gradient boosting methods show large errors from the XFS measurements particularly at larger RTT values. Comparing the performances

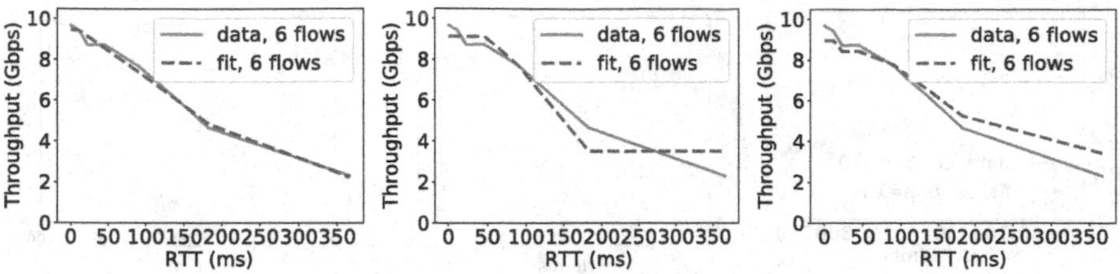

(a) artificial neural net (b) random forest (c) gradient boosting

Fig. 19. XDD XFS: Generic regression fits to file transfer measurements at 6 flows.

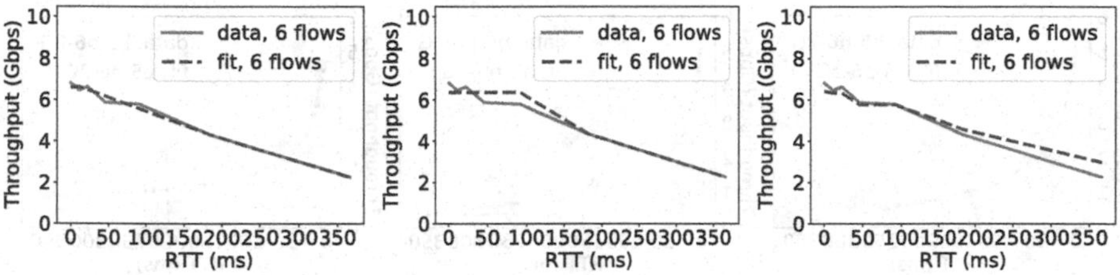

(a) artificial neural net (b) random forest (c) gradient boosting

Fig. 20. XDD Lustre: Generic regression fits to 8-striped direct I/O measurements at 6 flows.

between the XFS and Lustre throughputs, we notice that the fitting accuracy is little better with the Lustre datasets than XFS. However, from the concavity-convexity perspective, these generic fits do not provide as clear information as obtained via the sigmoid based regression approach.

GridFTP. The generic regression model fits to the GridFTP measurements are demonstrated in Fig. 21 for the CUBIC-Lustre-XFS configuration with the parameter values $cc = 5$ and $p = 10$, using ANN model from Python, and SVM and Gaussian kernel models from MATLAB. These plots are equivalent to Figs. 12(b) depicting the sigmoid regression fit. Overall, the characteristics of all the fitted models seems quite similar to each other. In particular, from the SVM fit in Fig. 21(b), we can observe the concave and convex portions of the fitted model respectively at the smaller and larger RTT values. However, exact specification of the transition-RTT value requires one more post-processing step on the fitted SVM model, which was not required in the sigmoid-regression method. The concave portion is also present in Gaussian kernel fit although less prominent compared to SVM fit, as shown in Fig. 21(c). Comparatively, the concave-convex shape is less discernable in neural network fit in Fig. 21(a), whereas the transition is apparent in the other two fits.

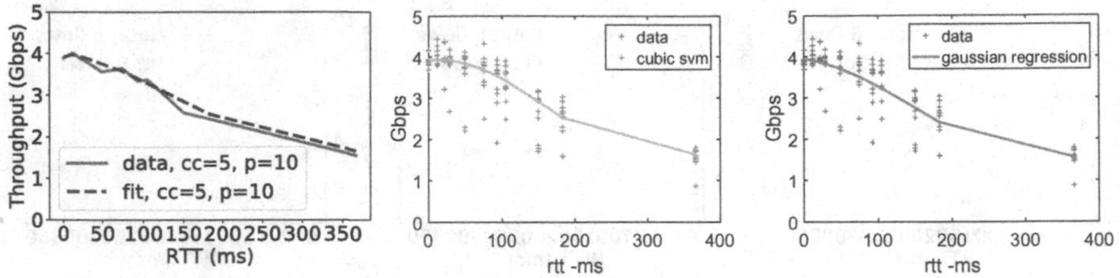

(a) artificial neural net (b) support vector machine (c) Gaussian kernel

Fig. 21. GridFTP: Generic regression fits to CUBIC-Lustre-XFS file transfer measurements.

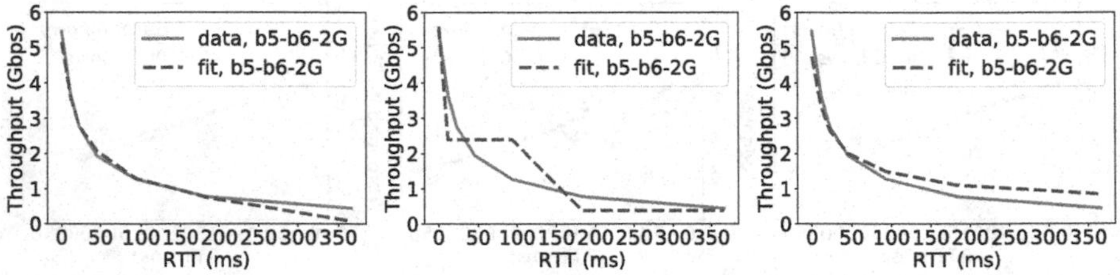

(a) artificial neural net (b) random forest (c) gradient boosting

Fig. 22. LNet: Generic regression fits to b5-b6-2G file transfer measurements.

LNet. The regression fits obtained using ANN, random forest, and gradient boosting techniques to the LNet throughput measurements are depicted in Fig. 22 for the b5-b6-2G configuration. Figure 16(a) depicts the equivalent sigmoid-regression fit. We do not include the generic regression fits to the other LNet datasets as they have similar trends. Among the three generic learning methods, the ANN model shows the best accuracy in fitting the measured LNet throughput data, followed by the gradient boosting and random forest models. Overall, these regression fits capture the entirely-convex trends of LNet throughput profiles, except the random forest fit, which is somewhat non-representative of convex profile.

6 Conclusions

We presented a machine learning approach that explicitly captures the dichotomous throughput profiles observed in throughput measurements collected in a number of data transport scenarios. Typically, it identifies concave regions at lower RTTs that indicate near-optimal performance, and convex regions at higher RTTs that indicate bottlenecks due to factors such as buffer or credit limits. This approach uses sigmoid functions to explicitly capture the concave and convex regions in a throughput profile, and also provides distribution-free confidence estimates for the closeness of the estimate's expected error to optimal error. We applied this method to throughput measurements of memory and file transfers over connections ranging from local to cross-country to round-the-earth distances. The convex-concave profile estimates enabled us to infer the near optimality of transfers in practical scenarios, such as XDD transfers between XFS filesystems, based on concave regions. In addition, this approach also enabled us to infer performance limits, such as in case of wide-area Lustre extensions using LNet routers, based on convex regions. A direct application of generic machine learning packages did not always highlight these critical performance regions nor provided as sharp confidence estimates, thereby making a case for customized machine learning methods such as the one proposed here.

In terms of future directions, characterizations of latency, jitter, and other dynamic properties will be of interest for data transport infrastructures, in addition to throughput profiles considered here. Additionally, it would be of future interest to derive confidence estimates for the transition-RTT computed using the measurements. Extensions of the proposed estimates and analyses to data transport over shared network connections would also be of future interest.

References

1. Allcock, W., et al.: The Globus striped gridFTP framework and server. In: Proceedings of the 2005 ACM/IEEE Conference on Supercomputing, SC 2005, pp. 54–64. IEEE Computer Society, Washington (2005)
2. Allen, B., Bresnahan, J., Childers, L., Foster, I., Kandaswamy, G., Kettimuthu, R., Kordas, J., Link, M., Martin, S., Pickett, K., Tuecke, S.: Software as a service for data scientists. Commun. ACM **55**(2), 81–88 (2012)
3. Aspera high-speed file transfer. http://www-03.ibm.com/software/products/en/high-speed-file-transfer
4. Cardwell, N., Cheng, Y., Gunn, C.S., Yeganeh, S.H., Jacobson, V.: BBR: congestion based congestion control. ACM Qeueue **14**(5), 50 (2016)
5. General Parallel File System. https://www.ibm.com/support/knowledgecenter/en/SSFKCN/gpfs_welcome.html
6. Gu, Y., Grossman, R.L.: UDT: UDP-based data transfer for high-speed wide area networks. Comput. Netw. **51**(7), 1777–1799 (2007)
7. Krzyzak, A., Linder, T., Lugosi, G.: Nonparametric estimation and classification using radial basis function nets and empirical risk minimization. IEEE Trans. Neural Netw. **7**(2), 475–487 (1996)
8. Liu, Q., Rao, N.S.V., Wu, C.Q., Yun, D., Kettimuthu, R., Foster, I.: Measurement-based performance profiles and dynamics of UDT over dedicated connections. In: International Conference on Network Protocols, Singapore, November 2016
9. Liu, Z., Kettimuthu, R., Foster, I., Ra, N.S.V.: Cross-geography scientific data transferring trends and behavior. In: ACM Symposium on High-Performance Parallel and Distributed Computing (HPDC), June 2018
10. Lustre Basics. https://www.olcf.ornl.gov/kb_articles/lustre-basics
11. Multi-core aware data transfer middleware. ndtm.fnal.gov
12. Pollard, D.: Convergence of Stochastic Processes. Springer, New York (1984). https://doi.org/10.1007/978-1-4612-5254-2
13. Rao, N.S.V.: Simple sample bound for feedforward sigmoid networks with bounded weights. Neurocomputing **29**, 115–122 (1999)

14. Rao, N.S.V.: Overlay networks of in-situ instruments for probabilistic guarantees on message delays in wide-area networks. IEEE J. Sel. Areas Commun. **22**(1), 79–90 (2004)
15. Rao, N.S.V.: Finite-sample generalization theory for machine learning practice for science. In: DOE ASCR Scientific Machine Learning Workshop (2018)
16. Rao, N.S.V., Imam, N., Hanley, J., Sarp, O.: Wide-area lustre file system using LNet routers. In: 12th Annual IEEE International Systems Conference (2018)
17. Rao, N.S.V., et al.: TCP throughput profiles using measurements over dedicated connections. In: ACM Symposium on High-Performance Parallel and Distributed Computing (HPDC), Washington, DC, July-August 2017
18. Rao, N.S.V., et al.: Experimental analysis of file transfer rates over wide-area dedicated connections. In: 18th IEEE International Conference on High Performance Computing and Communications (HPCC), Sydney, Australia, pp. 198–205, December 2016
19. Rao, N.S.V. et al.: Experiments and analyses of data transfers over wide-area dedicated connections. In: The 26th International Conference on Computer Communications and Network (ICCCN 2017) (2017)
20. Rhee, I., Xu, L.: CUBIC: a new TCP-friendly high-speed TCP variant. In: Proceedings of the Third International Workshop on Protocols for Fast Long-Distance Networks (2005)
21. Sen, S., Rao, N.S.V., Liu, Q., Imam, N., Foster, I.T., Kettimuthu, R.: On analytics of file transfer rates over dedicated wide-area connections. In: First International Workshop on Workflow Science (WOWS), Auckland, New Zealand, October 2017. in conjunction with 13th IEEE International Conference on e-Science
22. Settlemyer, B.W., Dobson, J.D., Hodson, S.W., Kuehn, J.A., Poole, S.W., Ruwart, T.M.: A technique for moving large data sets over high-performance long distance networks. In: IEEE 27th Symposium on Mass Storage Systems and Technologies (MSST), pp. 1–6, May 2011
23. Shorten, R.N., Leith, D.J.: H-TCP: TCP for high-speed and long-distance networks. In: Proceedings of the Third International Workshop on Protocols for Fast Long-Distance Networks (2004)
24. Vapnik, V.N.: Statistical Learning Theory. Wiley, New York (1998)
25. XDD - The eXtreme dd toolset. https://github.com/bws/xdd
26. XFS. http://xfs.org

Towards Analysing Cooperative Intelligent Transport System Security Data

Brice Leblanc, Emilien Bourdy, Hacène Fouchal(✉), Cyril de Runz, and Secil Ercan

CReSTIC, Universitè de Reims Champagne-Ardenne, Reims, France
{brice.leblanc,emilien.bourdy,hacene.fouchal,cyril.de-runz, secil.ercan}@univ-reims.fr

Abstract. C-ITS (Cooperative Intelligent Transport Systems) provide nowadays a very huge amounts of data from different sources: vehicles, roadside units, operator servers, smartphone applications. These amounts of data can be exploited and analysed in order to extract pertinent information as driver profiles, abnormal driving behaviours, etc. In this paper, we present a methodology for analysis of data provided by a real experimentation of a cooperative intelligent transport system (C-ITS). We have analysed mainly security issues as privacy, authenticity. We have used unsupervised machine learning approaches. The obtained results have shown interesting results in terms of latency, packet delivery ratio.

Keywords: C-ITS · Driving profile · Communication data · Data analysis · Machine learning · Clustering

1 Introduction

The deployment of connected vehicles becomes a hot challenge since many years. A dedicated WiFi has been designed for connected vehicles: IEEE 802.11p (denoted also ETSI ITS-G5). It is well suited to mobility compared to other WiFis. Over this WiFi a rich protocol stack layer has been proposed (by the ETSI, ISO and other standard institutes) and implemented by various providers. The target of these protocols is to ensure the efficiency of communications between vehicles in order to be sure that dangerous events, traffic jams will be disseminated precisely on time. The redundancy of sending messages is one of the main adopted principle. For these reasons, vehicles generate a lot of data when they respect the protocols defined within the previous stack. In the meantime many interested actors (road operators, telecom operators, car manufacturers)

Partially supported by The InterCor project number INEA/CEF/TRAN/M2015/1143833.

É. Renault et al. (Eds.): MLN 2018, LNCS 11407, pp. 23–32, 2019.
https://doi.org/10.1007/978-3-030-19945-6_2

need to explore these data for many purposes as traffic prediction, driver profile detection, alternative route search, *etc.*

The main idea of this paper is to use unsupervised data mining approaches on real dataset acquired during the InterCor Project TestFest held in Reims on April 2018. These data have been provided from different vehicles from more than 10 countries. Several scalable approaches from the literature have been tested on this dataset. This study allows us to build a technical evaluation methodology to be used for C-ITS. On all the data gathered during this event, we focus on analysing security issues currently used for C-ITS (authenticity and privacy).

The paper is structured as follows. Section 2 gives an overview of unsupervised data mining approaches and on C-ITS communications. Section 3 introduces the Test scenarios that provide driving data. Section 4 presents the results of the different unsupervised approaches over our real data. Section 5 gives the conclusion and perspectives.

2 Related Works

Machine learning area is very wide, there are different approaches to handle data and extract information from them. We will focus in this study on main clustering techniques and we will review some of the most popular algorithms in this field. We give a quick review about studies on security principles used for C-ITS.

2.1 Clustering Approaches

- Hierarchical-based algorithms
  Slink [1] and **Clink** [2] are two algorithms that optimize the basic principle of agglomerative clustering using respectively the single and the complete linkage (the linkage is the way hierarchical algorithms group or split clusters). They have an optimized complexity of $0(n^2)$ instead of $0(n^3)$ for the classic **Agnes**(**Ag**glomerative **Nes**ting) and **Diana** (**Di**visive **An**alysis) algorithms. **BIRCH** (Balanced Iterative Reducing and Clustering using Hierarchies) [3] is a scalable hierarchical algorithm used to find clusters in large data-sets using little memory and time. The big strength of BIRCH is the ability to create the clusters with only one scan of the data-set.
- Partitionning-based algorithms
  The **K-Means** [4] algorithm concept is to select at random k (k is a number defined by user) data points, to consider them as a cluster and to regroup the other points in the closest cluster until all points are inside a cluster. The center of the cluster is the mean of all the data points of the cluster and it is updated after a new point is added to the cluster. It is very simple and very efficient for many small data sets.
  The **K-Medoïds** [5] algorithm works as K-Means algorithm but does not rely on an average of a cluster center but on the most representative data point of the cluster.

- Density-based algorithms
  The **DBSCAN** (Density-based Algorithm for discovering clusters in large spatial databases with noise) [6] is a fast algorithm with a strong resistance to outliers. It is widely used for its effectiveness but it is very weak to cluster with varying density. This is the reason that led to the development of **OPTICS** (Ordering points to identify the clustering structure) [7]; it can handle the varying density of the cluster but at the cost of some speed performance.
- Artificial Neural Network (ANN) algorithms
  Algorithm using artificial neural network cannot handle a data set immediately but needs to be trained beforehand. Once trained they can efficiently handle larger amount of data than previous algorithm in a short time with a strong accuracy.
  SOM (Self Organized Map) [8] is an artificial neural network trained to produce a low-dimensional and discretized representation of the data. This makes it a strong visualisation tool for high dimensional data.
- Exemplar selection
  In [9], the authors propose a new approach of the classical *k*NN (*k* Nearest Neighbor) techniques to select exemplars, also called samples. Unlike the *k*NN techniques, this algorithm does not require training set to select the exemplars. The data with the highest neighborhood density is the first selected sample, and then, it is removed and its neighborhood from the dataset. The next samples are selected in the same way.

2.2 C-ITS Communications

For C-ITS, the ETSI standardisation institute in Europe [10] has proposed a protocol stack composed of:

- Physical and access layer based on the IEEE 802.11p.
- Network layer denoted Geonetworking. It is able to ensure multi-hop forwarding targeting geographical areas.
- Transport layer denoted Basic Transport Layer (BTP) which ensures the interface between Geonetworking protocol with a specific upper layer denoted Facilities layer.
- Facilities layer is interface between the application layer (close to the driver and the vehicle sensors) and the BTP protocol. This layer contains various protocols designed for specific needs as event notification, vehicle status, green light management.

We will focus in this study only on two main messages:

- CAM (Cooperative Awareness Message) [11]: A CAM message gives dynamic information about the vehicle (i.e. position, speed, heading, etc.). A vehicle sends CAMs to its neighbourhood using Vehicle-to-Vehicle (V2V) or Vehicle-to-Infrastructure (V2I) communications. The frequency of CAM messages varies from 1 Hz to 10 Hz depending on the average speed of the vehicle, on the congestion of the transmission channel.

- DENM (Decentralized Environmental Notification Message) [12]: DENM message is sent in order to notify any type of event (i.e. accident, traffic jam, road works, etc.). The event could be triggered automatically by a vehicle using cooperation between inside vehicles sensors or manually by drivers for more complicated situations as obstacle on the road.

Some issues studied in wireless sensor networks have similar features than C-ITS systems as routing and medium access. In [13] and [14] classification of routing protocols over WSNs is provided and gives an overview on the manner to choose a routing protocol regarding many parameters (topology, application, ...).

3 Test Environment

The general architecture of vehicular communications is presented on Figs. 1 and 2.

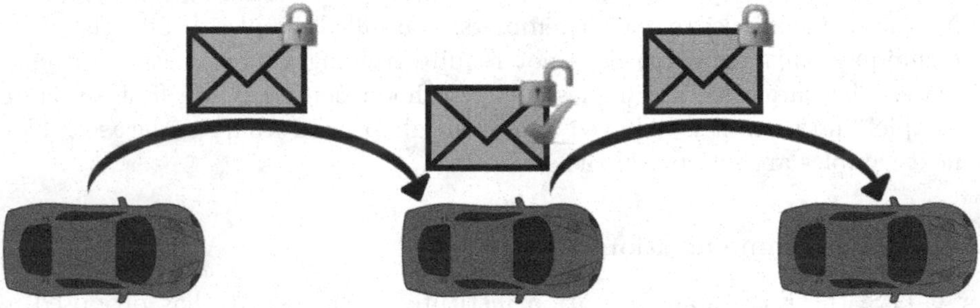

Fig. 1. A general scheme for Vehicle to vehicle communication

A vehicle is able to send a message through a G5 network in order to reach its neighbours. The message could reach other vehicles thanks to multi-hop forwarding.

An RSU (Road Side Unit) plays the same role as a vehicle for the forwarding aspect. In addition to that, the RSU handles all received messages from vehicles in order to run road operator's computations as traffic management, event recording, etc. In some cases the RSU disseminates events towards other RSUs within operator's network.

3.1 Security Mechanisms

In usual communication process, a node sends within a message its source identity (a MAC address for instance) and a destination identity. In vehicular communications, we often use a broadcast sending. Then the destination address is a generic broadcast address. But for the sender address, we should find a way to hide it in order to avoid driver tracking. The usual mechanism is to provide

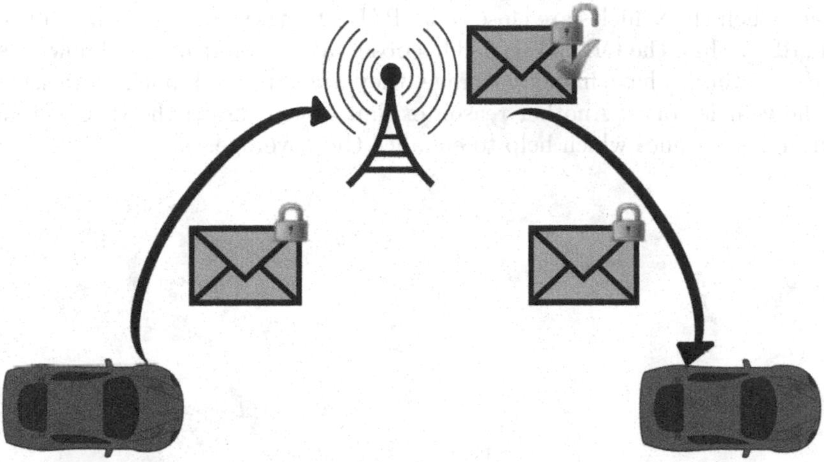

Fig. 2. A general scheme for Vehicle to infrastructure communication

drivers with a set of various identities. The driver will change his/her pseudonym according to a defined policy. In addition to privacy, we have to take care of the authenticity of sender in such eco-system. The adopted mechanism is to sign each sent message with a certificate of a driver. This certificate is provided by a common certificate authority. One certificate is not enough to avoid tracking, then a driver will be provided a set of certificates (denotes PC: pseudonym certificates) that he has to use during his/her travels. A vehicle has to send each message together with its signature and its certificate. When the certificate authority is common, vehicles are able to verify the sender authenticity but when vehicles are certified by different certificate authorities, vehicles have to verify trust chains which are implemented with trust lists. In the meantime, revocation lists have to be considered for misbehaving nodes.

3.2 Technical Evaluation

In order to evaluate the results of the experimentation, we have collected during four days of testing (on 30 km of urban road trajectory), 687.732705 Mb of data (Technical log files and PCAP files) provided by 16 vehicles and 4 Road Side Units. For each RSU an average of 10 Kb of Technical log is collected for each day. Moreover, 9 tablets were provided to 9 vehicles, to get there positions for the different tracks.

Network Communication Evaluation. The collected data have been analysed in order to measure some key performance indicators such as the packet delivery ratio for the OBUs (shown on Fig. 4) and for the RSUs (shown on Fig. 5). The trajectory of test field is shown on Fig. 5 where 4 RSUs have been involved in the event. The main observation on these maps is the delivery ratio is higher when the vehicles are close to an RSU and these ratios are much higher for the RSUs than the OBUs Fig. 3. This observation confirms well known issue: the RSUs antennas have higher coverage, they are supposed much more sensible than the vehicles ones. Another reason explains such result: the RSU antennas are on higher pylones which help to enhance the coverage.

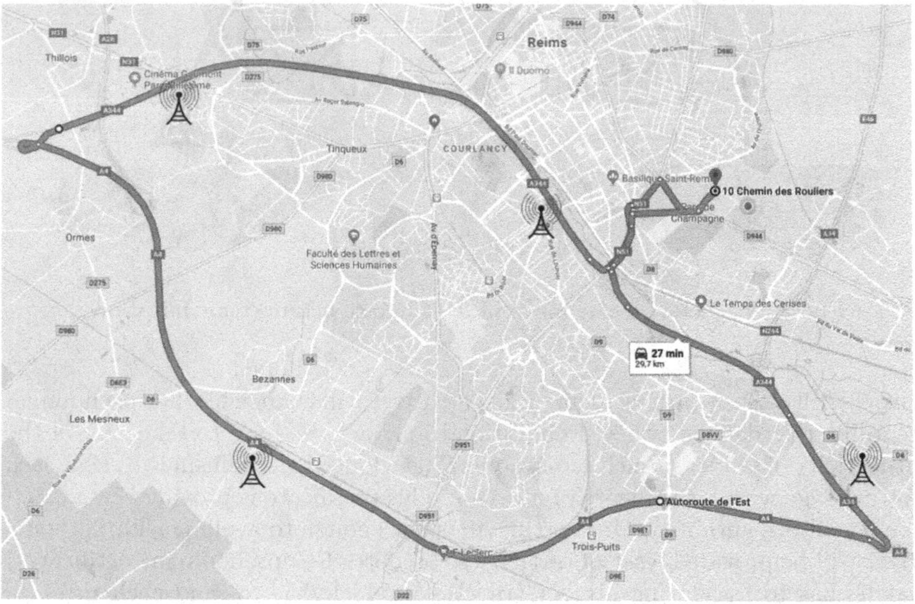

Fig. 3. The track of the Reims TestFest

					0	0	0	0.125	0	0.200	0.250	0.100	0	0	0.004
			0	0	0									0	
		0.083									0	0	0		
		0.167								0.046	0.001				
0.001	0.083	0.125							0.1		0.047	0.063	0.141		
0.006	0.009	0	0	0	0.167	0	0	0							

Fig. 4. A map representation of the OBU message received ratio

Network Security Evaluation. During this event, 133249 CAM messages have been received by RSUs. All these messages have been signed according to security schemes defined in security mechanisms section. On RSUs and OBUs specific log files on security failures have been recorded on most of the stations, they have been denoted "faulty message log files". The evaluation has observed that 20% of received messages are faulty messages.

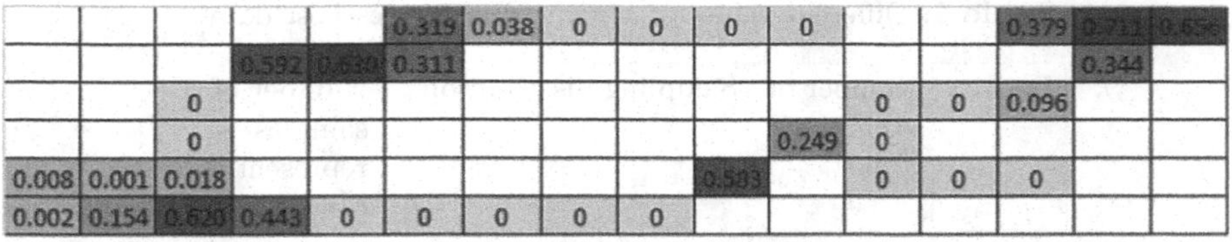

Fig. 5. A map representation of the RSU message received ratio

4 Data Analysis

From the data provided by the scenarios presented in the previous section we extracted the data generated by the vehicles equipped with tablets when they had to slow down or stop (red traffic light, stop sign, toll, etc.). These extracted observations created our data set of 58 observations of 95 variables for 9 different vehicles.

In order to reveal/extract driving (sub)profiles, some of the algorithms presented in the state of the art section were exploited. Selected approaches can generally be deployed using map/reduce paradigm and thus can be considered as scalable.

The Table 1 presents the results in terms of cluster purity index according to real driving profile using classical clustering algorithm.

Table 1. Purity table for classical algorithms

Algorithm	Kmeans	Agnes	Diana	Kmedoïd	Fanny	DBSCAN
Purity	0.3275862	0.3448276	0.3448276	0.3448276	0.3103448	0.2241379

On the other hand, the approach of Bourdy *et al.* [9] was applied. The results are presented in Table 2. According to that, when the density (k value in the table) is less than or equal to 24%, the exemplars are grouping observations (trajectories) of their own driver, and with a density of 82%, only 7 of the exemplars have neighbors from another driver, i.e. 87.9% of observations are grouping with an exemplar from the same driver. The Fig. 6 graphically illustrates those results. Each node is an observation, and each color/shape is a class. There are some groups of observations. Each group represents the sample at the center (with a big number) and its neighborhood (i.e. the observations represented by it). When no observation is linked together, it means that they are samples which do not represent other observations.

Indeed, the produced exemplars may be viewed as driving profile representatives. Those results can reveal a privacy issue. Indeed, even though no driver information is used, by grouping data, according to *Bourdy et al.* approach, we can identify a driving sub-profile. Nevertheless, the number of exemplars should moderate the conclusion.

Table 2. Different values of *k* with Intercor TestFest data

k	Density	Number of exemplars	Sampling distribution	Number of samples representing other classes
7	12%	57	(6, 8, 5, 2, 4, 9, 7, 3, 13)	0
10	17%			
14	24%		(7, 7, 5, 2, 4, 9, 7, 3, 13)	
19	33%	53	(6, 7, 5, 2, 4, 9, 7, 2, 11)	4
29	50%	44	(6, 7, 4, 2, 3, 7, 5, 1, 9)	7
38	66%	40	(6, 5, 4, 2, 3, 6, 5, 1, 8)	
48	82%	34	(7, 4, 3, 0, 3, 5, 4, 2, 6)	

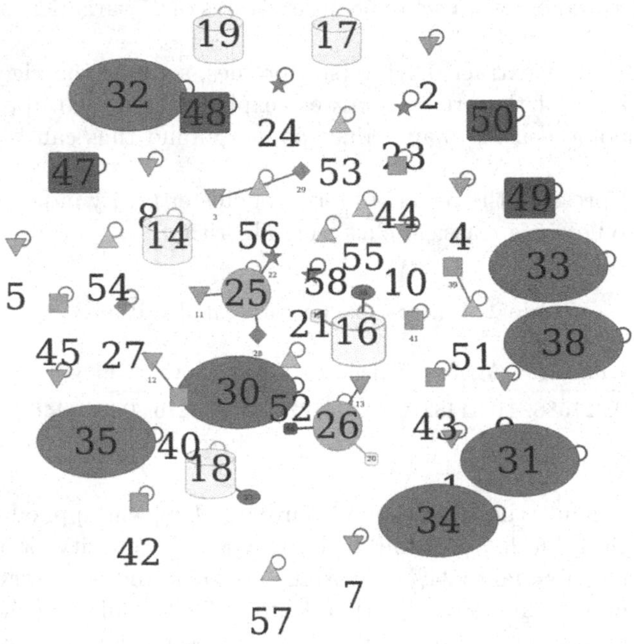

Fig. 6. Exemplar neighborhood with a *k* value of 48

5 Conclusion

In this paper we proposed to use some common clustering algorithms in order to analyse data provided by connected vehicles. We have analyzed some of the results obtained from our experimentations with 20 C-ITS stations. The obtained ratios are acceptable in the domain of C-ITS since redundancy is common over C-ITSs and is supposed to recover messages which are not received.

From this base, we extracted 58 trajectories from 9 vehicles, and tried to see with clustering approaches if we can identify interesting patterns in terms of driving profiles. Currently the density approach such as the method presented in [9] seems to give better results. This approach is scalable and portable using map/reduce.

As a future work, in terms of data analysis, we will keep exploring the density approaches with more evolved algorithms and still take a look at grid and artificial neural network methods to check if they can provide better results. We believe that combination of density approaches with ANN and Grid or Graph can provide significant results. We also need to keep an eye on streaming algorithms since our data set is quite small compared to the enormous amount of data the ITS are able to produce.

We intend to test better the scalability of our system by launching simulations. In the meantime we intend to analyze some other indicators related to the security.

References

1. Sibson, R.: SLINK: an optimally efficient algorithm for the single-link cluster method. Comput. J. **16**, 30–34 (1973)
2. Defays, D.: An efficient algorithm for a complete link method. Comput. J. **20**, 364–366 (1977)
3. Zhang, T., Ramakrishnan, R., Livny, M.: BIRCH: an efficient data clustering method for very large databases. In: Proceedings of the 1996 ACM SIGMOD International Conference on Management of Data, SIGMOD 1996, pp. 103–114. ACM, New York (1996)
4. MacQueen, J.: Some methods for classification and analysis of multivariate observations. In: Proceedings of the Fifth Berkeley Symposium on Mathematical Statistics and Probability, Volume 1: Statistics, pp. 281–297. University of California Press, Berkeley (1967)
5. Kaufman, L., Rousseeuw, P.: Clustering by means of medoids. In: Statistical Data Analysis Based on the L1 Norm and Related Methods, North-Holland, Amsterdam, pp. 405–416 (1987)
6. Ester, M., Kriegel, H.P., Sander, J., Xu, X., et al.: A density-based algorithm for discovering clusters in large spatial databases with noise, vol. 96, pp. 226–231 (1996)
7. Ankerst, M., Breunig, M.M., Kriegel, H.P., Sander, J.: Optics: ordering points to identify the clustering structure. ACM Sigmod Rec. **28**, 49–60 (1999)
8. Kohonen, T.: Self-organized formation of topologically correct feature maps. Biol. Cybern. **43**, 59–69 (1982)
9. Bourdy, E., Piamrat, K., Herbin, M., Fouchal, H.: New method for selecting exemplars application to roadway experimentation. In: Hodoň, M., Eichler, G., Erfurth, C., Fahrnberger, G. (eds.) I4CS 2018. CCIS, vol. 863, pp. 75–84. Springer, Cham (2018). https://doi.org/10.1007/978-3-319-93408-2_6
10. https://www.etsi.org/
11. ETSI EN 302 637-2: Intelligent Transport Systems (ITS); vehicular communications; basic set of applications; part 2: specification of cooperative awareness basic service. Number 302 637-2, ETSI (2014)

12. ETSI EN 302 637-3: Intelligent Transport Systems (ITS); vehicular communications; basic set of application; part 3: specifications of decentralized environmental notification basic service. Number 302 697-3, ETSI (2014)
13. Ramassamy, C., Fouchal, H., Hunel, P.: Classification of usual protocols over wireless sensor networks. In: Proceedings of IEEE International Conference on Communications, ICC 2012, Ottawa, ON, Canada, 10–15 June 2012, pp. 622–626. IEEE (2012)
14. Bennis, I., Fouchal, H., Zytoune, O., Aboutajdine, D.: An evaluation of the TPGF protocol implementation over NS-2. In: IEEE International Conference on Communications, ICC 2014, Sydney, Australia, 10–14 June 2014, pp. 428–433. IEEE (2014)

Towards a Statistical Approach for User Classification in Twitter

Kheir Eddine Daouadi[1(✉)], Rim Zghal Rebaï[2(✉)], and Ikram Amous[2(✉)]

[1] MIRACL-FSEGS, Sfax University, 3000 Sfax, Tunisia
khairi.informatique@gmail.com

[2] MIRACL-ISIMS, Sfax University, Tunis Road Km 10, 3021 Sfax, Tunisia
rim.zghal@gmail.com, ikram.amous@enetcom.usf.tn

Abstract. In this paper, we propose a novel technique for classifying user accounts on online social networks. The main purpose of our classification is to distinguish the patterns of users from those of organizations and individuals. The ability of distinguishing between the two account types is needed for developing recommendation engines, consumer products opinion mining tools, and information dissemination platforms. However, such a task is non-trivial. Classic and consolidated approaches of text mining use textual features from natural language processing for classification. Nevertheless, such approaches still have some drawbacks like the computational cost and time consumption. In this work, we propose a statistical approach based on post frequency, metadata of user profile, and popularity of posts so as to recognize the type of users without textual content. We performed a set of experiments over a twitter dataset and learn-based algorithms in classification task. Several supervised learning algorithms were tested. We achieved high f-measure results of 96.2% using imbalanced datasets and (GBRT), 1.9% were gains when we used imbalanced datasets with Synthetic Minority Oversampling technique and (RF), this yields 98.1%.

Keywords: Content-based approaches · Statistical-based approaches · Hybrid approaches · User classification · Individual vs. Organization

1 Introduction

Nowadays, social networks attract the attention of several parties in many communities, such as researchers, organizations, politicians and even governmental institutions. This gain of interests is due to the increasing number of applications targeted by the aforementioned actors. This growing phenomenon has become an important part of our everyday life; millions of people exchange a tremendous amount of data on such social networks which allow people to register through the creation of a profile account, sharing content and following other members. The emergence of social media has motivated a large number of recent research. Several research axes, such as user type classification [1–3], event detection [4], user recommendation [5] etc.

É. Renault et al. (Eds.): MLN 2018, LNCS 11407, pp. 33–43, 2019.
https://doi.org/10.1007/978-3-030-19945-6_3

Twitter is one of the leading social networks that allow users to post 280 characters, known as tweets. Users can generate three main types of tweets. These are mainly 'tweet', which means an original posting from a user, 're-tweet', which means a reposting of a tweet posted by another user, and 'reply', which means a comment for another tweet. This free micro blogging has been very popular not only by individuals to create, publish and share content, but also by organizations to spread information and engage users. We define an individual account as one having a non-organizational nature and usually managed by an individual whereas the organization account is one managed by a non-profit organization, an institution, an association, a corporation or any group of common interests. In [6], authors showed that the presence of organizations makes up 9.4% of the accounts in twitter. The manual judgment of the account types can be done easily by examining their profile names, their recent tweets, and their profile description. But such kind of intelligent judgment is non-trivial task.

In this paper, we attempt to distinguish between accounts belonging to individuals and organizations in twitter. The ability to classify the patterns from both users' types is needed for developing of many applications such as products opinion mining tools, recommendation engines, information dissemination platforms etc. Classic and consolidated approaches of text mining use textual features from natural language processing (NLP) for classification. Nevertheless, such approaches still have some drawbacks like the computational cost and time consumption. In this context, the main contribution of this work is to demonstrate the importance of using parameters from post frequency and metadata of user profile in the Individual-vs-Organization (IvO) classification task. In addition, we illustrated the benefits from using statistical features so as to enhance the performance the IvO classification users in a shorter time and with a lower computational cost. We build on the hypothesis that the post frequency and popularity of posts are a decisive factor in the process of IvO classification.

The remainder of this paper is organized as follows. In the Second Section, we will discuss the related works and justify the choice of the used approach. In Sect. 3, we will show the line of our proposed method: we will present the different phases of our proposed framework. In Sect. 4, we will discuss the results of our experiment. Finally, we will summarize the main contributions of our proposed method.

2 Related Works

Social networks user types can be analyzed across different dimensions and in different granularities. Classifying account types is considered as a type of latent attribute inferences, in which their main purpose is to infer various properties of online accounts. Much work has been done on a single and specific perspective such as fraud detection [7], age prediction [8], political orientation [9], organization and individual classification [3, 6, 10]. We divide the existing literature of twitter users classification into three main classes depending on the features used in the classification task. We distinguish mainly content-based approaches, statistical-based approaches, and hybrid approaches.

Content-based approaches focus solely on the textual features from (NLP) such as sentiments and emotion expressions, n-grams, informal language use, tweet style etc. In [10], authors proposed an approach to classify tweets according to their author type (individual and organization users). They used a rich set of features that exploit different linguistic aspects of tweet content using a supervised classifier framework. In [9], authors proposed an approach for political orientation detection. They examine users' political ideology using the textual feature only. In [8], authors used a deep convolution neural network with the content of the tweets so as to infer the age groups of users from those of adults and teenagers.

Statistical-based approaches used statistic features in the classification task such as the social network structure, and the post frequency. In [7], authors proposed a set of features from post frequency with a supervised learning framework so as to classify users into three main classes, mainly human, bots, or cyborgs. In [11], authors used solely the time distribution between posts in order to distinguish between three types of users including human, bot and organization.

Apart from content-based and statistical-based approaches, the third commonly used type of approaches are the hybrid approaches which tend to combine textual and statistical features in the classification task. In order to distinguish between individuals, organizations and celebrities, [2] used the content from the description of the user profile and the social connectivity as features in their classification task. In [3], authors proposed an approach which used the content, social, and temporal features so as to classify users as individuals or organizations. In [6], authors used post content features, stylistic features, structural and behavioral features based on how the account interacts with others. In [1], authors classify users who tweet about e-cigarette–related topics from individuals, vaper enthusiasts, informed agency marketers, and spammers. They used metadata of user profile with the derived tweeting behavior features.

Although content-based and hybrid approaches are good techniques and they were heavily used in the literature with respect to their specific aspect, these approaches are usually time-consuming and have an elevated computational cost. Such limitations stand as an obstacle to obtaining a real-time response. Moreover, they needed external resources of (NLP). Furthermore, their results depend on the accuracy of the external resources of natural languages processing (for example, the results of the works that use the sentiment features depend on the accuracy of the tools used in this task). In contrast, statistical approaches take a shorter time and have a lower computational cost than the other approaches. In addition, they do not use any external resources. Therefore, we proposed a statistical-based approach in order to perform an IvO classification of twitter users. Our work is similar to that of [3, 6, 10], but these previous works used content features in the classification task. This latter implies the use of external resources, which in turn still has some drawbacks such as computational cost and time consumption. Our proposed approach may offer an important advantage by removing noise from human accounts which do not have organization attributes, with a lower computational cost, less time consumption and a higher accuracy of classification than the previous works in the literature.

3 Proposed Method

In this section, we will discuss the progress of our proposed framework. As shown in Fig. 1, the proposed framework contains three main phases namely data collection, features extraction and classification. These phases are discussed in the following subsections.

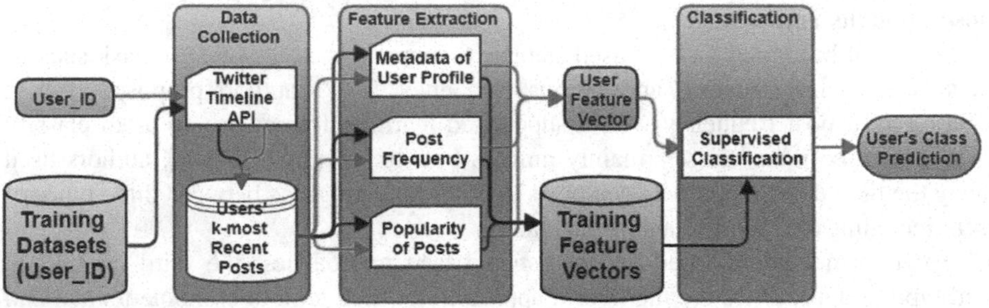

Fig. 1. Proposed framework.

3.1 Data Collection

In order to create our datasets, we went through two steps. First, we used twitter streaming (API) between 12 September 2017 and 30 October 2017, this latter allows to collect tweets in real time with specific keyword. Six names of the most popular products are used as a keywords, namely Pepsi, Coca-Cola, Galaxy Note 8, Iphone 8, Ryzen and I9 which are the last processors designed by Amd and Intel corporations. We achieved more than 1 million tweets provided by 971079 unique users. Based on our definition of the individual and organization users (in Sect. 1). Three coders were trained using names of users and twitter platform to distinguish the user types manually. Each user account was explored in the twitter platform, its behaviors and its sample of recent tweets on their timeline were verified, and then a label was attributed to each user account. Coding disagreements were resolved by an adjudicator coder. We achieved 2222 manual labeled users in which there are 170 organization users and 2052 individual users. Second, we used twitter timeline (API), this later allows to collect the k-most recent post of such user. Previous research have demonstrated that 200 posts are typically sufficient and are enough for predicting Twitter user characteristics [6]. The posts came with a few attributes. They were: User, Timestamp, and Tweet. The User attribute contained metadata of user profile such as the number of followers, number of followings, number of public lists that this user is a member of etc. The Timestamp attribute contained the date and time when the post was submitted by the user in the twitter microblog (for example, “05/04/2017 08:58:19”). Finally, the Tweet attribute contained the tweet metadata attributes such as the favorites count that this post has, the number of retweets that this post has, ‘is there a tweet, a retweet, or a reply?’ etc. the 200 recent posts for each user have been collected. This way, the next step of the method could follow.

3.2 Features Extraction

This phase aims to build out the feature vector space for the 2222 labeled users, extracting the metadata provided by twitter timeline (API) and engineering our proposed features from the 200 posts.

Metadata of User Profile Features. We exploited the different parameters from metadata of the user profile. As shown in Table 1, 10 features were obtained for each user. These types of metadata features have a demonstrated utility in classifying different type of users [1, 2]. We used the User attribute described in (Sect. 3.1) in order to get features from metadata of user profile. The numerical metadata features include number of followers, number of followings, the ratio of followers to followings, the number of lists that the user is a member of, the length of the description of the account, and the number of tweets that the user has liked. The binary metadata features include whether a user has enabled the possibility of geo-tagging their Tweets, whether the user has provided a URL in association with their profile, whether the user has not altered the background or theme of their user profile, and whether the user has a verified account.

Table 1. Metadata of user profile features.

Feature label	Description
Followers	The number of followers that the user has
Followings	The number of followings that the user has
Ratio	The ratio of number of followers to the number of followings
List	The number of public list that the user is a member of
Description	The length of the description of user profile
Liked	The number of posts liked by the users
Geo	If the user has enabled the possibility of geo-tagging their posts
url	If the user has provided an URL in association with their profile
DP	If the user has a default profile
Verified	If th user has a verified account

Statistical Features. In addition to the metadata of user profile features, we proposed parameters from of post frequency features and popularity of posts features. 130 features were calculated using the metadata of posts collected for each user (116 post frequency features, and 14 popularity of posts features. Our proposed post frequency features illustrate the activity of the users, their interaction with their social connectivity and provides an information about their social network usage. We used the both Timestamp and Tweet attributes described in (Sect. 3.1) so as to acquire post frequency features. Table 2 represent our proposed post frequency features. These features include the number of different types of post per day, per day of week, per hours of day and per week. The number of different type of post. The number of posts that contains a mention to another user etc. (see Table 2).

Table 2. Post frequency features.

Feature label	Description
P, T, R, T/R	The number of replies, tweets, and retweets. Ratio of T to R
APW	Average number T, P, R per week
APD	Average number T, P, R, T+P, T+R, and R+P per day
NPH	The number of T, R, P per Hour of day
NPD	The number of T, P, R per Day of week
I MIN	Minimum Interval between T, P, R, and T+P+R
I MAX	Maximum Interval between (posts, tweets, retweets, replies)
AIP	Average Interval between T, P, and R
Quoted	The number of posts in form of citation (quoted tweets)
Mention	The number of posts that contains a mention to another user

In addition to the post frequency features, we proposed 14 popularity of posts features. We used both Tweet and User attributes described in (Sect. 3.1) in order to acquire the popularity of posts features. We focused to the social media metric in order to capture the popularity of posts, two metric of social media are used (retweet and favorite metrics). These features are calculated using Eqs. (1), (2), (3), and (4).

$$fav(Posts) = \frac{\sum_{k=0}^{n} number\ of\ favorites}{number\ of\ following * number\ of\ (Posts)} \tag{1}$$

$$ret(Posts) = \frac{\sum_{k=0}^{n} number\ of\ retweets}{number\ of\ following * number\ of\ (Posts)} \tag{2}$$

$$all\,fav(P) = \left(\sum\nolimits_{k=0}^{n} number\ of\ favorites(\mathrm{P})\right) \tag{3}$$

$$all\ ret(P) = \left(\sum\nolimits_{k=0}^{n} number\ of\ retweets(P)\right) \tag{4}$$

Where: n is the number of posts. P may be (tweets, retweets, and replies). Posts may be (tweets, retweets, replies and tweets + retweets + replies). Number of retweets is the number of retweets to the posts done by the user. Number of favorites is the number of favorites to the posts done by the user.

Since there are several features, one feature selection method was used in order to filter the most relevant features. We used information gain method, which outputs each features from 0 to 1 in importance. The main purposes of using features selection method is to identify the non-relevant features and eliminate them from the classification process, which usually leads to better learning performance, lower computational cost and better model interpretability. We will discuss features importance in (Sect. 5).

3.3 Classification

The third phase of our proposed framework represent the classification task. Using a supervised machine-learning approach, a set of classifiers were used with the proposed features. Then, a comparison of the result of each classifier was performed so as to make a decision as to which classifier performs better with our proposed features. After choosing the classifier that give us the best results of accuracy, a final predictive model were constructed in order to use it in the process of the intelligent judgment of user classification task. The experimental setting and the results are discussed in the following section.

4 Experiment and Evaluation

In order to verify the validity of our proposed method, we carried out several experiments. Five most popular supervised classifiers of machine learning were used, namely Simple Logistic (SL), Logit Boost (LB), Gradient Boosting Regression Tree (GBRT), AdaBoost (AB) and Random Forest (RF). To test the performance measurement of each classifier, a 10-fold cross validation was performed. The tests were implemented using the Waikato Environment for Knowledge Analysis (WEKA), which is the one most commonly used by similar research. To evaluate our hypothesis that the post frequency and the popularity of posts features would vary between the two types of users, we first built three models: one based on metadata of user features, one based on post frequency and popularity of posts features, and the last one based on all features.

In order to evaluate the performance of our proposed method, we used the three performance metrics. In our context, average precision is presented by (Eq. 5), average recall is presented by (Eq. 6), and average F-measure can be defined as (Eq. 7).

$$AvgPrecision = \frac{1}{2}\left(\frac{TP_{org}}{TP_{org} + FP_{org}} + \frac{TP_{indv}}{TP_{indv} + FP_{indv}}\right) \tag{5}$$

$$AvgRecall = \frac{1}{2}\left(\frac{TP_{org}}{TP_{org} + FN_{org}} + \frac{TP_{indv}}{TP_{indv} + FN_{indv}}\right) \tag{6}$$

$$AvgF - measure = \frac{2 \times AvgPrecision \times AvgRecall}{AvgPrecision + AvgRecall} \tag{7}$$

Where: TP_{org} and TP_{indv} are the total number of users that are correctly classified as organizations and individuals respectively. FP_{org} is the number of individuals that are incorrectly classified as organizations and the FP_{indv} is the number of organizations that are incorrectly classified as individuals. The FN_{org} and FN_{indv} are the number of users that the classifier has classified as individuals and organizations respectively.

Table 3 show the average F-measure scores from the three models. These models reach a high average F-measure of 96.2% using all features and (GBRT), 95.8% with our proposed features and (GBRT), and 93.7% using metadata features and (SL).

Table 3. 10-fold cross validation average F-measure % with different features and using each classifier (2222 instances).

Features	SL	LB	GBRT	AB	RF
Metadata(m)	**93.7**	93.5	93.2	93.1	93.4
Proposed features(p)	95.2	95.5	**95.8**	95.6	95.7
(m)+(p)	95.3	95.8	**96.2**	95.6	95.8

Table 4. 10-fold cross validation confusions matrix (C: classified, L: labelled, I: individual, O: organization).

Features	(m)+ +SL		(p) +GBRT		(m)+(p)+GBRT	
	C as I	C as O	C as I	C as O	C as I	C as O
L as I	2018	**34**	2043	**10**	2040	**12**
L as O	**94**	76	**75**	96	**65**	105

As shown in Table 4, when we used the metadata features 34 individual users were classified as organizations and most of the organizations were classified as individuals. Using GBRT and the statistical features, 10 individual users were classified as organizations and the 74 organizations were classified as individuals. But when we used the combination of both metadata and post frequency features, only 65 organizations were classified as individuals. Tables 2 and 3 shows the impact of the proposed features, and their contribution in the enhancement of the overall performance result of the final prediction model.

Given the skewed distribution of the user account types, we used Synthetic Minority Oversampling Technique (SMOTE) [12] in order to balance the datasets. SMOTE algorithm allows to generate a new sample based on feature vector of the minority (in our case organization class), and is a powerful method that has seen successfully in many domains [13]. Significant performance gains were observed on balancing datasets. As shown in Table 5, when we used our proposed method with SMOTENN technique the average f-measure reach 98.1% using (RF).

Table 5. Classification performance of both imbalanced and balanced datasets. (Imbalanced: imbalanced datasets. Balanced: imbalanced datasets +SMOTEENN. P: AvgPrecision. R: AvgRecall. F: average F-Measure).

	Imbalanced			Balanced		
	P	R	F	P	R	F
AB	95.5	95.8	95.6	95.4	95.3	95.3
LB	95.7	96.0	95.8	95.7	95.7	95.7
RF	95.7	96.0	95.8	**98.2**	**98.1**	**98.1**
GBRT	**96.1**	**96.3**	**96.2**	98.0	98.0	98.0
SL	95.2	95.5	95.3	96.3	96.2	96.2

Table 6. Feature importance ranking using information gain method.

Rank	Score importance	Features
1	0.67711	Number of tweets posted in the 5th day of week
2	0.65695	Number of tweets posted in the 1st day of week
3	0.62995	Number of tweets posted at the 14th hour of day
4	0.61905	Number of tweets posted at the 15th hour of day
5	0.61756	Number of tweets posted in the 2nd day of week
6	0.6155	Number of tweets posted in the 7th day of week
7	0.61363	Number of tweets posted at the 17th hour of day
8	0.61224	Number of tweets posted at the 16th hour of day
9	0.60796	Using ret(tweet + retweet + reply) see (Eq. 2)
10	0.59518	Number of tweets posted at the 18th hour of day
11	0.57965	Number of tweets posted in the 6th day of week
12	0.56607	Number of retweets reposted in the 1th day of week
13	0.56175	Number of tweets posted in the 3rd day of week
14	0.55747	Minimum interval between all posts
15	0.55158	Number of tweets posted in the 4th day of week
16	0.54601	Using ret(retweets) see (Eq. 2)
17	0.54188	Number of tweets posted at the 13th hour of day
18	0.5337	Number of retweets reposted in the 7th day of week
19	0.52795	Number of tweets posted at the 12th hour of day
20	0.52237	Number of tweets posted at the 19th hour of day

5 Features Importance Analysis

In this section, we will discuss features importance. As described (in Sect. 3.2), we used information gain method, this outputs each features in range between 0 and 1 in importance. Table 6 shows the top 20 important features. It obvious that the posts frequency features and popularity of post was the most important features, they occupy all positions from 1 to 20. Metadata of user profile features were not included in the twentieth most important features according to information gain, meaning that those are not relevant enough to describe the both classes. Only two features that were ranked as 0 in their importance, these features are the number of replies that are posted at the 1st hour of day, and average interval between replies. These features were ignored for the classification process.

6 Comparison with Prior Work

Our proposed framework differ from the previous one in term of the features used in the classification task, previous work used the textual content of post as features, whereas some drawbacks such as time consumption, computational cost…etc. However, previous framework work well only with the user who posts only textual content, and fails with the user who posts multimedia posts (video, image…etc.). But our framework work well with all users because we not used the content of posts as features. Also we used the technique based on the usage of synthetic minority oversampling [12] to enhance existing datasets by generating additional labeled examples. This allow us to gain an average F-measure of 98.1 using (RF) in the account-level organization detection task.

7 Conclusion

In this paper, we presented an alternative approach for user classification in social networks using a set of rich statistical features, unlike the traditional approaches which used textual content to make a user classification task. This type of approach highlights a new field in the classification task of individual and organization users. Our proposed method differs from the prior works in terms of used features in the classification task. Previous work used features from (NLP). Nevertheless, such approaches still have some drawbacks like the computational cost, time consumption, and the need of external resources. Our work outperforms the previous one in three dimensions: first, the requirement of real time response of this type of application, previous work used features from (NLP) which used external resources to extract textual features, which in turn has some drawbacks such as computational cost and time consumption while our proposed method used statistical features. Second, in previous work, feature extraction needs multilingual and multidialectal resources, but in our proposed approach feature extraction is performed without any external resources and classifies users without taking into account the user's language. Third, the user can post textual data, images, or videos, previous work fails with the user who posts multimedia posts while the proposed method does well with the user who uses a multimedia content since it uses only the metadata of the user profile and post frequency features. Although we have chosen only the twitter platform but our proposed framework is generic and can be used for any social network such as Facebook and Instagram. Our predictive model achieved a high accuracy for both types of users using (RF) classifier. Examining the proposed post frequency features and popularity of posts was critical in improving the overall performance of IvO classification task. As future work, testing the other datasets should be performed. We also plan to make our system open source and implement a web service (ex an API) to allow the research community to perform user level organization detection using it.

References

1. Kim, A., Miano, T., Chew, R., Eggers, M., Nonnemaker, J.: Classification of Twitter users who tweet about E-cigarettes. JMIR Public Health Surveill. **3**(3), e63 (2017)
2. Nagpal, C., Singhal, K.: Twitter user classification using ambient metadata, arXiv preprint arXiv:1407.8499 (2014)

3. Oentaryo, R.J., Low, J.-W., Lim, E.-P.: Chalk and cheese in Twitter: discriminating personal and organization accounts. In: Hanbury, A., Kazai, G., Rauber, A., Fuhr, N. (eds.) ECIR 2015. LNCS, vol. 9022, pp. 465–476. Springer, Cham (2015). https://doi.org/10.1007/978-3-319-16354-3_51
4. Troudi, A., Zayani, C.A., Jamoussi, S., Amous, I.: A new social media mashup approach. In: Madureira, A.M., Abraham, A., Gamboa, D., Novais, P. (eds.) ISDA 2016. AISC, vol. 557, pp. 677–686. Springer, Cham (2017). https://doi.org/10.1007/978-3-319-53480-0_67
5. Kalaï, A., Wafa, A., Zayani, C.A., Amous, I.: LoTrust: A social Trust Level model based on time-aware social interactions and interests similarity. In: 14th Annual Conference on Privacy, Security and Trust (PST), pp. 428–436. IEEE, New Zealand (2016)
6. McCorriston, J., Jurgens, D., Ruths, D.: Organizations are users too: characterizing and detecting the presence of organizations on Twitter. In: 9th International Conference on Web and Social Media ICWSM, pp. 650–653. The AAAI Press, UK (2015)
7. Tavares, G.M., Mastelini, S.M., Barbon Jr., S.: User classification on online social networks by post frequency. In: CEP, vol. 86057, pp. 970–977 (2017)
8. Guimaraes, R.G., Rosa, R.L., De Gaetano, D., Rodriguez, D.Z., Bressan, G.: Age groups classification in social network using deep learning. IEEE Access **5**, 1–11 (2017)
9. Preoţiuc-Pietro, D., Liu, Y., Hopkins, D., Ungar, L.: Beyond binary labels: political ideology prediction of twitter users. In: Proceedings of the 55th Annual Meeting of the Association for Computational Linguistics, Canada, pp. 729–740 (2017)
10. De Silva, L., Riloff, E.: User type classification of Tweets with implications for event recognition. In: ACL, pp. 98–108 (2014)
11. Tavares, G., Faisal, A.: Scaling-laws of human broadcast communication enable distinction between human, corporate and robot Twitter users. PLoS One **8**(7), e65774 (2013)
12. Chawla, N.V., Bowyer, K.W., Hall, L.O., Kegelmeyer, W.P.: SMOTE: synthetic minority over-sampling technique. J. Artif. Intell. Res. **16**, 321–357 (2002)
13. He, H., Garcia, E.A.: Learning from imbalanced data. IEEE Trans. Knowl. Data Eng. **21**(9), 1263–1284 (2009)

RILNET: A Reinforcement Learning Based Load Balancing Approach for Datacenter Networks

Qinliang Lin(✉), Zhibo Gong, Qiaoling Wang, and Jinlong Li

Network Technology Laboratory, 2012 Labs, Huawei Technologies Co., Ltd., Shenzhen, China
{linqinliang,gongzhibo,wangqiaoling,lijinlong767}@huawei.com

Abstract. Modern datacenter networks are facing various challenges, e.g., highly dynamic workloads, congestion, topology asymmetry. ECMP, as a traditional load balancing mechanism which is widely used in today's datacenters, can balance load poorly and lead to congestion. Variety of load balancing schemes are proposed to address the problems of ECMP. However, these traditional schemes usually make load balancing decision only based on network knowledge for a snapshot or a short time past. In this paper, we propose a Reinforcement Learning (RL) based approach, called RILNET (ReInforcement Learning NETworking), aiming at load balancing for datacenter networks. RILNET employs RL to learn a network and control it based on the learned experience. To achieve a higher granularity of control, RILNET is constructed to route flowlet rather than flows. Moreover, RILNET makes routing decisions for aggregation flows (an aggregation flow is a flow set that includes all flows flowing from the same source edge switch to the same destination edge switch) instead of a single flow. In order to test performance of RILNET, we propose a flow-level simulation and a packet-level simulation, and the both results show that RILNET can balance traffic load much more effectively than ECMP and another load balancing solution, i.e., DRILL. Compared with DRILL, RILNET outperforms DRILL in data loss and maximal link delay. Specifically, the maximal link data loss and the maximal link delay of RILNET are 44.4% and 25.4% smaller than DRILL, respectively.

Keywords: Load balancing · Datacenter network · Flowlet · Reinforcement learning · Machine learning

1 Introduction

Modern datacenter networks are worldwide deployed across various applications, e.g., databases, web services, data analytics, etc. Unlike traditional networks, datacenter networks usually leverage Clos topologies (e.g., fat-tree [1], BCube [9], etc.) to achieve large bisection bandwidth and high network performance. Clos topologies provide multiple paths between host pairs, and therefore this can

É. Renault et al. (Eds.): MLN 2018, LNCS 11407, pp. 44–55, 2019.
https://doi.org/10.1007/978-3-030-19945-6_4

enable load balancing in datacenter networks for high performance. The standard load balancing routing strategy in today's datacenter networks is called Equal Cost Multi-Path (ECMP), in which traffic flows are routed based on hashing flow-related fields in the packet header. However, it is well known that ECMP can balance traffic loads poorly [3,20], and the reasons are because

(1) ECMP can do nothing about elephant flows because they may be routed in same paths due to hash collisions;
(2) ECMP performs poorly in asymmetric topologies.

To address ECMP's problems, variety of load balancing solutions have been proposed (see Sect. 4 for more details). Solutions in Hedera [2] and Planck [17] try to reroute elephant flows when congestions occur. However, these elephant flows oriented approaches are coarse-grained for load balancing, because elephant flows are usually too large in size [4], and therefore it is hard to achieve load balancing just by rerouting them. Local decision oriented solutions (e.g., MPTCP [16], FLARE [12], Presto [11], etc.) make routing decisions only based on local information rather than global network knowledge, which can lead to congestion due to highly complicated and dynamic traffic in datacenter networks. In CONGA [3], the solution cannot support topologies with more than two layers while three layers are common used in modern datacenters. Moreover, specialized networking hardware is required for all switches, which makes it complicated and expensive for deploying.

Broadly speaking, in existing load balancing solutions,

(1) routing decisions are usually made only based on a snapshot of network knowledge or historical traffic variation for a short time past, without considering the variation pattern of network congestion;
(2) local-decision based solutions can only provide suboptimal routing decisions due to the lack of global congestion information, and moreover, they cannot perform well with asymmetric topology which is common in datacenters due to link failures [7,10];
(3) flow-oriented solutions are coarse-grained, because elegant flows are small in number, but yield the majority of traffic data.

In this paper, we propose a novel load balancing approach, called RILNET (ReInforcement Learning Networking). Our contributions in RILNET are as follows:

(1) Unlike traditional solutions, RILNET leverages reinforcement learning (RL) to learn the knowledge and experience for a network, and therefore, is able to learn how to control the network optimally[1]. Moreover, we propose a center-controlled architecture in RILNET to sense, learn and control a network.

[1] Note that RILNET can be used for multiple purposes, including load balancing, reducing data loss, reducing flow completion time, etc. In this paper, we focus on load balancing and leave the other purposes in our future work.

(2) In order to achieve a higher granularity of control, RILNET is constructed to routing flowlet rather than TCP flows.
(3) RILNET makes routing decisions for aggregation flows instead of a single flow. Note that an aggregation flow is defined as a set for all flows flowing from the same source router/switch to the same destination router/switch. This is because there are tens of thousands of flows in a data center [4], and it could be a heavy burden for a controller to control every single flow in real time.

2 Design of RILNET

RILNET leverages a center-controlled architecture (e.g., software-defined networking) to manage a network, and uses RL to learn the hidden congestion pattern based on the knowledge collected by the controller. In this paper, the reinforcement learning algorithm we use in RILNET is Deep DPG (DDPG) [14], which is a model-free, off-policy actor-critic algorithm and uses deep function approximators that can learn policies in continuous action spaces.

2.1 Methodology

Reinforcement learning, which is an area of machine learning, is a framework for decision-making in which an agent aims to take actions in an environment so as to maximize cumulative reward. The typical framing of a reinforcement learning consists of

(1) an environment, which is typically formulated as a Markov decision process;
(2) an agent, which takes actions in the environment;
(3) an interpreter, which reads states from the environment and computes rewards for the agent.

Actor-critic methods [8], which are a category of reinforcement learning, consist of an actor for outputting actions and a critic for value function approximation. DDPG [14] is an actor-critic approach, which is a deterministic policy gradient based algorithm and can learn policies in high-dimensional, continuous action spaces. In RILNET, we employ DDPG to be the brain of a network.

DDPG consists of four deep neural networks: (1) an actor network and a target actor network, denoted as μ and μ', respectively, and (2) a critic network and a target critic network, denoted as Q and Q', respectively. The critic network, parameterized as θ^Q, can output a Q-Value denoted as $Q(s_t, a_t)$ where s_t and a_t is a state and an action at time t,respectively, and it can be updated by standard temporal difference update, i.e.,

$$Q^{new}(s_t, a_t|\theta^Q) \leftarrow Q(s_t, a_t|\theta^Q) + \alpha(r + \gamma Q(s'_{t+1}, \mu(s_{t+1}|\theta^\mu)|\theta^Q)), \tag{1}$$

where α is known as learning rate, r is known as the reward value given state s_t and action a_t, and γ is known as discount factor, while $\mu(s_{t+1}|\theta^\mu)$ denotes the output of the actor network. The loss function for the critic network can defined as follows:

$$L_{\theta^Q}(s_t, a_t) = \left(Q(s_t, a_t|\theta^Q) - (r + \gamma Q'(s_{t+1}, \mu'(s_{t+1}|\theta^{\mu'})|\theta^{Q'}))\right)^2. \quad (2)$$

Then, the critic network can be updated by gradient descent via this loss function.

The actor network updates its weights for lager Q-values. Therefore, we can update the actor network via the gradients which are provided by the critic network and point at the direction to larger Q-values. In this context, for state s_t, the gradient for updating the actor policy is as follows:

$$\nabla_{\theta^\mu} \mu(s_t) = \nabla_a Q(s_t, a|\theta^Q) \nabla_{\theta^\mu} \mu(s_t|\theta^\mu) \quad (3)$$

We can see in Eqs. (2) and (3) that updating the critic's policy is also affecting the actor's update, and this will lead to divergence for both networks. This is the reason why two target networks are proposed in order to make the updating stable and convergent. DDPG updates the target networks by slowly tracking the learned networks, which can change the target network's weights slowly and improve the stability of learning. More specifically, the target networks are updated as following (see [14] for more details):

$$\theta^{u'} \leftarrow \tau\theta^u + (1-\tau)\theta^{u'}, \quad (4)$$

$$\theta^{Q'} \leftarrow \tau\theta^Q + (1-\tau)\theta^{Q'}. \quad (5)$$

2.2 State Space and Action Space

2.2.1 Input

We define a network state as a snapshot of all throughput between all pairs of edge switches, i.e., $s_t = \{T^t_{e_1 e_2} : e_1, e_2 \in \{edge\ nodes\}\}$, where $T^t_{e_1 e_2}$ is denoted as the throughput from edge node A to edge node B at time t. Here, we define an aggregation flow $e_1 e_2$ as a flow set which includes all flows flowing from edge node e_1 to node e_2.

When we input historical network states into a RL agent, the agent is able to automatically learn how to predict implicitly by maximizing the cumulative reward. In RILNET, its input is a time array of throughput for all aggregation flows. To make this clear, we take the topology shown in Fig. 1 as an instance. The input of RILNET for this topology is the past K aggregation flow, i.e.,

$$s_t = \{T^{t-K+1}_{AB}, T^{t-K+2}_{AB}, \cdots, T^t_{AB}\}. \quad (6)$$

2.2.2 Output

The output of RILNET is the routing decisions for all aggregation flows. Note that, RILNET does not control an individual flow's routing, and the reason is because there are a large number of flows in a datacenter network (tens of thousands [4]), and control the routing for an individual flow can lead to huge

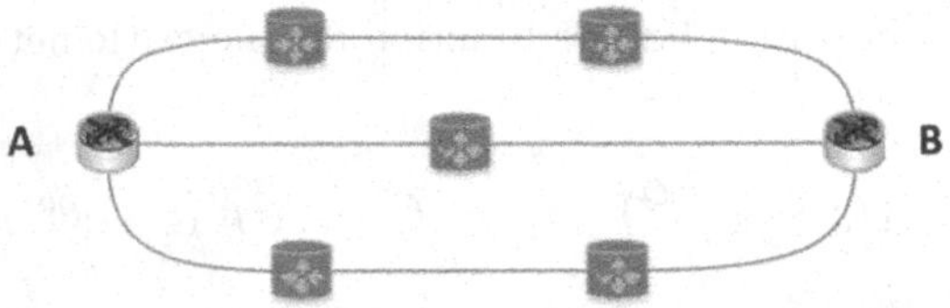

Fig. 1. A topology with three paths between edge node A and B.

computational cost and slow convergence or even divergence for RL. In RILNET, we define the output as the routing proportions for all pairs of edge nodes. In this context, as an instance topology shown in Fig. 1, the output is a routing proportion (denoted as P_{AB}) for node A and B. For P_{AB} (suppose $P_{AB} = P_1\% : P_2\% : P_3\%$, where $P_1\% + P_2\% + P_3\% = 1$), it means that, for all the flowlets flowing from node A to B, $P_1\%$ of them are routed through path 1, P_2 of them are routed through path 2%, and the others are routed through path 3. Note that we assume all paths are precomputed in RILNET.

2.3 Congestion-Aware Reward Design

The aim of RILNET is load balancing for a network. To achieve this, we design a congestion-aware reward function for this aim.

For the topology shown in Fig. 1 as an instance, the optimal load balancing solution for this network is that the utilizations of three paths are the same, i.e., $u(P_i) = u(P_j)$ $(i, j \in [1, 3])$. Note that we define a path P's utilization (denoted as $u(P)$) as the maximal link utilization of the links belonging to the path P, i.e.,

$$u(P) = \max(u(l)), \quad l \in P, \tag{7}$$

where l denotes a link and $u(l)$ denotes link l's utilization. In this context, we design a reward function as a variance vector for RILNET, i.e.,

$$\overrightarrow{R_t} = [r_{ab} : a, b \in E], \tag{8}$$

where E denotes a set containing all edge nodes in the network, and r_{ab} denotes the reward between edge node a and b, and,

$$r_{ab} = 1 - \frac{\sum_{P \in P_{ab}} |u(P) - \overline{u}|}{|P_{ab}|}, \tag{9}$$

where P_{ab} denotes a path set containing all paths between a and b, and $\overline{u}$ denotes the mean value of utilization of all paths between a and b.

2.4 Reinforcement Learning Architecture

As mentioned above, RILNET uses DDPG to model the network, which consists of two actor networks and two critic networks. In all networks, the time series of states are preprocessed by a convolutional neural network (CNN) layer and a pooling layer, which can learn the feature of the variation of traffic loads. In the CNN layer, we set filter size as 128 and kernel size as 5. We set the pooling layer with pooling size equal to 4. After states are preprocessed, the networks employ three fully connected layers consisting of 512-512-512 units. We use an ADAM optimizer [13] and both actor's and critic's learning rate are set to 10^{-4}. Target networks are updated by slowly tracking the learned actor network and critic network respectively, using a learning rate $\tau = 10^{-4}$.

2.5 Architecture of RILNET

RILNET uses a center-controlled architecture for collecting traffic data, configuration and routing control. As shown in Fig. 2, the main components of RILNET include:

(1) Edge switches, which upload states to the controllers, receive routing proportions and route flowlets based on the received proportions.
(2) Data collector, which receives states sent by edge switches, and formats and analyzes the data.
(3) AI controller, which updates the parameters for RL after receiving states, and computes the routing decision based on the updated RL.

Note that the AI controllers and the data collector are the brain of a network whose jobs are to sense, learn and control a network.

Specifically, for every time step Δt, the brain collects network states from all edge nodes. Based on the collected states, the brain updates RILNET, and then computes new routing proportions and sends these policies to all edge nodes.

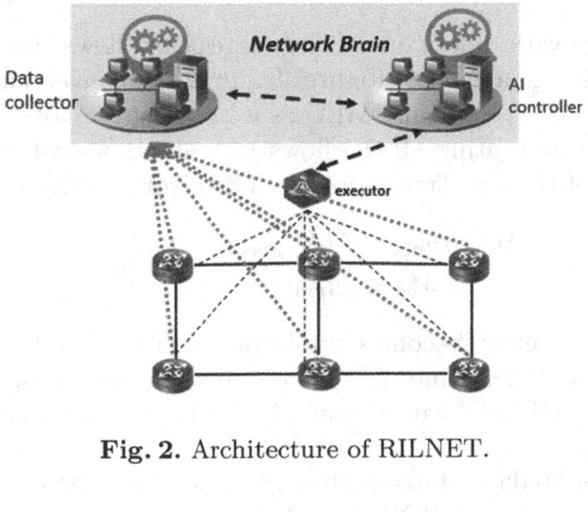

Fig. 2. Architecture of RILNET.

3 Evaluation

In order to test the performance of our solution, we propose a flow-level simulation and a packet-level simulation to test the performance of RILNET comparing with ECMP and DRILL [6]. DRILL is a load balancing solution for datacenter networks. When a packet arrives, DRILL randomly choose a port among $d+m$ ports, where m is the number of ports which have the smallest queue length in last time slot, and d is the number of ports which are chosen randomly. In this simulation, we set $d = 0$ and $m = 1$. The control interval Δt is 100 ms. The topology we employ in this evaluation is fat-tree [1] with $k = 4$, where k is the number of a switch's ports.

The packet-level simulation in this paper is based on OMNeT++ [19]. However, OMNeT++ is very slow in speed (about tens of minutes for simulating 1 s). Since RILNET usually needs hundreds of thousands of time steps to converge, it will spend unacceptable convergence time in OMNeT++. Therefore, we propose a flow-level based on Python to accelerate the training of RILNET.

In this section, we employ MLU (maximal link utilization) to show how load-balance a network is. The smaller the MLU is, the more load-balance a network is, and vice versa.

3.1 Flow-Level

We generate flow traffic based on real datasets which are collected from WIDE ISP networks by the MAWI (Measurement and Analysis on the WIDE Internet) working group [18]. We generate flows for 56 edge switch pairs by using 56 days' dataset for training RILNET. Then we use another 56 days' dataset to test the performance of RILNET. All links' capacities are set as 100 Gbps and K is set as 8. Note that in this simulation, we assume flows can be split under continuous proportions.

Figure 3a shows variation of one of the aggregation flows, and we can see that the traffic loads change abruptly. Figure 3b shows the convergence of RILNET's training. Here, we use the optimal MLU as a benchmark. Note that the optimal MLU is computed by routing all the flows into all equal-cost paths evenly due to the symmetry of the Fat-Tree topology. The error is computed by

$$\frac{MLU_{rilnet} - MLU_{optimal}}{MLU_{optimal}} (100\%).$$

We can see that the error becomes small and stable since 220,000 time steps. Moreover, the error is less than 3% since 500,000 time steps. In other words, after convergence, RILNET can output MLU which is at most 3% larger than the optimal MLU.

We use another 56 days' data (each aggregation flow lasts 15 min) to test the performance of the trained RILNET, and the result is shown in Fig. 3c. We can see that all the errors are below 0.92% for the test dataset.

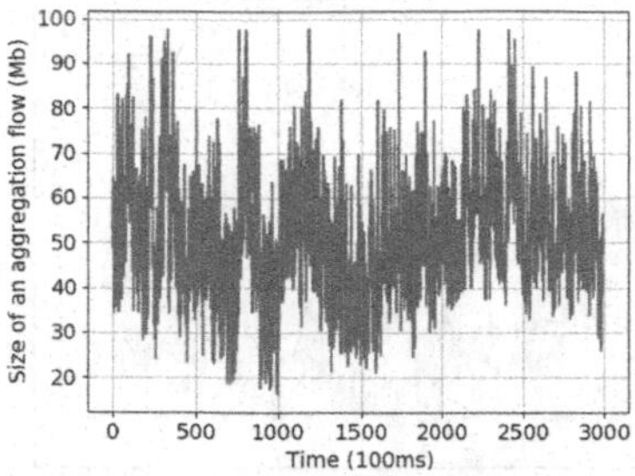

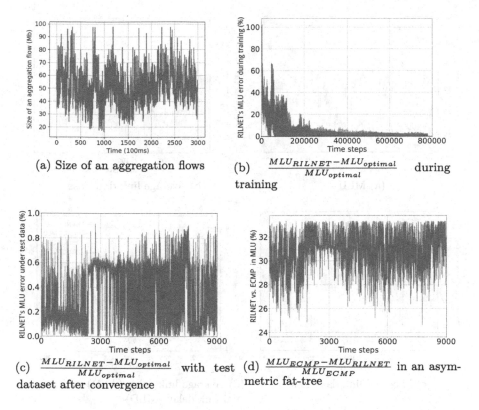

(a) Size of an aggregation flows

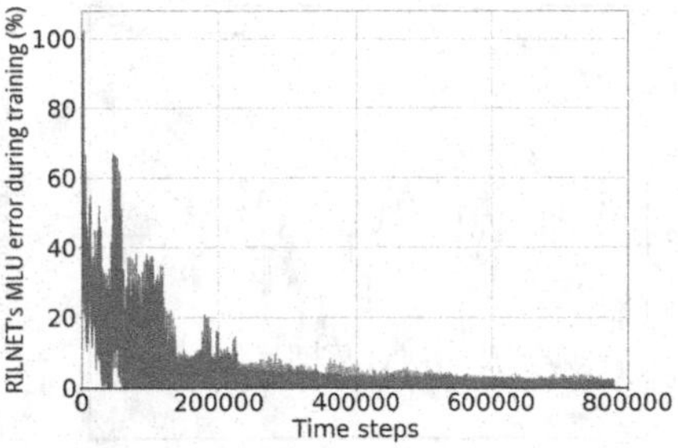

(b) $\frac{MLU_{RILNET}-MLU_{optimal}}{MLU_{optimal}}$ during training

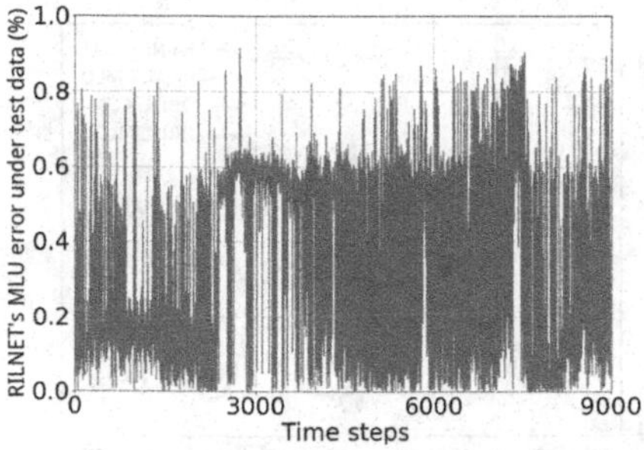

(c) $\frac{MLU_{RILNET}-MLU_{optimal}}{MLU_{optimal}}$ with test dataset after convergence

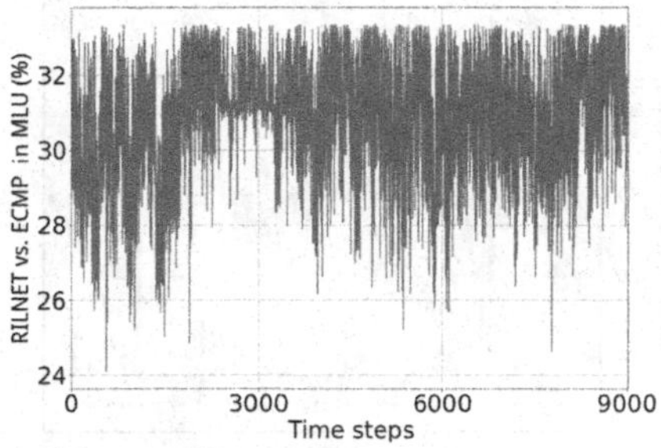

(d) $\frac{MLU_{ECMP}-MLU_{RILNET}}{MLU_{ECMP}}$ in an asymmetric fat-tree

Fig. 3. Results of the flow-level simulations.

Figure 3d shows how much smaller the MLU of RILNET is compared with ECMP in an asymmetric fat-tree (we halve the capacity of 4 links chosen from the network randomly), i.e.,

$$\frac{MLU_{ecmp} - MLU_{rilnet}}{MLU_{ecmp}} \ (100\%).$$

We can see that RILNET's MLU is smaller than ECMP, and the number is 30.9% on average. In a word, RILNET can balance the network efficiently with a close performance comparing with the optimal solution and outperforms ECMP.

3.2 Packet-Level

We use OMNeT++ [19] to build a packet level simulation to test the performance of RILNET compared with DRILL. Note that, in RILNET, data packets can be routed based on flowlets with low probability of out-of-order [3], while DRILL routes individual packet directly which can therefore lead to packet reordering.

Note that the RL model used in OMNeT++ has been already trained in the flow-level simulation. We generate aggregation flows by using eight WIDE

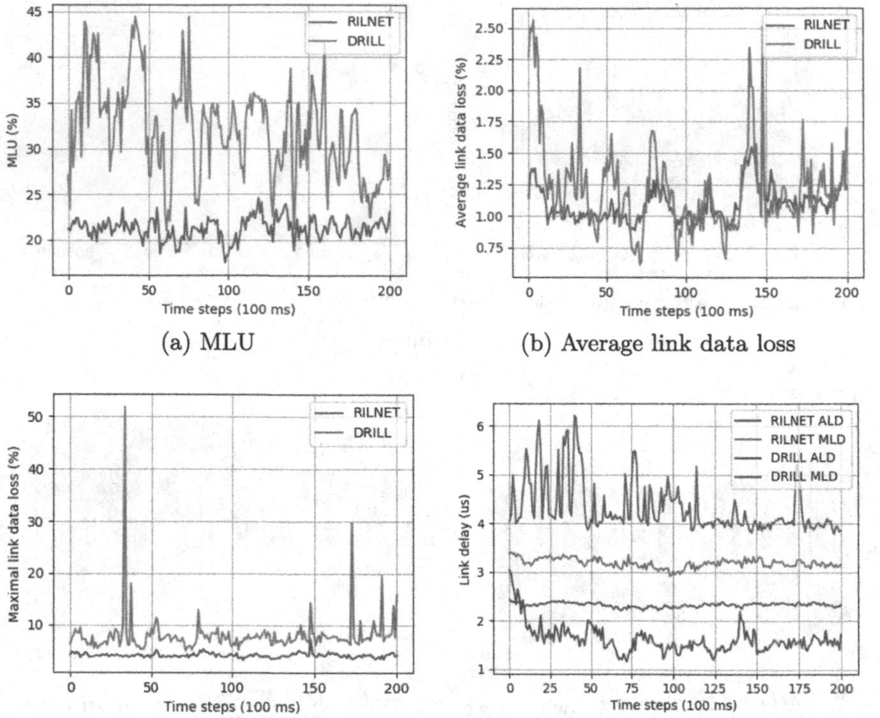

(a) MLU

(b) Average link data loss

(c) Maximal link data loss

(d) Average link delay (ALD) and maximal link delay (MLD)

Fig. 4. Results of the packet-level simulations.

datasets. More specifically, we generate eight aggregation flow for eight edge switches, respectively, and the destination of each aggregation flow is randomly chosen from the other seven edge switches. The simulation results are shown in Fig. 4.

Figure 4a shows that RILNET have much smaller MLU than DRILL. RILNET' MLU is 31.9% smaller than DRILL on average. Therefore, we can see that RILNET can load the network more balance than DRILL.

Figure 4b shows the results of the average data loss (ADL) in all links. We can see that RILNET has smaller ADL than DRILL in most of time steps. On average, RILNET's ADL is 5.4% smaller than DRILL.

Figure 4c shows results of the maximal data loss (MDL) in all links. We can see that RILNET has a smaller and smoother MDL than DRILL, and it is 44.4% smaller than DRILL on average.

Figure 4d shows the results of the average link delay (ALD) and the maximal link delay (MLD) among all links. We can see that DRILL has a smaller ALD than RILNET because it chooses ports with the smallest queue length for routing. However, RILNET has a much smaller MLD, which is 25.4% smaller than DRILL on average. In a word, compared with DRILL, RILNET can improve the performance of the links with worst performance.

4 Related Work

In this section, we just briefly discuss some representative related work. Hedera [2] focuses on rerouting elephant flows rather than all flows. In Planck [17], a novel network measurement architecture is presented and based on it, a traffic engineering application that can reroute congested flows in milliseconds is proposed. MPTCP [16] is a transport protocol which is deployed on hosts, and it can routes several subflows over multiple paths concurrently. However, MPTCP is a host-oriented solution, which relies on local information rather than global network knowledge, and requires host stack changing when deploying. An edge-based solution is proposed in Hermes [20], which reroute flows with low sending rate for link failures or congestions by detecting path conditions.

In FLARE [12], a traffic splitting algorithm is proposed with a new routing unit, called flowlet, which is fine-grained and can avoid causing packet reordering. Alizadeh *et al.* [3] propose a distributed load balancing scheme, i.e., CONGA, which employs congestion-aware flowlet switching to load balance the network. The load balancing scheme proposed in Presto [11] uses edge vSwitches to split flows into flowcells, and distributes them evenly to multi paths. However, routing flowcells is based on local decision and suffers in asymmetry topology.

The solution proposed in [5] employs Q-learning to maximize throughput and reduce latency for the UbuntuNet network. Lin *et al.* [15] introduce a distributed hierarchical control plane architecture and propose a QoS-aware adaptive routing based on Q-learning. However, Q-learning

(1) can only support discrete action with small dimensions;
(2) is slow in convergence.

Therefore, Q-learning is not capable for large scale network such as datacenter networks.

5 Conclusion and Future Work

In this paper, we propose a load balancing approach based on reinforcement learning, called RILNET. RILNET leverages DDPG to learn a network and control it. To test the performance of RILNET, we propose a flow-level simulation and a packet-level simulation, and compare it with an optimal solution, ECMP and DRILL. The results show that RILNET can balance a network's load with a close performance with the optimal solution, and more effectively than ECMP and DRILL. Moreover, RILNET has a smaller data loss and a comparative data delay compared with DRILL.

As aforementioned, RILNET can be used for multiple purposes, including but not limited to load balancing, reducing data loss/delay and reducing flow completion time. In our future work, we will try to include more information (e.g., data loss, delay, flow completion time, etc.) into the input state for RILNET, so that RILNET can learn the network better and control it more effectively.

References

1. Al-Fares, M., Loukissas, A., Vahdat, A.: A scalable, commodity data center network architecture. In: ACM SIGCOMM Computer Communication Review, vol. 38, pp. 63–74. ACM (2008)
2. Al-Fares, M., Radhakrishnan, S., Raghavan, B., Huang, N., Vahdat, A.: Hedera: dynamic flow scheduling for data center networks. In: NSDI, vol. 10, p. 19 (2010)
3. Alizadeh, M., et al.: CONGA: distributed congestion-aware load balancing for datacenters. In: ACM SIGCOMM Computer Communication Review, vol. 44, pp. 503–514. ACM (2014)
4. Benson, T., Akella, A., Maltz, D.A.: Network traffic characteristics of data centers in the wild. In: Proceedings of the 10th ACM SIGCOMM Conference on Internet Measurement, pp. 267–280. ACM (2010)
5. Chavula, J., Densmore, M., Suleman, H.: Using SDN and reinforcement learning for traffic engineering in UbuntuNet alliance. In: 2016 International Conference on Advances in Computing and Communication Engineering (ICACCE), pp. 349–355. IEEE (2016)
6. Ghorbani, S., Godfrey, B., Ganjali, Y., Firoozshahian, A.: micro load balancing in data centers with drill. In: Proceedings of the 14th ACM Workshop on Hot Topics in Networks, p. 17. ACM (2015)
7. Gill, P., Jain, N., Nagappan, N.: Understanding network failures in data centers: measurement, analysis, and implications. In: ACM SIGCOMM Computer Communication Review, vol. 41, pp. 350–361. ACM (2011)
8. Grondman, I., Busoniu, L., Lopes, G.A., Babuska, R.: A survey of actor-critic reinforcement learning: standard and natural policy gradients. IEEE Trans. Syst. Man Cybern. Part C (Appl. Rev.) **42**(6), 1291–1307 (2012)
9. Guo, C., et al.: BCube: a high performance, server-centric network architecture for modular data centers. ACM SIGCOMM Comput. Commun. Rev. **39**(4), 63–74 (2009)
10. Guo, C., et al.: Pingmesh: a large-scale system for data center network latency measurement and analysis. ACM SIGCOMM Comput. Commun. Rev. **45**(4), 139–152 (2015)
11. He, K., Rozner, E., Agarwal, K., Felter, W., Carter, J., Akella, A.: Presto: edge-based load balancing for fast datacenter networks. ACM SIGCOMM Comput. Commun. Rev. **45**(4), 465–478 (2015)
12. Kandula, S., Katabi, D., Sinha, S., Berger, A.: Dynamic load balancing without packet reordering. ACM SIGCOMM Comput. Commun. Rev. **37**(2), 51–62 (2007)
13. Kingma, D.P., Ba, J.: Adam: a method for stochastic optimization. arXiv preprint arXiv:1412.6980 (2014)
14. Lillicrap, T.P., et al.: Continuous control with deep reinforcement learning. arXiv preprint arXiv:1509.02971 (2015)
15. Lin, S.C., Akyildiz, I.F., Wang, P., Luo, M.: QoS-aware adaptive routing in multi-layer hierarchical software defined networks: a reinforcement learning approach. In: 2016 IEEE International Conference on Services Computing (SCC), pp. 25–33. IEEE (2016)
16. Popa, L., Kumar, G., Chowdhury, M., Krishnamurthy, A., Ratnasamy, S., Stoica, I.: FairCloud: sharing the network in cloud computing. In: Proceedings of the ACM SIGCOMM 2012 Conference on Applications, Technologies, Architectures, and Protocols for Computer Communication, pp. 187–198. ACM (2012)

17. Rasley, J., et al.: Planck: millisecond-scale monitoring and control for commodity networks. In: ACM SIGCOMM Computer Communication Review, vol. 44, pp. 407–418. ACM (2014)
18. The MAWI Working Group: MAWI working group traffic archive. http://mawi.wide.ad.jp/mawi/. Accessed 21 June 2018
19. Varga, A.: OMNeT++ user manual version 4.6. OpenSim Ltd (2014)
20. Zhang, H., Zhang, J., Bai, W., Chen, K., Chowdhury, M.: Resilient datacenter load balancing in the wild. In: Proceedings of the Conference of the ACM Special Interest Group on Data Communication, pp. 253–266. ACM (2017)

Building a Wide-Area File Transfer Performance Predictor: An Empirical Study

Zhengchun Liu[1,2](✉), Rajkumar Kettimuthu[1,2], Prasanna Balaprakash[1], Nageswara S. V. Rao[3], and Ian Foster[1,2]

[1] Argonne National Laboratory, 9700 Cass Avenue, Lemont, IL 60439, USA
{zhengchun.liu,kettimut,pbalapra,foster}@anl.gov
[2] University of Chicago, Chicago, IL 60637, USA
[3] Oak Ridge National Laboratory, Oak Ridge, TN 37830, USA
raons@ornl.gov

Abstract. Wide-area data transfer is central to geographically distributed scientific workflows. Faster delivery of data is important for these workflows. Predictability is equally (or even more) important. With the goal of providing a reasonably accurate estimate of data transfer time to improve resource allocation & scheduling for workflows and enable end-to-end data transfer optimization, we apply machine learning methods to develop predictive models for data transfer times over a variety of wide area networks. To build and evaluate these models, we use 201,388 transfers, involving 759 million files totaling 9 PB transferred, over 115 heavily used source-destination pairs ("edges") between 135 unique endpoints. We evaluate the models for different retraining frequencies and different window size of history data. In the best case, the resulting models have a median prediction error of ≤21% for 50% of the edges, and ≤32% for 75% of the edges. We present a detailed analysis of these results that provides insights into the cause of some of the high errors. We envision that the performance predictor will be informative for scheduling geo-distributed workflows. The insights also suggest obvious directions for both further analysis and transfer service optimization.

1 Introduction

With increasing data generation rates, wide-area data transfer is becoming inevitable for many science communities [1–3]. Wide-area data transfer is an important aspect of distributed science. An accurate estimation of data transfer time will significantly help both resource selection and task scheduling of geographically distributed workflows (e.g., one the cloud) [4–8]. Typically, wide-area transfers involve a number of shared resources including the storage systems and data transfer nodes (DTNs) [9] at both the source and destination, and the multi-domain network in between. Even for transfers between a user's machine and an experimental, analysis, or archival facility, the resources at the facility

É. Renault et al. (Eds.): MLN 2018, LNCS 11407, pp. 56–78, 2019.
https://doi.org/10.1007/978-3-030-19945-6_5

and the network connection are shared. Because of the fluctuation in the load on various shared resources, the data transfer rates can fluctuate significantly [10]. A number of attempts have been made to model and predict wide-area data transfer performance [11–17]. Many of these studies use a data-driven approach; some have used analytical models of individual components to predict transfer performance in dedicated environments [18]. Building an end-to-end model using the analytical models of individual components in a shared environments is challenging because the behavior of individual components can vary in an unpredictable fashion, due to competing load.

While a variety of tools are used for high-performance wide-area data transfers, such as GridFTP [19], bbcp [20], FDT [21], and XDD [22], the features that impact transfer performance are similar in most cases (e.g., number of TCP connections, number of network and disk I/O threads or processes, data transfer size, number of files). A data-driven model built using the data obtained from one high-speed data transfer tool should be generally applicable for all high-performance wide-area data transfers.

In this work, we apply a number of machine learning algorithms to develop models that can be used to predict wide-area data transfer times. We use the logs from the Globus transfer service [23] for our study. While this service orchestrates GridFTP or HTTP transfers between storage systems, we focus here on GridFTP transfers. GridFTP extends FTP to provide features such as high-speed, fault tolerance, and strong security. A number of features that are known to impact the transfer times, such as number of TCP connections, number of network and disk I/O threads, transfer size, and number of files are used to build machine-learning models.

The rest of this paper is organized as follows. First, in Sect. 2, we present background and motivation for building the predictor as well as the data we used in this paper. Next, in Sect. 3 we introduce the features we used to build the machine learning based predictor. A brief introduction of the machine learning algorithms as well as their the hyperparameters tuning, and model selection are given in Sect. 4. Then, we describe experiments, analyze results, and study feature importance in Sect. 5. Further insights into prediction errors are given in Sect. 6. Finally, we review related work in Sect. 7, and summarize our conclusions and briefly discuss future work in Sect. 8.

2 Background, Motivation, and Data

2.1 Background

To understand the characteristics of wide-area data transfer throughput, we analyzed the Globus transfer logs of the 1,000 source-destination pairs ("edges") that transferred the most bytes. Figure 1 shows the relative standard deviation, also known as coefficient of variation—a measure of variability relative to the mean—of the throughput for these edges. We observe a large variation in throughput for many of these edges, which reiterates the fact that predicting wide-area transfer performance is a challenging task.

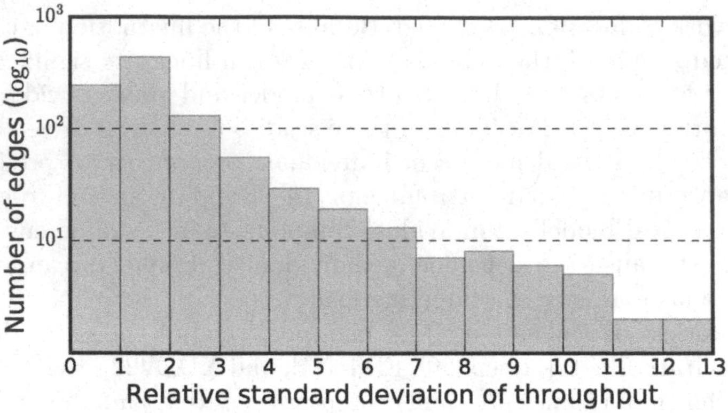

Fig. 1. Relative standard deviation (standard deviation divided by the mean) for the top 1,000 heavily used edges in Globus

In previous work [18], we characterized the features that impact wide-area data transfers and used linear and nonlinear models to validate the explainability of the features. Our goal in that work was to *explain* the performance of wide-area transfers through interpretable models. Here, we focus on applying machine learning algorithms to *predict* transfer performance.

2.2 Motivation

The primary motivation of this work is to provide a reasonably accurate estimate of data transfer time so that workflow management systems (e.g., pegasus [24]) can allocate appropriate resources for distributed science workflows and schedule the tasks more efficiently. In general, an accurate data transfer performance prediction model can be useful for many purposes, including the following:

1. It can be used for resource selection (when computing resource are available on multiple HPC centers or regions of cloud), and scheduling (e.g., reserve computing resource beforehand) of distributed workflows [4,5].
2. Choosing appropriate tunable parameters of data transfer tools plays a crucial role in data transfer time [3,25,26]. A model that captures the influence of tunable parameters under different circumstances can be used to identify optimal tunable parameters of the transfer tools under a given condition (i.e., dataset characteristics and external load), e.g., by allowing for exhaustive search over parameters without adding real load to the infrastructure [26–29].
3. They can be used to improve user satisfaction, by setting expectations at the start of the transfer. Users can also plan their job accordingly based on the prediction.

2.3 Data

For this study, we use Globus transfer logs from 2016/01/01 to 12/31/2017, which contains 2.7 million records. We discard transfers with elapsed time shorter than 10 s, because we do not see it as useful to predict these small transfers in practice; transfers that were canceled by users; and transfers from Globus tutorial endpoints, leaving 0.9 million transfers, which involve 24,951 unique source-destination pairs. The top 10% of the 24,951 edges contribute 83.6% of the transfers. Only 607 edges have more than 100 transfers. Figure 2 shows the distribution of the number of transfers over edges. We consider edges with more than 150 transfers as *heavily used*.

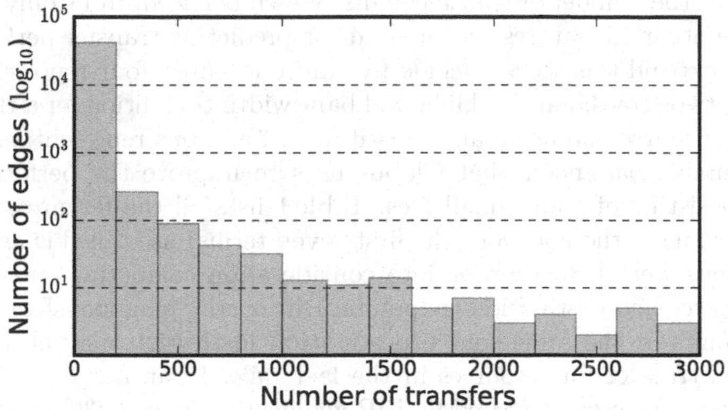

Fig. 2. Distribution of the number of transfer over edges.

To evaluate the models for online use, we set aside transfers from May 1, 2017 onwards for testing, and use only the transfers before this date for tuning the hyperparameters of machine learning algorithms, model selection, and training selected machine learning models: 201,388 transfers over 115 heavily used source destination pairs (involving 135 unique endpoints), totaling 9 PB in 759 million files. In terms of round trip time (RTT, a key difference among edges) between source and destination endpoints, the 50th and 75th percentile of these 115 edges are 2.46 ms and 18.63 ms respectively, compared with 2.37 ms and 19.40 ms for all edges, respectively. We believe that these 115 edges are sufficiently representative to capture the potential behavior of our models on other edges that are heavily used but lack sufficient transfers before and after May 1, 2017 to be considered in this paper.

3 Transfer Features

Feature engineering is a key data preparation and processing step in machine learning. Often it is a manual process of transforming raw features into derived features that better represent the problem structure to the predictive models, resulting in improved model accuracy on testing data. This process involves constructing new features from raw features by combining or splitting existing features into new ones, often by the application of domain knowledge.

In our previous work [18], we employed 16 features for explaining wide-area data transfer performance, in three groups: tunable parameters specified by users, transfer characteristics that describe file properties, and competing load. We extracted the features in the first two groups directly from Globus transfer logs, but constructed those in the third group to capture resource load conditions on endpoints. Specifically, we created profiles for endpoint CPU load and network interface card load via feature engineering and showed that these profiles can be used to explain a large fraction of transfer performance in many cases. Besides the number of transfer faults, which is known to us only after the transfer, the other 15 features can be used for predicting transfer performance.

Here we expand this set to include five more features: four representing the wide-area network outbound and inbound bandwidth contention for a transfer at the transfer's source and destination institutions, and one representing pipeline depth, a tunable parameter that Globus uses to improve the performance of transfers consisting of many small files. Table 1 lists all the features (lower 20 terms) and some of the notation (the first seven terms) used in this article.

In previous work [18], we took into consideration competing transfers at a transfer's source and destination endpoints. But overlapping transfers involving other endpoints at the same source/destination institution may also compete with a given transfer for resources in the last mile. In an attempt to capture this contention, we used the MaxMind IP geolocation service [30] to obtain the approximate location of endpoints and used that information to estimate competing load based on proximity. We considered the throughput of simultaneous transfers within eight kilometers as competing transfers to the transfer of interest.

We introduce four features to quantify the wide-area network contention from the same institution. Assume that there is a transfer k from institution I_{src} to institution I_{dst}. The Globus **contending transfer rate** for a transfer k at that transfer's source institution (I_{src}) and destination institution (I_{dst}) endpoints is as follows:

$$Q^{x\in\{sout,sin,dout,din\}}(k) = \sum_{i\in A_x} \frac{\mathcal{O}(i,k)}{Te_k - Ts_k} R_i, \qquad (1)$$

where *sout* and *sin* denote outgoing and incoming at institution I_{src}, respectively, and *dout* and *din* represent outgoing and incoming at institution I_{dst}, respectively; A_x is the set of transfers (excluding k) with I_{src} as source, when $x = sout$; I_{src} as destination, when $x = sin$; I_{dst} as source, when $x = dout$; and I_{dst} as destination when $x = din$; and R_i is the throughput of transfer i, and $\mathcal{O}(i,k)$ is the overlap time for the two transfers:

$$\mathcal{O}(i,k) = \max(0, \quad \min(Te_i, Te_k) - \max(Ts_i, Ts_k)).$$

Table 1. Notation used in this article. The lower 20 terms are used as features in our machine learning algorithms, of which the first 15 are from Liu et al. [18] and the remaining five are developed in this paper.

src_k	Source endpoint of transfer k
dst_k	Destination endpoint of transfer k
I_k^{src}	Institution of the source endpoint of transfer k
I_k^{dst}	Institution of the destination endpoint of transfer k
Ts_k	Start time of transfer k
Te_k	End time of transfer k
R_k	Average transfer rate of transfer k
K^{sin}	Contending incoming transfer rate on src_k
K^{sout}	Contending outgoing transfer rate on src_k
K^{din}	Contending incoming transfer rate on dst_k
K^{dout}	Contending outgoing transfer rate on dst_k
C	Concurrency: Number of GridFTP processes
P	Parallelism: Number of TCP channels per process
S^{sin}	Number of incoming TCP streams on src_k
S^{sout}	Number of outgoing TCP streams on src_k
S^{din}	Number of incoming TCP streams on dst_k
S^{dout}	Number of outgoing TCP streams on dst_k
G^{src}	GridFTP instance count on src_k
G^{dst}	GridFTP instance count on dst_k
Nf	Number of files transferred
Nd	Number of directories transferred
Nb	Total number of bytes transferred
Q^{sin}	Contending incoming transfer rate on I_k^{src}
Q^{sout}	Contending outgoing transfer rate on I_k^{src}
Q^{din}	Contending incoming transfer rate on I_k^{dst}
Q^{dout}	Contending outgoing transfer rate on I_k^{dst}
D	Pipeline depth

Pipelining, D, speeds transfers involving many small files by dispatching up to D FTP commands over the same control channel, back to back, without waiting for the first command's response [3]. This method reduces latency and keeps the GridFTP server constantly busy since it is never idle waiting for the next command.

Although we performed this work using Globus data, we believe that our methods and conclusions are applicable to all wide-area data transfers. The reason is that the features we used (number of TCP connections, number of network and disk I/O threads/processes, data transfer size, number of files, competing load) are generic features that impact the performance of any wide-area data transfer, irrespective of the tool employed. The pipeline feature is applicable to all tools that involve a command response protocol. The data used in this paper are publicly available at https://github.com/ramsesproject/wide-area-data-transfer-logs.

4 Machine-Learning-Based Modeling

We use supervised learning, a class of machine learning (ML) approach, to model throughput as a function of input features. The success of the supervised-learning approach depends on several factors such as identifying appropriate data-preprocessing techniques, selecting the best supervised-learning algorithm that performs well for the given training data, and tuning the algorithm's hyper-parameters. Figure 3 illustrates the process we use to obtain the best model for each edge. We detail these steps next.

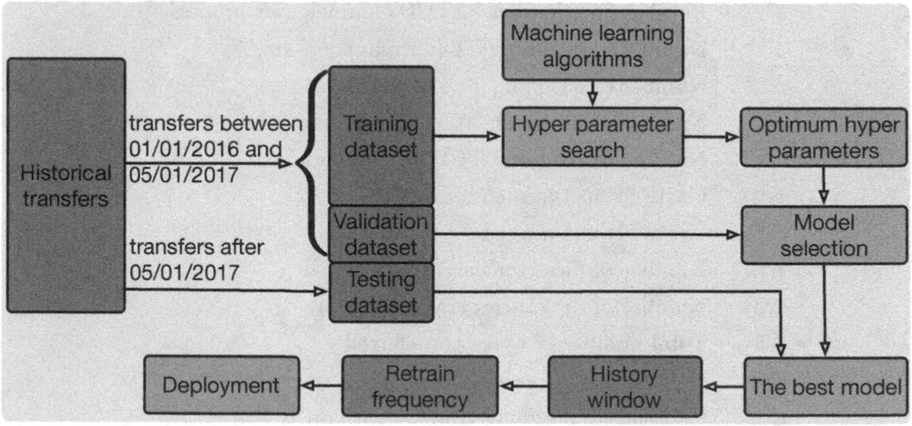

Fig. 3. Model selection process

4.1 Preprocessing

Many ML algorithms are more effective if the features have the same scale and the training data set is standardized. We apply scale transformation, which involves computing the standard deviation for each feature and dividing each value in that feature by that standard deviation. We also observed that the variance of the measured throughput is large for several endpoints. Since several ML algorithms minimize the mean squared error on the training set, if we use raw throughput, then the model fitting is biased towards large throughputs values but not the small ones. To avoid this, we applied logarithm transformation for the throughput.

4.2 Supervised Learning Algorithms

Many supervised learning algorithms exist in the ML literature. They can be categorized as regularization, instance-based, kernel-based, bagging, and boosting algorithms. Typically, the best algorithmic choice depends on the type of relationship between the input features and the throughput. Instead of focusing on a single ML algorithm for our study, we selected a set of ML algorithms of increasing complexity. We start from linear regression, the most basic ML algorithm. We adopt ridge regression that addresses overfitting problem of linear regression using regularization. Ensemble methods build a number of simple models and combine them to obtain better predictive performance and they are shown to obtain state-of-the-art results in a wide range of predictive modeling problems [31]. Two families of ensemble methods are (1) *bagging* that build several simple models independently and then average their predictions—Bagging, ExtraTrees, and Random Forest algorithms are selected from this class; (2) *boosting* that build simple models incrementally to decrease the training error in a sequential way—GradientBoosting, AdaBoost, and eXtreme Gradient Boosting algorithms are selected from this class. We also evaluated support vector machines as a candidate for a kernel-based regression method, but found that it did not perform well and so we omitted it from the study. In the rest of the section, we provide a high-level overview of the algorithms considered in our study.

Multivariate linear regression tries to find $h(x)$ by fitting a linear equation $h(x) = c + \sum_{i=1}^{M} \alpha^i \times x^i$, where c is a constant and α^i is the coefficient of the input x^i. Appropriate values of (c, α) are obtained by the method of least-squares, which minimizes the sum of the squared deviations between the y_i and $h(x_i)$ in $\mathcal{D}$.

Ridge regression (RR) [32] is a regularization algorithm that addresses the overfitting problem faced by linear regression. Overfitting occurs when the model becomes sensitive to even small variations in the training set output and lose prediction accuracy on the testing and validation set. To avoid this problem, RR, in addition to minimizing the error between predicted and actual observations, penalizes the objective by a factor α of the input coefficients. Here, $\alpha \geq 0$ is a user-defined parameter that controls the tradeoff between minimizing error and minimizing sum of square of coefficients.

Bagging trees (BR) [33] builds `nt` decision trees. Each decision tree is obtained on a random subsample of the original dataset by recursively splitting the multi-dimensional input space into a number of hyper-rectangles such that inputs with similar outputs fall within the same rectangle. Given a new point x^* to predict, each tree follows if-else rule and returns the constant value at the leaf as the predicted value. The mean of all the predicted values will be returned as the final prediction value.

Random forest (RFR) [34,35] is also a bagging approach that builds `nt` decision trees similar to BR but at each split, RFR considers only a random subset of inputs and selects the best one for split based on mean squared error.

Extremely randomized trees (ETR) [36] differs from RFR in the way in which the splits are computed. As in RFR, a random subset of inputs is considered for the split, but instead of looking for the best mean squared error, the values are drawn at random for input and the best of these randomly-generated values is picked as the splitting rule.

Adaptive boost regressor (ABR) [37] is a tree-based boosting approach that differs from tree-based bagging in the way it builds the trees and aggregates the results: `nt` trees are built sequentially such that the mth tree is adjusted to correct the prediction error of the $(m-1)^{th}$ tree. Each sample point in the training set is assigned a weight proportional to the current error on that point so that subsequent tree focuses on points with relatively higher training errors. Given a new test point x^*, each tree calculates a predicted value. These predicted values are weighted by each tree's weight, and their sum gives the prediction for the ensemble model.

Gradient boosting trees (GBR) [38] are similar to ABR, in which `nt` trees are built sequentially. The key difference is that GBR tries to minimize squared error loss function by using the residuals of the $(m-1)^{th}$ model as the negative gradient. It generates a new model at the mth iteration by adding the mth tree that fits the negative gradient to the $(m-1)^{th}$ model.

eXtreme Gradient Boosting (XGB) [39] is a high-performing gradient boosting optimized software framework and implementation, which is used widely by data scientists to achieve state-of-the-art results on many ML challenges such as Kaggle and KDDCup. XGB is similar to GBR but the former adopts a more regularized model formalization than the latter which gives it better performance. Moreover, XGB uses several novel algorithmic optimizations—a novel approximate tree learning algorithm, an efficient procedure to handle training point weights in approximate tree learning, and system level optimization—out-of-core computation to process data that is too large to fit in main memory and cache-aware learning to improve speed and scalability.

4.3 Hyperparameter Tuning

Many ML algorithms require user-defined values for hyperparameters that strongly influence the performance of the algorithm. These parameters include not only categorical but also numerical parameters such as α, number of trees, and maximum tree depth. Choosing appropriate parameter settings can have a significant impact on the performance. Unfortunately, finding performance-optimizing parameter settings is data dependent and difficult task.

This task has traditionally been tackled by hand, using a trial-and-error process and/or grid search. In the past few years, new algorithmic methods have begun to emerge to address these serious issues. In particular, a random search algorithm has shown to be more effective than grid search for tuning the hyperparameters of ML algorithms [40]. We consider a random search over the feasible domain D without replacement. This consists in sampling n_s random parameter configurations for a given supervised learning algorithm, evaluate each of them on the training set by cross validation, and selecting the best according to a user-defined metric (e.g., the median absolute percentage error).

We tuned the hyperparameters for each learning algorithm by using a random search that samples $n_s = 200$ configurations and selected the best based on cross validation on the training data. The parameters of each algorithm considered for tuning are as follows. **XGB**: The number of boosted trees to fit, the maximum tree depth for base learners, the minimum loss reduction required to make a further partition on a leaf node of the tree, the subsample ratio of the training instance, and the subsample ratio of columns when constructing each tree. **BR**: The number of base estimators in the ensemble, the number of samples to train each base estimator, and whether samples are drawn with replacement. **GBR**: The number of boosting stages to perform, the maximum depth of the individual regression estimators, and the learning rate. **RFR** and **ETR**: The number of trees in the forest, the maximum depth of the tree, and whether bootstrap samples are used when building trees. **ABR**: The maximum number of estimators at which boosting is terminated, and the learning rate. **RR**: The regularization strength. Note that linear regression does not have tunable parameters.

4.4 Model Selection

For each edge, we select the best model as follows. First, as shown in Fig. 3, we split transfers before 05/01/2017 as two sets: the first 70% of the transfers as training set and the remaining 30% as validation set. For each supervised learning algorithm, we run hyperparameter optimization with $n_s = 200$, 10-fold cross validation on the training set, and root mean squared error as the metric that needs to be minimized. This is repeated for each algorithm. Once we obtain the best hyperparameters from each of the seven ML algorithms, we predict the throughput on the validation set. The best model for the edge is the one that has the minimal prediction error on the validation set.

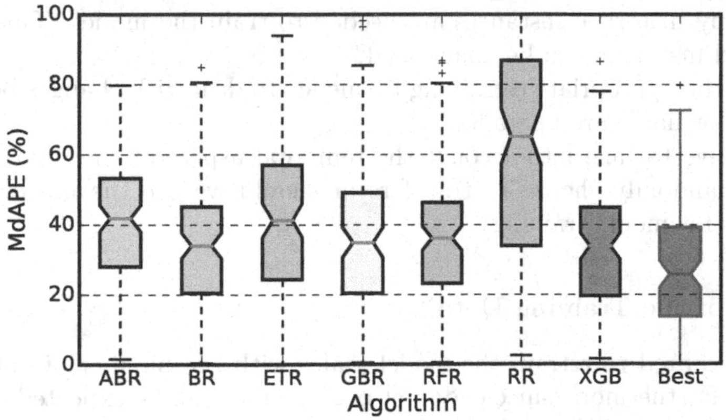

Fig. 4. Validation results. *Best* represent validation errors when the best algorithm is used for each edges.

Figure 4 shows the overall accuracy of each algorithm on the validation set, in which the median absolute percentage error (MdAPE) is used to represent model accuracy. The **Best** in the plot is the accuracy when each edge uses the best algorithm.

From the results, we can observe that **XGB** works the best for most number of edges; **GBR** and **BR** seem to have similar accuracy. However, the best case obviously outperforms any single algorithm. Table 2 presents the statistics on the winning algorithms.

Table 2. Statistics of the best model selection

Algorithm	Pairs
GradientBoostingRegressor	23
Ridge	6
XGBRegressor	31
BaggingRegressor	18
AdaBoostRegressor	9
RandomForestRegressor	14
ExtraTreesRegressor	14

5 Experimental Results

After identifying the best model for each edge, we search hyperparameters again and retrain the model with the combined training and validation data sets. Then, we evaluate model accuracy on the test data set (i.e., transfers between 05/01/2017 and 12//31/2017). From a practical deployment perspective, we focus on the following three issues:

1. How many historical instances are needed to train the model? That is, how many old instances can be abandoned?
2. What is the appropriate retraining frequency to deal with changes because of software or hardware upgrades?
3. Do we have to take into account the temporal aspect of the data? That is, can we randomly choose K transfers or should we use the most recent K transfers for model training.

5.1 Size of the Training Data

The time required to retrain the model scales with the number of data points considered. Furthermore, since endpoint performance can be expected to change over time, for example because of software or hardware upgrades, it is necessary to remove the data points before such changes. But identifying the point at which an upgrade or other change occurs may not always be straightforward in the online setting. Therefore, we empirically evaluate the volume of historic data required to make a reasonable prediction. Here, we use the most recent w ($w \in \{25, 50, 100, 500, 1000, 1500, 2000\}$) transfers before May 1, 2017, to train the model, and we test each trained model on all transfers from May 1, 2017 to December 31, 2017. Figure 5 compares quantile distribution of MdAPE under different window sizes for training and the corresponding model training time. In the plot, *Q25*, *Q50* and *Q75* represent 25th, 50th and 75th quantile of MdAPE separately. *RCT* and *RND* denotes most recent N_{train} transfers and randomly chosen N_{train} transfers, respectively. *All* means that we used all transfers before May 1, 2017, to train the model.

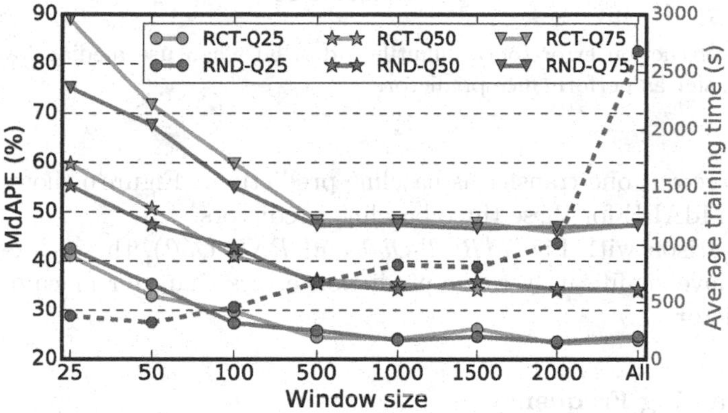

Fig. 5. Prediction errors (solid lines) and model training time (dotted blue line) as a function of the number of transfers (N_{train}) used to train the model. **Q25**, **Q50** and **Q75** represent 25th, 50th and 75th quantile of MdAPE separately. **RCT** and **RND** denotes most recent N_{train} transfers and randomly chosen N_{train} transfers individually. **All** means that we used all transfers before May 1, 2017, to train the model. (Color figure online)

Models trained with random data points are slightly better than models trained with the most recent data points; in most cases this performance improvement is only a little (except the $w = 25$ case). The results show that increasing the values of w decreases the error but only until $w = 1000$. The observed improvements become insignificant for $w > 1000$. However, the model training time increases significantly when the window size is increased (this is true for all cases except when w is increased from 25 to 50 and from 1000 to 1500). These results indicate that most recent 500 to 1,000 data points is sufficient for training the model.

We use the methods that users would use in practice, specifically, *average*, *median* and the performance of the *previous* transfer, as the baselines. Specifically, for a given new transfer, there are w historical transfers over the same edge, we consider the median performance, average performance and the performance

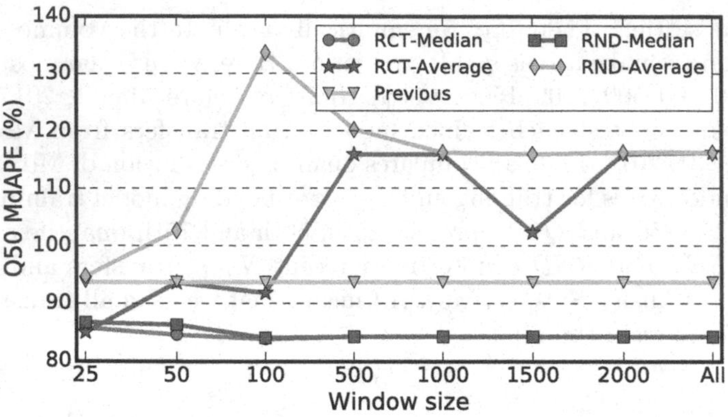

Fig. 6. The prediction error (50th quantile MdAPE) when use median, average and previous transfer as performance predictor.

of the most recent one transfer as baseline predictions. Figure 6 shows the 50th quantile of MdAPE for these three baseline predictors.

In comparison with Fig. 5 (*RCT-Q50* and *RND-Q50*), these three baseline predictors have significantly worse prediction errors than our machine learning based predictor.

5.2 Retraining Frequency

We evaluated and compared the results obtained when retraining models every 3 months, 1 month, 2 weeks, 1 week, 2 days, and online. To this end, we split the test dataset into multiple datasets based on the frequency. For the 3 month frequency, we train the model using all transfers before May 1, 2017, and evaluate it with transfers from May 1, 2017, to July 31, 2017. Then, we retrain model using data before August 1, 2017 and evaluate it with the transfer from August 1, 2017 to October 31, 2017, and so on. Similarly, for the monthly frequency, we evaluate the model trained using all transfers before May 1, 2017 with the transfers in May, 2017. Then we retrain the model using all the data before June 1, 2017, and evaluate it with transfers in June, 2017, and so on. We follow a similar procedure for other retraining frequencies. For online retraining, we monitor the prediction error of new transfers and retrain the model with the most recent w transfers if the MdAPE of the recent k transfers is larger than a certain threshold. In this study, we set w, k, and threshold to 100, 4, and 15%, respectively. We also experimented with w, k, and threshold of 500, 10, and 15%, respectively, but those values did not result in a significant difference.

Figure 7 compares the MdAPE distribution on all edges for different retraining frequencies. We observe that more frequent retraining results in lower errors and that online retraining achieves the lowest error.

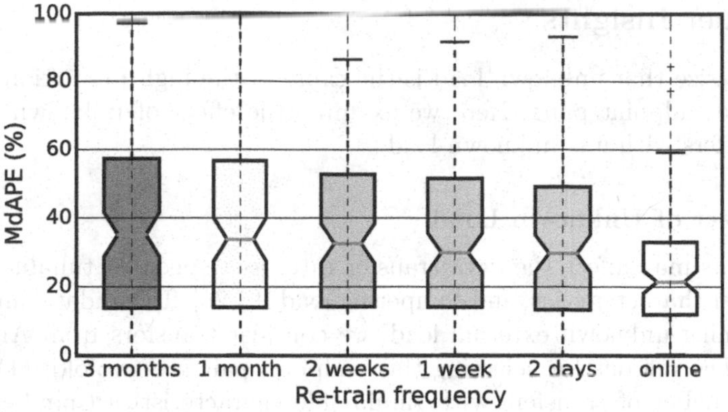

Fig. 7. Retrain frequency versus prediction error on testing dataset.

5.3 Feature Importance

To gain insights into the impact of the features listed in Table 1 on the throughput, we analyze their importance using the relative feature importance capability of XGB [39]. First, we compute the relative importance of the features on each edge and select the five most important ones. Then we tabulate the frequency of the top five in all the edges in Fig. 8.

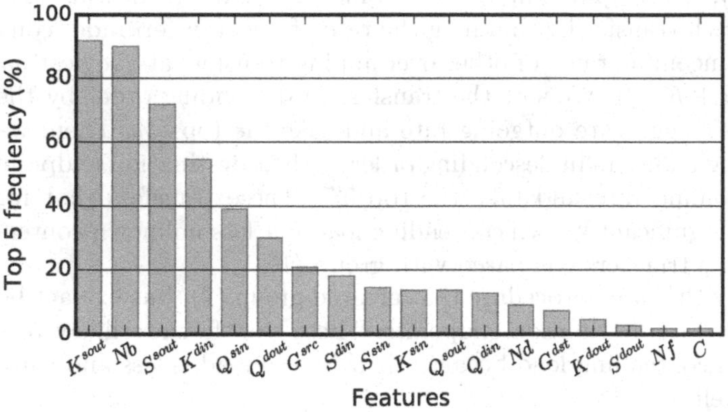

Fig. 8. Frequency of features that appear in the top 5 most important features of each edge model.

As one can see from Fig. 8, transfer size Nb and K^{sout}, K^{din} and S^{sout} which have contention in the same direction with the transfer of interest, are important for most of the endpoint pairs. The four new introduced features in this paper (Q^{sout}, Q^{din}, Q^{sin} and Q^{dout}), which quantify contention from simultaneous transfers from the same institution (within eight kilometers), are also important.

6 Further Insights

We hypothesize that unknown load is the cause of the high prediction error for some of the endpoint pairs. Here we examine the effect of unknown load and study transfers with less unknown load.

6.1 Impact of Unknown Load

Three factors may affect the data transfer rate: user-specified tunable parameters, transfer characteristics, and competing load [18,26]. To validate our hypothesis concerning unknown external load, we consider transfers from Argonne to Fermilab. The reasons for selecting this endpoint pair are twofold: (1) it has a sufficient number of transfers with similar file characteristics (number of files, directories, and average file size) and the same tunable Globus parameters; and (2) we observe quite different rates for these transfers. Since transfer parameters and characteristics are identical, the transfer rate should be affected only by competing load.

We split the transfers over this edge into three groups:

1. Transfers with rate greater than 50% of the maximum rate observed over this endpoint pair. These transfers are likely to have less contending load.
2. After filtering out the transfers in group (1), for each remaining transfer, we compute the aggregate outgoing rate on source endpoint during the transfer by summing the rate of the transfer under consideration and the outgoing rates for other overlapping transfers at the source endpoint ($R + K^{sout}$). Similarly, we compute the aggregate incoming rate on destination endpoint during each transfer by summing the rate of the transfer under consideration and the incoming rates of other overlapping transfers at the destination endpoint ($R + K^{din}$). We sort the transfers in descending order by their source endpoint's aggregate outgoing rate and take the top 5%. Then, we sort the remaining transfers in descending order of their destination endpoint's aggregate incoming rate and take the top 5%. These transfers that form group (2) have significant known contending load but less unknown contending load among the transfers compared with group (1).
3. We apply the same procedure that created group (2), but extract bottom 5% on source and destination endpoints. These transfers are likely to have high unknown contending load because their known load is less and throughput is low as well.

As a result, we get 161 transfers in group (1), 175 in group (2), and 175 in group (3) (shown in Fig. 9). We split each group by 70% and 30% for training and testing respectively. Table 3 shows the prediction errors with different algorithms.

Group 1 and 2 have low prediction error with nonlinear models (the second and third model in Table 3) and group 3 always have higher error than the first two groups. The linear model works well for group 1 but not for group 2 or 3. The poor prediction accuracy of linear model and the high prediction accuracy of machine learning models in group 2 indicate nonlinear relationship between 'known load' and 'transfer rate', which is consistent with findings in [18].

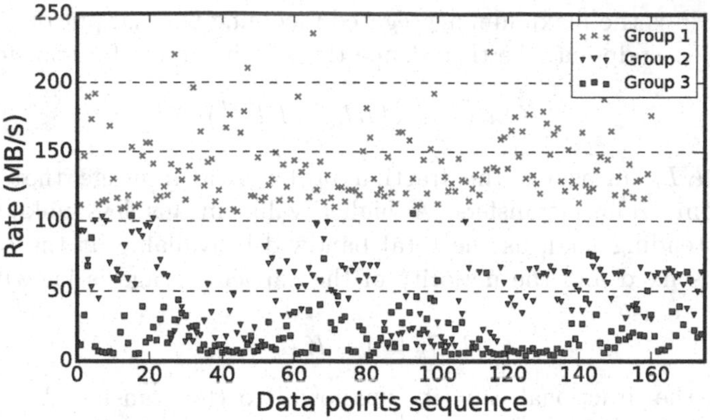

Fig. 9. Group transfers by their rate and known load.

Table 3. Prediction error with different machine learning algorithm on the three groups.

Algorithm	Group	Q50(%)	Q75(%)	Q90 (%)
Ridge regression	1	11.24	18.19	22.63
	2	20.04	33.08	64.33
	3	35.37	126.54	223.29
XGBRegressor	1	11.85	22.91	25.20
	2	8.20	18.06	29.36
	3	27.16	51.02	72.49
BaggingRegressor	1	9.54	18.83	25.02
	2	9.46	14.81	32.64
	3	29.85	51.27	133.48

6.2 Transfers with Fewer Unknowns

Since transfers with high unknown load is the source of noise when we use them to train the machine learning model, here we trained the models using transfers with less unknown load and study how they perform. We compute the relative 'known load (KL)' for a transfer as the ratio of aggregate transfer rate to maximum observed aggregated throughput. For the source endpoint of the transfer of interest, this is given by:

$$KL_k^{src} = \frac{K^{sout} + R_k}{DR^{max}}, \tag{2}$$

where DR^{max} is the maximum aggregated outgoing throughput observed from the source endpoint. Similarly, for the destination endpoint of the transfer of interest, we have:

$$KL_k^{dst} = \frac{K^{din} + R_k}{DW^{max}}. \tag{3}$$

where DW^{max} is the maximum aggregated incoming throughput observed from the destination endpoint. We then define the relative load of a transfer k as:

$$KL_k = \max\left(RL_k^{src}, RL_k^{dst}\right). \tag{4}$$

Intuitively, KL_k measures the fraction of bandwidth usage that has been observed from Globus transfers. A higher value means less unknown (non-Globus) contending load, as the total bandwidth available at the source and destination is fixed and the majority of the capacity usage is known. Thus in this section, we use

$$UC_k = 1 - KL_k \tag{5}$$

to represent the 'fractional capacity unknown' to the transfer. A larger value means that the transfer is more likely to have experienced unknown load.

In order to verify the influence of unknown load on the model being trained, we use only transfers with $UC_k \leq 0.5$ to train the model. Specifically, we ignore transfers with $UC_k > 0.5$ in the training data set as these transfers are more likely to have high unknown contending load.

Then, we use transfers with $UC_k \leq 0.5$ to train the model and test the trained model on the *whole* testing set (i.e., transfers between 05/01/2017 and 12/31/2017. We infer that the model trained with these clean data is more representative of transfers with less unknown load because there is less noise in the training set. Thus, we expect this model to make better predictions for transfers with less unknown load.

Figure 10 compares the relationship between 'fractional capacity unknown' and the absolute percentage error. The green cross points are prediction errors for the model that was trained with all transfers before 05/01/2017 (irrespective of the 'fractional capacity unknown') and red star points represent prediction errors for the model that was trained by using transfers before 05/01/2017 and with $UC_k \leq 0.5$.

As shown in Fig. 10, when we look at the prediction errors for either model i.e., the model trained with all transfers (green cross points) or the model trained only by using transfers with $UC_k \leq 0.5$ (red star points), we observe that the transfers with lower 'fractional capacity unknown' get more accurate prediction. The reason is that these transfers are less likely to have high unknown load. When we compare the prediction error of the same transfer, i.e. a green point and a red point with the same UC_k, model that was trained with cleaner data (marked by red star points) make more accurate predictions than the model that was trained with all transfers (marked by green cross points), when UC_k is less than 0.5. However, we cannot use this method to improve the model accuracy because the relative known load is not known to us before the transfers are done. We use this analysis to show the influence of external (unknown) load in training as well as in prediction. To improve the prediction, we need to develop methods that can measure the external load online.

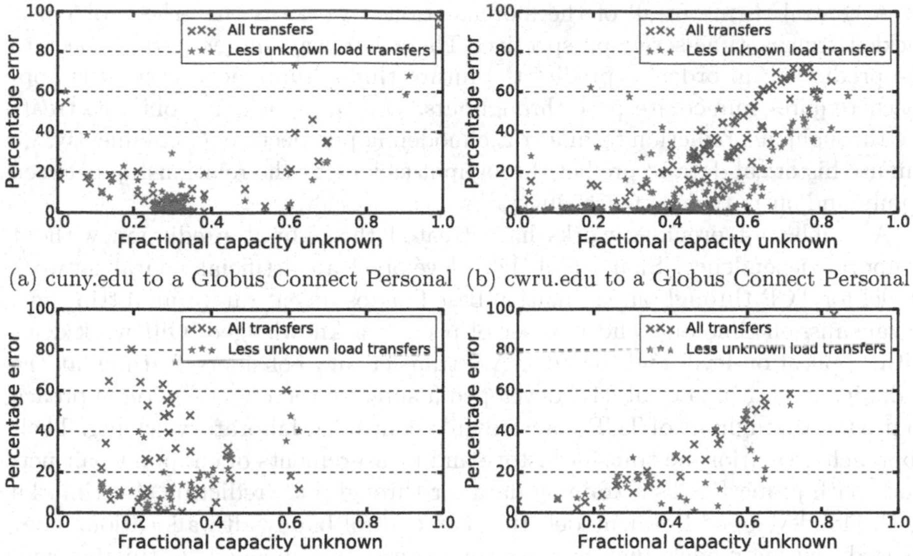

(a) cuny.edu to a Globus Connect Personal (b) cwru.edu to a Globus Connect Personal

(c) harvard.edu to a Globus Connect Personal (d) westgrid.ca to a Globus Connect Personal

Fig. 10. Comparison of the 'fractional capacity unknown' of a transfer versus absolute performance prediction error (%), for two models: one trained only with less unknown load transfers and the other trained with all transfers. We show results for four different source endpoints. We note that the destination endpoints in the figures are also different from each other. (Color figure online)

7 Related Work

Several researchers have developed regression-based throughput prediction models using historical data. Vazhkudai et al. [41] developed mean, median, autoregressive, sliding window, and rule-based approaches to predict the performance of wide area data transfers. Swany et al. [42] developed a multivariate time-series forecasting technique to enable Network Weather Service, a TCP throughput prediction framework, to automatically correlate monitoring data from different sources, and to exploit that observed correlation to make better, longer-ranged forecasts. Lu et al. [43] developed TCP throughput predictions based on an exponentially weighted moving average of bandwidth probes. He et al. [44] studied analytical and empirical models for predicting the throughput of large TCP transfers. Their empirical approach used standard time series forecasting techniques to predict throughput based on a history of throughput measurements from previous TCP transfers on the same path. Huang et al. [45] developed time series of windows of segments arriving at the receiver, and predicted future throughput by exploiting knowledge of how TCP manages transfer window size. When the file transfer time series resembles a known TCP pattern, this information is used for prediction, otherwise simple heuristics are used.

A central theme in all of the aforementioned works is that the problem is tackled from a time-series perspective. There is always a temporal aspect to the prediction: in order to predict the future throughputs accurately, the approach requires immediate past throughputs. Our approach, in contrast, treats the throughput prediction from a static modeling perspective. Consequently, we can use historical data to predict throughput as long as the relationship between inputs and outputs does not change.

A number of previous works have treated throughput prediction without temporal dependency. Shah et al. [46] developed an artificial neural network model for TCP throughput estimation based on loss event rate, round trip time, retransmission time out, and number of packets acknowledged. Our work deals with application-level end-to-end WAN transfer and considers features at the endpoint level. Mirza et al. [47] developed a support vector regression approach to predict throughput of TCP transfers in a controlled laboratory setting. Their approach uses prior file transfer history and measurements of endpoint pair network path properties as a training data for throughput prediction. Kettimuthu et al. [48] developed linear models that can control bandwidth allocation. They showed that log models that combine total source concurrency, destination concurrency, and a measure of external load are effective. Nine et al. [49] analyzed historical data consisting of 70 million file transfers and applied data mining techniques to extract the hidden relations among the parameters and the optimal throughput. They used neural networks and support vector machines to develop predictive models and to find optimal parameter settings. Our approach considers a broad range of features, several high performing supervised learning algorithms, and adopts a rigorous model selection methodology in which the best is automatically selected based on the training data.

Arslan et al. [25] developed the HARP auto-tuning system, which uses historical data to derive endpoint-pair-specific models based on protocol parameters. By running probing transfers, HARP captures the current load on the network, which is then augmented with a polynomial regression model to increase the predictive accuracy. Moreover, we consider several models and choose the best automatically to hedge against the poor performance of any single method. Hours et al. [50] introduced the use of copula, a statistical approach to model the dependence between random variables, to model the relationship between network parameters and throughput as multidimensional conditional densities. While this method is orthogonal to the approach we used, it is limited by the data assumptions and presence of outliers.

Liu et al. [51] conducted a systematic examination of a large set of data transfer logs to characterize transfer characteristics, including the nature of the datasets transferred, achieved throughput, user behavior, and resource usage. Their analysis yielded new insights that can help design better data transfer tools, optimize networking and edge resources used for transfers, and improve the performance and experience for end users. Liu et al. [52,53] analyze various logs pertaining to wide area data transfers in and out of a large scientific facility.

That comprehensive analysis yielded valuable insights on data transfer characteristics and behavior, and revealed inefficiencies in a state-of-the-art data movement tool.

8 Discussion and Conclusion

Wide-area data transfer is an important aspect of distributed science. Faster delivery of data is important but predictable transfers are equally (or even more) important. We developed machine learning models to predict the performance of wide area data transfer. We also evaluated them for practical deployment. We believe that our evaluation strategy will serve as a reference to evaluate machine learning models for practical deployments not just for predicting the performance of wide area data transfers but also for other prediction use cases. We showed that our models perform well for many transfers, with a median prediction error of $\leq$21% for 50% of source-destination pairs ("edges"), and $\leq$32% for 75% of the edges. For other edges, errors are high, an observation that we attribute to unknown load on the network or on endpoint storage systems. We also showed that unknown load can interfere with model training and that eliminating transfers with high unknown load from training data can improve prediction accuracy for transfers with less unknown load. Thus, collecting more information about endpoint load can further improve the prediction accuracy.

Acknowledgments. This material is based upon work supported by the U.S. Department of Energy, Office of Science, under contract DE-AC02-06CH11357. We gratefully acknowledge the computing resources provided and operated by the Joint Laboratory for System Evaluation (JLSE) at Argonne National Laboratory.

References

1. Kettimuthu, R., Agrawal, G., Sadayappan, P., Foster, I.: Differentiated scheduling of response-critical and best-effort wide-area data transfers. In: 2016 IEEE International Parallel and Distributed Processing Symposium, pp. 1113–1122, May 2016
2. Allcock, W., et al.: Data management and transfer in high-performance computational grid environments. Parallel Comput. **28**(5), 749–771 (2002). https://doi.org/10.1016/S0167-8191(02)00094-7
3. Kettimuthu, R., Liu, Z., Wheeler, D., Foster, I., Heitmann, K., Cappello, F.: Transferring a petabyte in a day. Future Gener. Comput. Syst. **88**, 191–198 (2018). https://doi.org/10.1016/j.future.2018.05.051
4. Stavrinides, G.L., Duro, F.R., Karatza, H.D., Blas, J.G., Carretero, J.: Different aspects of workflow scheduling in large-scale distributed systems. Simul. Model. Pract. Theory **70**, 120–134 (2017). https://doi.org/10.1016/j.simpat.2016.10.009
5. Liu, Z., Kettimuthu, R., Leyffer, S., Palkar, P., Foster, I.: A mathematical programming- and simulation-based framework to evaluate cyberinfrastructure design choices. In: IEEE 13th International Conference on e-Science, October 2017, pp. 148–157 (2017). https://doi.org/10.1109/eScience.2017.27

6. Bicer, T., Gürsoy, D., Kettimuthu, R., De Carlo, F., Foster, I.T.: Optimization of tomographic reconstruction workflows on geographically distributed resources. J. Synchrotron Radiat. **23**(4), 997–1005 (2016)
7. Kettimuthu, R., et al.: Toward autonomic science infrastructure: architecture, limitations, and open issues. In: The 1st Autonomous Infrastructure for Science Workshop, AI-Science 2018. ACM, New York (2018). https://doi.org/10.1145/3217197.3217205
8. Rao, N.S.V., Liu, Q., Liu, Z., Kettimuthu, R., Foster, I.: Throughput analytics of data transfer infrastructures. In: Gao, H., Yin, Y., Yang, X., Miao, H. (eds.) TridentCom 2018. LNICST, vol. 270, pp. 20–40. Springer, Cham (2019). https://doi.org/10.1007/978-3-030-12971-2_2
9. Kettimuthu, R., Vardoyan, G., Agrawal, G., Sadayappan, P., Foster, I.: An elegant sufficiency: load-aware differentiated scheduling of data transfers. In: SC15: International Conference for High Performance Computing, Networking, Storage and Analysis, pp. 1–12, November 2015
10. Vazhkudai, S.: Enabling the co-allocation of grid data transfers. In: Proceedings of First Latin American Web Congress, pp. 44–51, November 2003
11. Wei, D.X., Jin, C., Low, S.H., Hegde, S.: FAST TCP: motivation, architecture, algorithms, performance. IEEE/ACM Trans. Netw. **14**(6), 1246–1259 (2006)
12. Tierney, B., Johnston, W., Crowley, B., Hoo, G., Brooks, C., Gunter, D.: The NetLogger methodology for high performance distributed systems performance analysis. In: 7th International Symposium on High Performance Distributed Computing, pp. 260–267. IEEE (1998)
13. Kosar, T., Kola, G., Livny, M.: Data pipelines: enabling large scale multi-protocol data transfers. In: 2nd Workshop on Middleware for Grid Computing, pp. 63–68 (2004)
14. Kelly, T.: Scalable TCP: improving performance in highspeed wide area networks. ACM SIGCOMM Comput. Commun. Rev. **33**(2), 83–91 (2003)
15. Wolski, R.: Forecasting network performance to support dynamic scheduling using the Network Weather Service. In: 6th IEEE Symposium on High Performance Distributed Computing, Portland, Oregon (1997)
16. Hacker, T.J., Athey, B.D., Noble, B.: The end-to-end performance effects of parallel TCP sockets on a lossy wide-area network. In: 16th International Parallel and Distributed Processing Symposium, IPDPS 2002, p. 314. IEEE Computer Society, Washington, DC (2002). http://dl.acm.org/citation.cfm?id=645610.661894
17. Rao, N.S.V., Sen, S., Liu, Z., Kettimuthu, R., Foster, I.: Learning concave-convex profiles of data transport over dedicated connections. In: Renault, É., Mühlethaler, P., Boumerdassi, S. (eds.) MLN 2018. LNCS, vol. 11407, pp. 1–22. Springer, Cham (2019)
18. Liu, Z., Balaprakash, P., Kettimuthu, R., Foster, I.: Explaining wide area data transfer performance. In: 26th ACM Symposium on High-Performance Parallel and Distributed Computing (2017)
19. Allcock, W., et al.: The Globus striped GridFTP framework and server. In: SC, Washington, DC, USA, pp. 54–61 (2005)
20. www.slac.stanford.edu/abh/bbcp/, BBCP (2017). http://www.slac.stanford.edu/~abh/bbcp/. Accessed 3 Jan 2017
21. FDT: FDT - Fast Data Transfer. http://monalisa.cern.ch/FDT/. Accessed Apr 2017
22. Settlemyer, B.W., Dobson, J.D., Hodson, S.W., Kuehn, J.A., Poole, S.W., Ruwart, T.M.: A technique for moving large data sets over high-performance long distance networks. In: 2011 IEEE 27th Symposium on Mass Storage Systems and Technologies (MSST), pp. 1–6, May 2011

23. Chard, K., Tuecke, S., Foster, I.: Globus: recent enhancements and future plans. In: XSEDE 2016 Conference on Diversity, Big Data, and Science at Scale, p. 27. ACM (2016)
24. Deelman, E., et al.: Pegasus: a workflow management system for science automation. Future Gener. Comput. Syst. **46**, 17–35 (2015)
25. Arslan, E., Guner, K., Kosar, T.: Harp: predictive transfer optimization based on historical analysis and real-time probing. In: Proceedings of the International Conference for High Performance Computing, Networking, Storage and Analysis, SC 2016, pp. 288–299, November 2016
26. Liu, Z., Kettimuthu, R., Foster, I., Beckman, P.H.: Towards a smart data transfer node. Future Gener. Comput. Syst. **89**, 10–18 (2018)
27. Arslan, E., Guner, K., Kosar, T.: HARP: predictive transfer optimization based on historical analysis and real-time probing. In: SC, Piscataway, NJ, USA, pp. 25:1–25:12 (2016). http://dl.acm.org/citation.cfm?id=3014904.3014938
28. Arslan, E., Kosar, T.: A heuristic approach to protocol tuning for high performance data transfers, ArXiv e-prints, August 2017
29. Kim, J., Yildirim, E., Kosar, T.: A highly-accurate and low-overhead prediction model for transfer throughput optimization. Clust. Comput. **18**(1), 41–59 (2015)
30. www.maxmind.com: MaxMind: IP Geolocation and Online Fraud Prevention (2017). https://www.maxmind.com. Accessed 3 Apr 2017
31. Maclin, R., Opitz, D.W.: Popular ensemble methods: an empirical study, CoRR, vol. abs/1106.0257 (2011). http://arxiv.org/abs/1106.0257
32. Hoerl, A.E., Kennard, R.W.: Ridge regression: biased estimation for nonorthogonal problems. Technometrics **12**(1), 55–67 (1970)
33. Breiman, L.: Bagging predictors. Mach. Learn. **24**(2), 123–140 (1996)
34. Ho, T.K.: Random decision forests. In: 3rd International Conference on Document Analysis and Recognition, ICDAR 1995, pp. 278–282. IEEE (1995). http://dl.acm.org/citation.cfm?id=844379.844681
35. Breiman, L.: Random forests. Mach. Learn. **45**(1), 5–32 (2001). https://doi.org/10.1023/A:1010933404324
36. Geurts, P., Ernst, D., Wehenkel, L.: Extremely randomized trees. Mach. Learn. **63**(1), 3–42 (2006)
37. Freund, Y., Schapire, R.E.: A desicion-theoretic generalization of on-line learning and an application to boosting. In: Vitányi, P. (ed.) EuroCOLT 1995. LNCS, vol. 904, pp. 23–37. Springer, Heidelberg (1995). https://doi.org/10.1007/3-540-59119-2_166
38. Friedman, J.H.: Greedy function approximation: a gradient boosting machine. Ann. Stat. **29**, 1189–1232 (2001)
39. Chen, T., Guestrin, C.: XGBoost: a scalable tree boosting system, arXiv preprint arXiv:1603.02754 (2016)
40. Bergstra, J., Bengio, Y.: Random search for hyper-parameter optimization. J. Mach. Learn. Res. **13**, 281–305 (2012). http://dl.acm.org/citation.cfm?id=2188385.2188395
41. Vazhkudai, S., Schopf, J.M., Foster, I.: Predicting the performance of wide area data transfers. In: International Parallel and Distributed Processing Symposium, 10-pp. IEEE (2001)
42. Swany, M., Wolski, R.: Multivariate resource performance forecasting in the Network Weather Service. In: Supercomputing Conference, p. 11. IEEE (2002)
43. Lu, D., Qiao, Y., Dinda, P.A., Bustamante, F.E.: Characterizing and predicting TCP throughput on the wide area network. In: 25th IEEE International Conference on Distributed Computing Systems, pp. 414–424. IEEE (2005)

44. He, Q., Dovrolis, C., Ammar, M.: On the predictability of large transfer TCP throughput. Comput. Netw. **51**(14), 3959–3977 (2007)
45. Huang, T.-i., Subhlok, J.: Fast pattern-based throughput prediction for TCP bulk transfers. In: International Symposium on Cluster Computing and the Grid, vol. 1, pp. 410–417. IEEE (2005)
46. Shah, S.M.H., ur Rehman, A., Khan, A.N., Shah, M.A.: TCP throughput estimation: a new neural networks model. In: International Conference on Emerging Technologies, pp. 94–98. IEEE (2007)
47. Mirza, M., Sommers, J., Barford, P., Zhu, X.: A machine learning approach to TCP throughput prediction. IEEE/ACM Trans. Netw. **18**(4), 1026–1039 (2010)
48. Kettimuthu, R., Vardoyan, G., Agrawal, G., Sadayappan, P.: Modeling and optimizing large-scale wide-area data transfers. In: 14th IEEE/ACM International Symposium on Cluster, Cloud and Grid Computing, pp. 196–205. IEEE (2014)
49. Nine, M., Guner, K., Kosar, T.: Hysteresis-based optimization of data transfer throughput. In: 5th International Workshop on Network-Aware Data Management, p. 5. ACM (2015)
50. Hours, H., Biersack, E., Loiseau, P.: A causal approach to the study of TCP performance. ACM Trans. Intell. Syst. Technol. (TIST) **7**(2), 25 (2016)
51. Liu, Z., Kettimuthu, R., Foster, I., Rao, N.S.V.: Cross-geography scientific data transferring trends and behavior. In: Proceedings of the 27th International Symposium on High-Performance Parallel and Distributed Computing, HPDC 2018, pp. 267–278. ACM, New York (2018). https://doi.org/10.1145/3208040.3208053
52. Liu, Z., Kettimuthu, R., Foster, I., Liu, Y.: A comprehensive study of wide area data movement at a scientific computing facility. In: IEEE International Conference on Distributed Computing Systems. Scalable Network Traffic Analytics. IEEE (2018)
53. Rao, N., Liu, Q., Sen, S., Liu, Z., Kettimuthu, R., Foster, I.: Measurements and analytics of wide-area file transfers over dedicated connections. In: 20th International Conference on Distributed Computing and Networking. ACM (2019)

Advanced Hybrid Technique in Detecting Cloud Web Application's Attacks

Meryem Amar(✉), Mouad Lemoudden, and Bouabid El Ouahidi

Mohammed-V University, I.P.S.S, Rabat, Morocco
Amar.meryem@gmail.com, mouad.lemoudden@gmail.com,
bouabid.ouahidi@gmail.com

Abstract. Recently cloud computing has emerged the IT world. It eventually promoted the acquisition of resources and services as needed, but it has also instilled fear and user's renunciations. However, Machine learning processing has proven high robustness in solving security flaws and reducing false alarm rates in detecting attacks. This paper, proposes a hybrid system that does not only labels behaviors based on machine learning algorithms using both misuse and anomaly-detection, but also highlights correlations between network relevant features, speeds up the updating of signatures dictionary and upgrades the analysis of user behavior.

Keywords: Attack-detection · Cloud · IDS · Machine learning · Security · Similarities

1 Introduction

Not long ago, software, technology and internet emerged our daily life. More than 54% of world population are using the web and the number continues to grow faster and faster [1]. As a result, of the big sizes of producing and requesting data, traditional technologies had become unable to provide resources and services as needed. Cloud computing overcomes these breakthroughs by providing on demand self-service, pay per use and multi-tenancy characteristics [2]. However, these properties are mainly based on virtualization and multi-tenant technologies which involves not only cloud risks such as putting sensitive data at stake and spreading unauthorized accesses but also traditional vulnerabilities [3, 4].

In fact, the simplicity of web application protocols and its increasing usage opens new security flaws that attackers easily make use of. They define new attack behaviors, steal and spoof identities. Traditional mechanisms are unable to provide an appropriate defense to protect both websites and their users.

According to OWASP top 10 vulnerabilities over the eleven years, more than 2/3 of the top 10 vulnerabilities in 2007 are still in the top 10 list of 2018 [5]. Most of these attacks are small variations of previously known cyber-attacks and preventing them is a priority for every user and company. Thus, it requires an immediate remediation and the first targeted server-side attacks are web applications.

Detecting attacks may be manually, by analyzing log files or automatically by providing automatic reactions and alerts when discovering an attack. Knowledge Based

É. Renault et al. (Eds.): MLN 2018, LNCS 11407, pp. 79–97, 2019.
https://doi.org/10.1007/978-3-030-19945-6_6

Intrusion Detection (KBID) [6] consists of comparing collected data to a dictionary of signatures, but this technique causes many false alarms, especially, when anomaly defined grounds are not updated or based on volatile characteristics. Not only that, but it is also static and does not calculate distance between a suspicious behavior and an attack and neither predicts new anomalies. The anomaly-based intrusion detection (ABID) overcomes theses drawbacks by building dynamically a normal profile and considers every unusual behavior as suspicious. Nevertheless, it does not consider attacks signatures in its process neither reduces the number of false positives.

Recent anomaly-based analysis is generally based on machine learning algorithms [7–10]. Machine Learning and deep learning analysis have shown a success in various big data fields and have drawn several interests in cyber security. Because of its high-level feature extraction capabilities and self-taught processing in uncovering hidden patterns and building profiles, Machine Learning techniques could be a resilient mechanism to small mutations in creating a novel attack.

The results of our previous works [11–13] has shown that applying the MapReduce algorithm over a big data of centralized log files enhances the system latency and reduces the resources acquisition. However, applying the FP-Growth mining approach [14] anticipates only at a level of reliability the following attack based on previously identified ones. Nevertheless, it does not predict new attacks or substitutes new features with their correlated ones.

In this paper, we proposed to address the detecting of attacks on a web application by providing a hybrid attack detection system. This system combines both a rule-based and an anomaly-based detection technique over centralized log files. The remainder of this paper is organized as follows. Section 2 describes the related works, Sect. 3 overviews basic concepts, Sect. 4 presents our training dataset, Sect. 5 gives a detailed encounter of our proposed solution, Sect. 6 provides a qualitative analysis of experimental results and final Sect. 7 summarizes the survey and highlight key motivations of our future work.

2 Related Works

Lots of research has been made in detecting malwares and especially IDS. This section overviews a number of researches made in the same concept.

Zahoor Rehman, Sidra Nasim and Sung Wook Baik have proposed a hybrid model to detect malware in Android Apps [15]. Their model is based on both signatures and heuristic analysis of the manifest.xml files. It compares, before each installation of new Android Package (APK), the constant strings of downloaded APK files to constant strings of malware applications and the manifest .xml files.

In the AI^2 solution Kaylan, Ignacio proposed an end-to-end active learning security system based on analyzing log files, applying both a supervised and an unsupervised learning algorithm to label new examples and updating the existing intrusion detection rules [16]. In this approach, the output of the unsupervised learning algorithm is validated by a human analyst. The system then takes into consideration the analyst interactions to update the labels of the supervised training data. In our case and in order to eliminate the risk of an intruder audit, the human analyst is replaced by the NSLKDD database which is constantly updated with the results of the previous analysis.

Yunaucheng Li, Rong Ma and Runhai Jiao have used deep learning approaches in detecting malicious codes, using autoencoders for feature extraction and Deep Belief Network (DBN) as a classifier [9]. Results showed that the proposed model is more accurate than using only DBN. However, its major limitation is the update of training dataset step. This process is basically different from ours. It updates the security system rules, merges different types of log files (audit, access, error, ssh error) to cover all relevant features, then analyze them together to get not only a network insight but also to learn each user's profile behavior at the application layer.

Lekaning Djionang and Gilber Tindo proposed a modular architecture to improve intrusion detection systems [17]. The first module is based on an ABID technique in detecting either the unknown behavior is legitimate or suspicious. If the input connection is suspicious it is communicated to the following modules where each module detects a unique type of attacks. To do so, the system starts by normalizing the NSLKDD dataset, and using a two classes neuron network. Then, it identifies the best number of neurons to use in hidden layer to have the best rate of recognition. Afterwards, each network is trained apart using backpropagation algorithm. Results showed the importance of feature selection in increasing/decreasing the performance rate. However, using a modular processing in the first may be of high false positive rates, especially when there is a new normal behavior that wasn't taken into consideration in the training phase. We overcome these limits by using a hybrid system that firstly updates the training dataset and calculates distance between a normal and the unknown behavior and compare it to a deviation measure when labeling it [13, 18].

3 Basic Concepts

3.1 Malware Detection in a Network

Several techniques have been proposed for detecting malware which can be divided into two classes: rule-based (misuse or static) and anomaly-based (dynamic) methods:

Rule-based analysis, is an approach in which a malicious behavior is detected based on static rules such as source code, serial number of developers, session fixation or the attack signatures. This strategy, compares activities to known signatures of attacks and has two models:

- Negative security model: also known as a blacklist approach has a default policy that considers everything as normal except for blacklisted connections flagged as attacks. Easy to implement and yields a very little number of false positives. Nevertheless, defined rules don't change and should be adapted to new findings, and new vectors in defining an attack.
- Positive Security model: In the of the Negative security model, the positive one denies all connections by defaults and allows only whitelist ones considered as normal and good traffic. Firewalls are usually configured this way. The whitelist is defined manually or during the learning phase.

Anomaly-based detection, known as behavioral-based analysis, uses ML algorithms, collects information at runtime and craft a model, then, it matches each new comportment to the crafted model. This strategy is based on dynamic rules and usually builds a profile or a baseline of "normal" traffic in a learning phase and considers every deviating behavior as suspicious. It is based on many features such as a number of command execution, a number of times that a specific server port has been opened, how many times they used or requested a root access.

Actually, static analysis is fast and effective, but various techniques can be used to dodge the system and enable it to detect polymorphic malware [19]. It may easily lack accuracy when attack's signatures are outdated or when the rule-based features don't select resilient ones, such as the Initial Sequence Number (ISN) in defining a Trojan that alters a lot or Zero-day attacks [14]. In the other hand, anomaly detections fail in a high number of false positives when the deviations from the model isn't fair enough (Fig. 1).

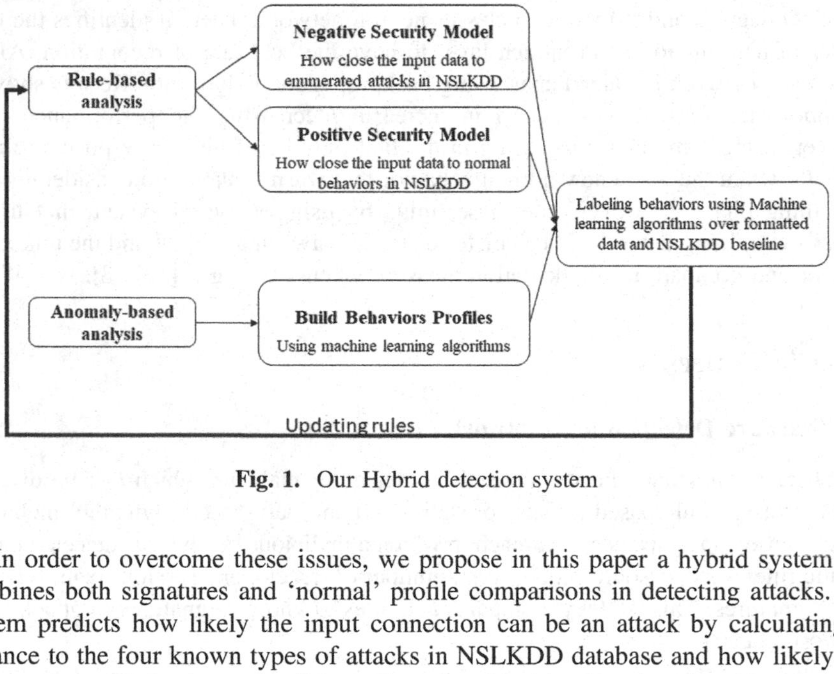

Fig. 1. Our Hybrid detection system

In order to overcome these issues, we propose in this paper a hybrid system that combines both signatures and 'normal' profile comparisons in detecting attacks. Our system predicts how likely the input connection can be an attack by calculating its distance to the four known types of attacks in NSLKDD database and how likely it is close to normal behavior and build profiles using Machine learning algorithm and then finds out the outliers.

3.2 Dynamic-Analysis and Performance Metrics

In cyber security, dynamic analysis is basically relying on machine learning algorithms in detecting malwares. The IDS, identifies the infection's patterns and controls the channels such as identifying usually used websites in simulating an attack and sorting out malicious executable from benign ones. In the opposite of static analysis that are very slow in detecting unknown attacks and requires less resources, dynamic analysis requires more resource in training the system at a runtime but are much better in identifying new unknown malwares. As a result, all recent contributions are on reducing the computational requirements and false positives key metrics [20].

Thus, to evaluate the truthfulness of each new system and decide whether to trust it or not, there are several performance metrics for attack detection such and that we have used to evaluate our solution.

Accuracy (ACC): It is the most common evaluation metrics for classification problem. It shows the overall effectiveness of a classifier. The accuracy, is the ratio of the number of correctly predicted behaviors to the overall behaviors [20].

$$ACC = \frac{TP + TN}{TP + TN + FP + FN} \tag{1}$$

Detection rate (DR): is the ratio of correctly predicted behaviors to the total number of perfect predictions and wrongly, negatively labeled behaviors.

$$DR = \frac{TP}{TP + FN} \tag{2}$$

False alarm rate (FAR): is the ratio of wrongly predicted behaviors, over the total number of true negatives and false positives, it generally calculates the erroneous of the detection system decision when it is caused by noise data.

$$FAR = \frac{FP}{TN + FP} \tag{3}$$

Precision: shows the agreement of the data labels with predicted labels given by the classifier. It is the ratio of correctly predicted positive behaviors to the total number of positive observations. The higher is the precision, the lower is the false positive rate and the better is the detection system.

$$\frac{TP}{TP + FP} \tag{4}$$

Recall: shows the effectiveness of a classifier in detecting positive labels. Also called sensitivity, is the ratio of correctly positive predicted behaviors to all actual class observations (TP and FN). The higher it is, the less are false negatives and true positives. Which make insufficient to decide whether a system is trusted or not.

$$Recall = \frac{TP}{TP + FN} \tag{5}$$

F1-Score: It shows the relation between data's positive labels and those given by the classifier. This score is more useful than accuracy when false positives and false negatives hasn't the same costs.

$$F1\ Measure = \frac{2.Recall.Precision}{Recall + Precision} \tag{6}$$

Receiver operating Characteristics (ROC) is a popular performance metric that evaluates and compares the robustness of different classifiers [21]. It is based on two basic evaluation measures: the FPR, FAR and sensitivity.

4 Training Dataset

The NSLKDD dataset is an improvement of KDDCUP'99 data set which is the most widely used training dataset in building profiles in the anomaly-detection based strategies. This data solves the inherent problems existing in the KDD'99 data by eradicating the number of redundant records, keeps the same scale of difficulties from selected rows to their original ones (KDD'99) [22]. Furthermore, the sizes of train and test dataset are reasonable which make them affordable for real time anomaly analysis.

Our training and test datasets contain 4 types of attacks (DOS, U2R, R2L, Probes) and a number of Normal signatures. The following figure resumes the number of different records in each type of behavior.

Our training dataset is composed of 41 features (listed in Table 1) [22], classified into four classes:

Table 1. Rules to calculate each NSLKDD feature

Features
Basic (B)
duration: The duration of a session
Protocol_type: Used protocol: tcp, udp, icmp
service: Destination network service: http, http_443, ftp, imap4…
src_bytes: Size of transferred data (in bytes) from source
dst_bytes: Size of downloaded data (in bytes)
flag: Status of the connection: reset, established, closed … For example: SF, RSTO, RSTR, S0, S1, S2, S3, REJ, SH…
land: It is a Boolean attribute that takes 1 if the source and destination IP addresses and port numbers are equals and 0 otherwise
wrong_fragment: The total number of wrong fragments in a connection
urgent: The number of urgent packets in the same connection
Content (C)
hot: Number of indicators considered as 'hot' in content, for example: entering a system directory, creating a program (XSS), executing a program (.exe)
num_failed_loggins: The number of failed attempts to log in
logged_in: 1 if the user is logged in and 0 otherwise
num_compromised: Number of compromised conditions in a connection request. This is usually, recorded in the 'error_log' file
root_shell: 1 if the user got the root shell permissions, 0 otherwise

(*continued*)

Table 1. (*continued*)

Features
su_attempted: 1 if 'su root' commands are attempted to get the root access permissions and 0 otherwise
num_root: Number of operations performed as a root user
num_file_creations: Number of file creations
num_shells: Number of shell prompts commands
num_access_files: Number of operations on access control files, 'htpasswd', 'passwd', 'htdocs', 'htaccess', 'httpd', '.conf'
num_outbound_cmds: Number of outbound commands in an ftp session
is_host_login: 1 if the login belongs to the 'hot' list (root or admin). Rule: if root_shell = 0 then is_host_login = 0 else if hot = 0 then is_host_login = 0 else if su_attempt! = 0 then is_host_login = 1
is_guest_login: 1 if the host address has a 'guest' request and 0 otherwise
Traffic (T)
count: Number of connections to the same host destination in the two seconds window time
Srv_count: Number of connections to the same service in the two seconds window time
serror_rate: The percentage of connections that have activated: S0, S1, S2 or S3 flags. Among connections in 'count' Rule: Card({S0, S1,S2,S3} in count)/count
rerror_rate: The percentage of connections that have activated REJ flag among connections in 'count'. Rule: Card(REJ in count)/count
Same_srv_rate: The percentage of connections that were to the same 'service', among connections in 'count' Rule: Card(Connections to same port in count)/count
Diff_srv_rate: The percentage of connections that were to different services among connections in 'count' Rule: 1-same_srv_rate
Srv_serror_rate: The percentage of connections that have activated S0, S1, S2 or S3 flags among connections in 'srv_count' Rule:Card({S0,S1,S2,S3} in srv_count)/srv_count
Srv_rerror_rate: The percentage of connections that have activated REJ flag among connections in 'srv_count' Rule: Card(REJ in srv_count)/srv_count
Srv_diff_host_rate: The percentage of connections that were to different destination machines among the connections in srv_count. Rule: Card({connections to different hosts in srv_count)/srv_count
Host (H)
Dst_host_count: Number of connections having the same destination IP address. Rule: Card({same host destination}
Dst_host_srv_count: Number of connections having the same port number. Rule: Card({same service})
Dst_host_same_srv_rate: The percentage of connections to the same service among connections in 'dst_host_count'

(*continued*)

Table 1. (*continued*)

Features
Rule: Card(same service and same host detination)/dst_host_count
Dst_host_diff_srv_rate: The percentage of connections to different services among connections in 'dst_host_count' Rule: 1-dst_host_same_srv_rate
Dst_host_same_src_port_rate: The percentage of connections that were to the same source port among connections in 'dst_host_srv_count'. Considering the fact that each new connection opens a new source port. Rule: 1/dst_host_srv_count
Dst_host_srv_diff_host_rate: The percentage of connections that were to different destination machines among connections in 'dst_host_srv_count' Rule:1-dst_host_same_src_port_rate
Dst_host_serror_rate: The percentage of connections that have activated one of the flags S0, S1, S2 or S3 among connections in 'dst_host_count' Rule:Card({S0,S1,S2,S3}) in dst_host_count/dst_host_count
Dst_host_srv_serror_rate: The percentage of connections that have activated one of the flags S0, S1, S2 or S3 among the connections in 'dst_host_srv_count' Rule: Card({S0, S1, S2, S3} in dst_host_srv_count)/dst_host_srv_count
Dst_host_rerror_rate: The percentage of connections that have activated REJ flag among connections in 'dst_host_count' Rule: Card(REJ in dst_host_count)/dst_host_count
Dst_host_srv_rerror_rate: The percentage of connections that have activated REJ flag among connections in dst_host_srv_count Rule: Card(REJ in dst_host_srv_count)/dst_host_srv_count

- Basic (B): where the features are related to the individual TCP connections, which means these attributes combines collected data in the network layer
- Content (C): are the attributes that define the domain knowledge (the level of user permissions when requesting a connection, the number of compromised rules, the number of attempts to access sensitive files…)
- Traffic (T): they are the features calculated using the two-second time window and the information are related to the B features in the same time window.
- Host (H): are features that identify attacks lasting for more than two seconds. These attributes are related to B and C ones.

5 Proposed Solution

To fill up existing gaps in dynamic and rule-based detection, we proposed in this paper a hybrid approach that do not only combines ABID and KBID but considers also relevant factors in detecting intrusions such as the performance, the false positives, the adaptability and scalability. The system generates also dynamically correlation's network features and updates constantly the training data to reduce the number of false positives.

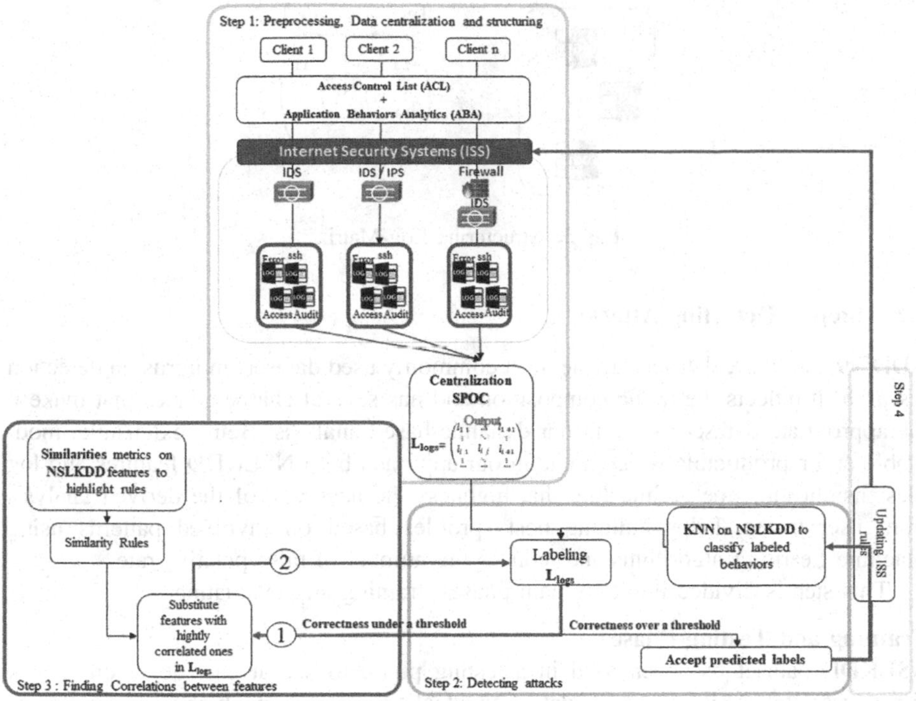

Fig. 2. Proposed Solution

5.1 Step 1: Preprocessing, Data Centralization and Structuring

In this step we start by taking away all generated log files into a Single Point of Contact (SPOC), a Cloud Provider [12, 13]. Centralization upgrades the security mechanism. It unable attackers from deleting their traces, helps deriving better insight from used machine learning algorithm in sub-Sect. 3 by putting together correlations in a variety of security measures deriving from different connected clients. Generated log files amount to a size that encourages the use of predictive algorithms suitable to a big data.

In order to get user, network and application information we use altogether different types of log files, ssh and error files bring information at the network layer level and access and audit files logs application and users' behaviors, structuring the input data is primeval. For these reasons, we map logged data into a single matrix of 41 features corresponding exactly to NSLKDD attributes having the variety of relevant features to detect attacks in network, transport and application layers.

- Let $E = (e_1\ldots, e_{41})$ be a set of NSLKDD attributes, e_j the jth attribute of NSLKDD dataset.

Let $L = \{l_1\ldots, l_n\}$ be a set of n log records, each with dimension 41. Where a_i is the ith projected log row on E and a_{ij} the corresponding value of e_j in the ith log row (Fig. 3).

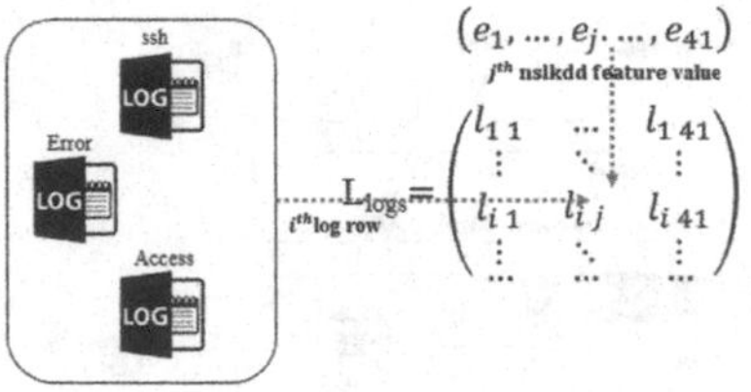

Fig. 3. Structuring Log Matrix

5.2 Step 2: Detecting Attacks

KDD'99 and ISCX datasets are the most commonly used datasets in intrusion detection research. It reflects the traffic composition and has several characteristics that make it the appropriate dataset to use in our dynamic-based analysis. Being extensible, modifiable and reproducible we combine in our approach both NSLKDD features and log files insight into one architecture that highness the accuracy of the derived analysis from user traces, helps building hosts profiles based on involved patterns using Machine Learning algorithms and reduces the number of false positive rate.

This step is divided into two main phases: training and evaluating.

Training and Testing Phases

NSLKDD dataset has been used in a testing phase to see at each level our attack signatures database is accurate in detecting attacks and to select which algorithm to use in labeling our Log structured matrix. This dataset is available in csv format.

In this step we considered four machine learning algorithms for traffic classification and behavior labeling. K-nearest neighbors, Naïve Bayes, Support Vector Machine (SVM) and Logistic Regression (LR) [23]. Classifications are made in a two class and five class behaviors cases.

Figure 4 presents confusions matrix for each applied algorithm and Table 2 explicit the precision values of all traffic categories with KNN, Naïve Bayes, SVM and LR. It can be seen that for our experimental dataset, the classification model based on KNN produces higher precision for all five behavioral classes and is more stable.

Results showed that KNN is the most appropriate algorithm in classifying traffic network data, with 98,80% accuracy, 99.80% precision, 98.80% of Recall and only 0.9% of false positive rate.

In order to identify the level at which the system keeps good accuracy, sensitivity and precision in detecting attacks when the class labels are reduced to only two (normal and attack categories), we made the same previous test on two classes training and testing dataset. Results in Table 3 showed that even with one attack class and normal behaviors, the KNN algorithm keeps a high accuracy and low false positive rate.

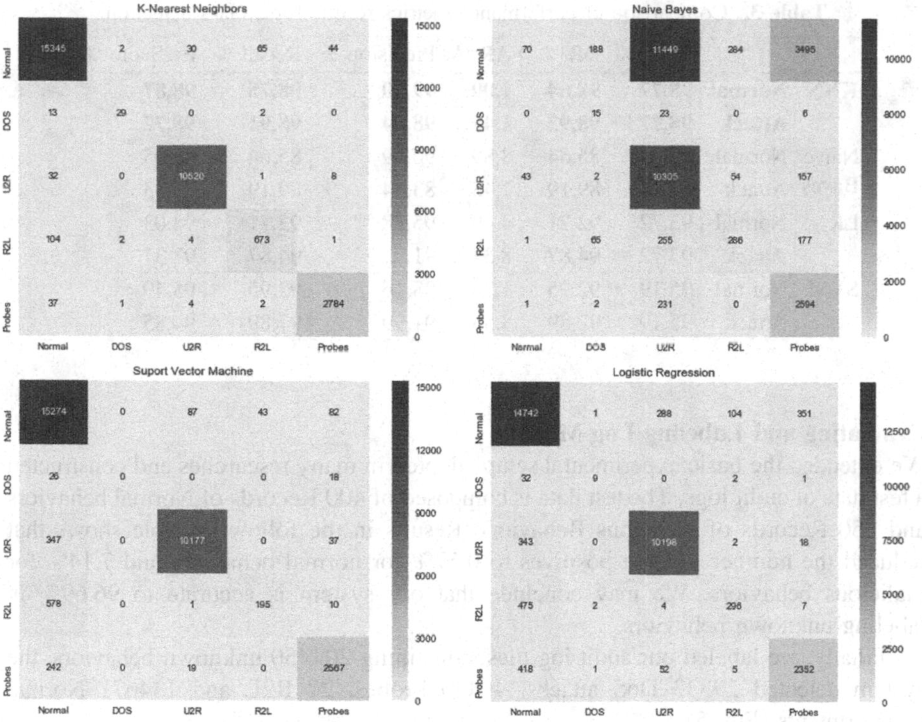

Fig. 4. Five behavior's confusion matrices for KNN, RF, SVM and LR

Table 2. Performance metrics results for 5 behaviors detection

		ACC %	DR %	FAR %	Precision %	Recall %	F1-Score %
KNN	**Normal**	**98,89**	**98,80**	**0,91**	**99,09**	**98,80**	**98,95**
	DOS	**99,93**	**85,29**	**44,12**	**65,91**	**85,29**	**74,36**
	U2R	**99,73**	**99,64**	**0,39**	**99,61**	**99,64**	**99,63**
	R2L	**99,39**	**90,58**	**14,94**	**85,84**	**90,58**	**88,15**
	Probes	**99,67**	**98,13**	**1,55**	**98,44**	**98,13**	**98,29**
Naïve Bayes	Normal	47,95	60,87	13405,22	0,45	60,87	0,90
	DOS	99,04	5,51	10,66	34,09	5,51	9,49
	U2R	58,88	46,29	1,15	97,58	46,29	62,79
	R2L	97,19	45,83	79,81	36,48	45,83	40,63
	Probes	86,30	40,35	3,64	91,73	40,35	56,04
LR	Normal	93,23	92,08	4,65	95,20	92,08	93,61
	DOS	99,87	69,23	269,23	20,45	69,23	31,58
	U2R	97,62	96,74	3,44	96,56	96,74	96,65
	R2L	97,98	72,37	119,32	37,76	72,37	49,62
	Probes	97,13	86,19	17,44	83,17	86,19	84,65
SVM	Normal	95,27	92,76	1,29	98,63	92,76	95,60
	DOS	99,85	-	-	0,00	-	0,00
	U2R	98,35	98,96	3,73	96,36	98,96	97,64
	R2L	97,87	81,93	247,48	24,87	81,93	38,16
	Probes	98,63	94,58	9,62	90,77	94,58	92,64

Table 3. Comparison of performance metrics results for 2 class detection

		ACC %	DR %	FAR %	Precision %	Recall %	F1-Score %
KNN	**Normal**	**98,77**	**98,64**	**1,00**	**99,00**	**98,75**	**98,87**
	Attack	**98,77**	**98,92**	**1,47**	**98,64**	**98,91**	**98,77**
Naïve Bayes	Normal	87,24	85,64	8,89	90,59	85,64	88,05
	Attack	87,24	89,19	17,45	83,64	89,19	86,33
LR	Normal	93,72	92,71	4,49	95,38	92,71	94,03
	Attack	93,72	94,87	8,34	91,92	94,87	93,37
SVM	Normal	95,19	92,95	1,74	98,16	92,95	95,49
	Attack	95,19	97,89	8,53	91,99	97,89	94,85

Evaluating and Labeling Log Matrix

We extended the basic experimental setup adopted in many researches and constructed a test data of audit logs. The test data is composed of 400 Records of Normal behaviors and 350 Records of malicious Behaviors. Results in the following table shows that reduced the number of false positives to 0.57% for normal behaviors and 7.14% for malicious behaviors. We may conclude that our system is accurate to 96.69% in labeling unknown behaviors.

Finally, we labeled our audit log files, containing 204450 unknown behaviors, the system detected 29333 DoS attacks, 40420 Probes, 27 R2L and 134671 Normal comportments (Fig. 5).

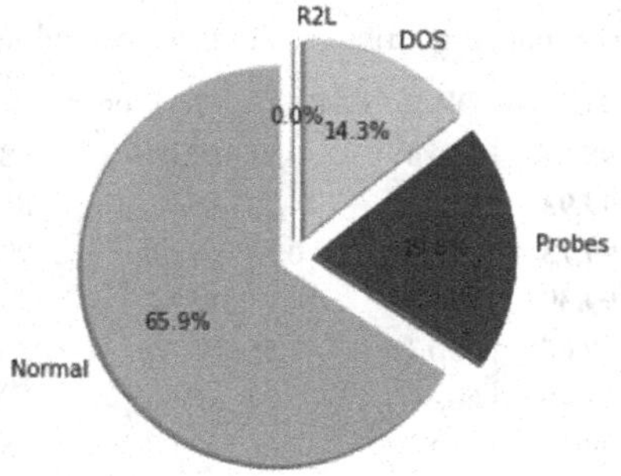

Fig. 5. Percentage of each detected behavior in web log

5.3 Finding Correlations Between Features

Finding correlations between our training dataset features enables the system to discover new malicious behaviors and identify new signatures in identifying an attack. Furthermore, the rest of unlabeled data from formatted audit logs matrix can be labeled by substituting their highly correlated features. The new substituted matrix is after that communicated to the previous step to be analyzed and labeled again (1 and 2 steps in Fig. 2), till the system converges.

In order to yield similarity scores between the different objects in our training data set, we have used several similarity metrics each one depending on the prototype of the features [24, 25].

Jaccard Coefficient: is used when dealing with data objects having binary attributes, the objects are generally unordered set of a collection of data. In our case it is taken into consideration when seeing the similarities of binary features such as logged_in, root_shell, is_host_login, is_guest_loggin…

$$J(X_i, X_j) = \frac{X_i \cap X_j}{X_i \cup X_j} \tag{8}$$

Our binary features are six, and from the results of Jaccard similarity matrix we can presume that whether the user is a guest login or not isn't important in defining a behavior and may be deleted from our dataset, and that the land of connection, the status of the connection, the root shell privileges and the attempts to get the root user permissions are highly correlated to each other and can be reduced (Fig. 6).

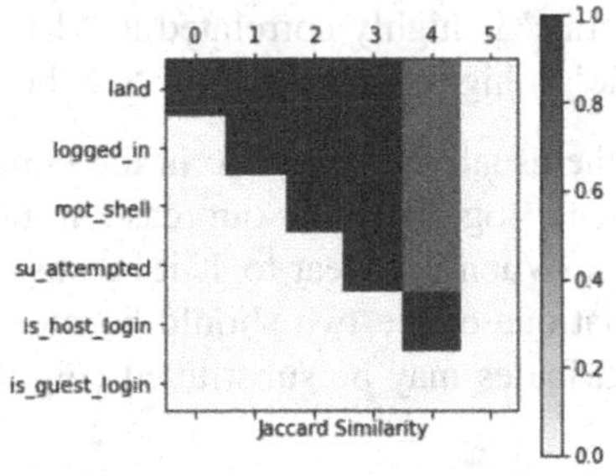

Fig. 6. Jaccard similarity results

Correlation metric: Is investigated when seeking for relationship between two quantitative, continuous variables, it measures the strength of the association between attack's features (Fig. 7).

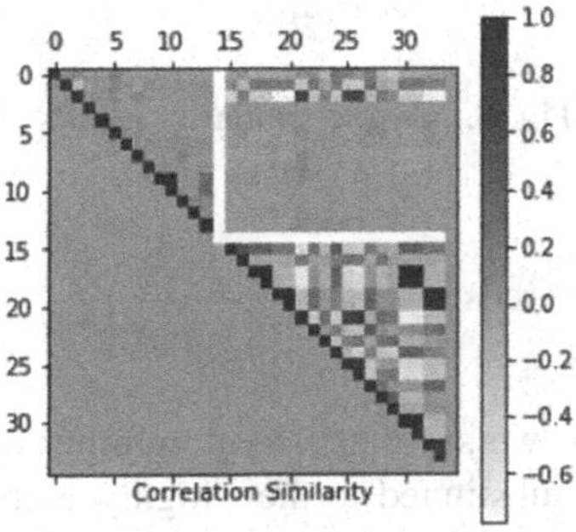

Fig. 7. Correlation similarity results

$$\rho_{X_iX_j} = \frac{cov(X_i, X_j)}{\sigma_{X_i}.\sigma_{X_j}} \tag{9}$$

- $cov(X_1, X_2)$: is the covariance measure
- σ_{X_i}: is the standard deviation measure.

From the Correlation measures, we see that:

- 'srv_count' is highly correlated to 'serror_rate', 'dst_host_srv_diff_host_rate', and 'dst_host_serror_rate'
- 'serror_rate' is highly correlated to 'dst_host_srv_diff_host_rate' and 'dst_host_serror_Rate'
- 'srv_serror_rate' is highly correlated to 'rerror_rate', 'dst_host_srv_serror_rate' and 'dst_host_rerror_rate'
- 'rerror_rate' is highly correlated to 'dst_host_srv_serror_Rate' and 'dst_host_rerror_rate'
- 'dst_host_count' is highly correlated to 'dst_host_srv_count'
- 'dst_host_srv_diff_host_rate' is highly correlated to 'dst_host_serror_rate'
- 'dst_host_srv_serror_rate' is highly correlated to 'dst_host_rerror_rate'

Cosine similarity: It is the usually used metric in the context of text mining, while comparing documents, spams, log files…in our case it underlines the normalized product of each two features. When it is near to 1, it means that the two features have the same orientation and that one of the two should be more than enough to find the other one. Thus, the two attributes may be substituted (Fig. 8).

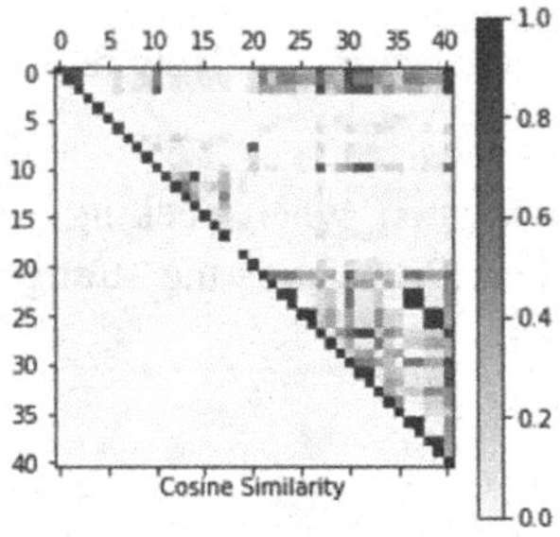

Fig. 8. Cosine similarity results

$$\cos(X_i, X_j) = \frac{X_i.X_j}{\|X_i\|\|X_j\|} \tag{10}$$

From similarities metrics, we see that a large number of features are correlated to each other, and thus can be substituted to their highly correlated ones and drastically decrease the number of features following some rules:

- Rule 1: Duration is highly correlated to: protocol type, service and the percentage of connections to different hosts having the same service.
- Rule 2: Protocol type is highly correlated to the percentage of connections to different hosts into the same port destination.
- Rule 3: the number of connections to the same service in the same two seconds window time is highly correlated to the number of connections to the same port outside the time window and to percentage of connections that activated one of the flags S0, S1, S2 or S3 in the same period of time.
- Rule 4: the number of connections to the same destination host and have activated the S0, S1, S2 or S3 flags in a 2 s time window is mainly correlated to the its number when it is to the same destination and overpasses the range time and the number of connections that were to different port destination outside the time window.
- Rule 5: the number of connections to the same service and have activated the S flags is proportional to the number of rejected connections to the same host destination in the same time window and to the number of rejected connections in a different time window to the same host destination. It is also related to the number of connections the same service that have activated the S flags outside the time window.
- Rule 6: the percentage of rejected connections that were to same host destination in the 2 s is highly related to its percentage when it is still to the same destination but outside the time window. It is also related the number of connections to the same service that have activated the S flags outside the time window.
- Rule 7: the number of connections that were to the same host destination are strongly related to the number of connections that were in the same service.
- Rule 8: the percentage of connections that were to different destination machines, having the same service and outside the time window is proportional to the percentage of connections that have activated the S flags when requesting the same destination host.
- Rule 9: the percentage of connections outside the time window to the same service and have activated the S flags is proportional to the percentage of rejected connections to the same host machine that were outside the time window.

5.4 Step 4: Updating the AppISS Rules

In this step, the cloud auditor updates the tracking security system rules. It adds eligible signatures to the training dataset and evaluates constantly that a high accuracy and precision are maintained to get very slow false positive rates.

Given that our model robustness is based on the AUC calculation too, which tends to be influenced by the imbalance of positive and negative data. We repeated our basic labeling experiment, adding normal and malicious audit logs samples to training dataset.

Those changes to the training/testing set showed obvious differences in the accuracy of the unknown behaviors identification, inferior results with the low update rate, and a continuous increasing of the system robustness by updates.

Finally, from AUC surfaces in Fig. 9, results show that the more our training dataset is complete and accurate the better are our prediction rate's, accuracies and ROC surfaces.

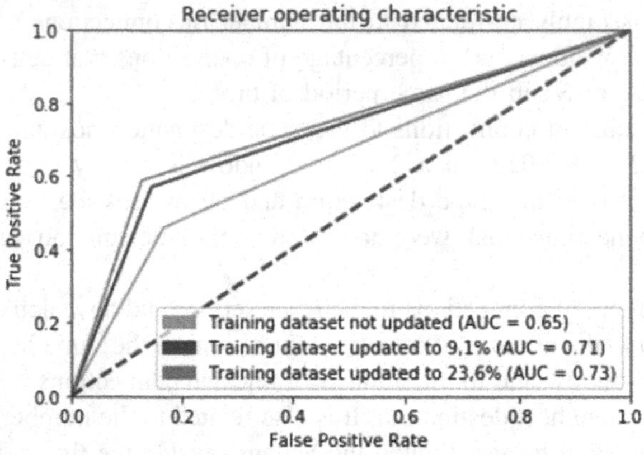

Fig. 9. ROC comparison of Update effect

6 Experiment Results

Our experiment has many objectives. Enhancing the velocity of our system using Big data technique, Selecting the best classifier over audit logs and NSLKDD datasets, finding correlation between network feature to finally detect attacks based on web application log files and our hybrid attack detection system.

NSLKDD dataset is used as a training dataset and log files as testing, evaluating and tracking database that helps us secure the system. It contains 61685 signatures of normal behaviors, 42706 rows of DOS attacks, 197 records of U2R attacks, 3006 of R2L different signatures and 11217 data rows of Probes.

The audit logs used in our experiment is from The Honeynet Project, challenge scan 31 [26]. It traces 204450 unknown behaviors that we aim to label. As the fragments in audit logs were structured in Preprocessing step, both NSLKDD and L matrix have four categories of features, basic (B) ones related to TCP connections, domain knowledge (D) containing both connection's status and user attempts, the third category has information about a traffic network in a 2 s window and long-term attacks.

The experimental result has demonstrated several standards. The first one is the detection rate in Table 3. It exhibits that KNN is better than Naïve Bayes, Random Forest and Support Vector Machine for both binary and multi-classes. As it can be seen, KNN is 98,64% to 98,77% accurate in labeling behaviors, data agree with positive labels to 99% and a very slow false positive rate around 1% to 1,47%.

We also evaluated the robustness of the proposed architecture, applying a second test phase on formatted audit logs matrix (L). The detection rate in Table 4 also shows that KNN is the best classifier with an accuracy of 96,36%, an overall precision factor

Table 4. Performance metrics for audit logs labeling

		ACC %	DR %	FAR %	Precision %	Recall %	F1-Score %
KNN	Normal	**96,69**	**94,89**	**0,57**	**99,40**	**94,89**	**97,09**
	Attack	**96,69**	**99,21**	**7,14**	**93,28**	**99,21**	**96,15**
Naïve Bayes	Normal	45,36	53,19	304,26	14,88	53,19	23,26
	Attack	45,36	43,92	8,63	83,58	43,92	57,58
LR	Normal	59,93	75,27	105,38	41,67	75,27	53,64
	Attack	59,93	53,11	11,00	82,84	53,11	64,72
SVM	Normal	79,14	88,89	35,56	71,43	88,89	79,21
	Attack	79,14	71,26	8,98	88,81	71,26	79,07

around 93,28% to over 99,4%. It has also a low score of false positives that varies from 0,57% to 7,14%.

The experiment also evaluated the importance of centralization in increasing the network insight, selecting new attack signatures and finally updating the training dataset to enhance the system robustness.

7 Conclusion and Perspectives

This paper proposed a novel hybrid approach in detecting attacks named MLSecure to dynamically create a pattern and profiles of network behaviors based on different log categories. It provides discussions about different techniques in detecting attacks and a detailed encounter of our proposed solution.

In this study, we centralized different categories of log files and combined the insight of each in a unique formatted matrix. This first step disables malicious users from modifying or deleting their traces. In the Cloud Provider we applied a basic MapReduce programming to handle the big volume of the matrix and assign a weight to each behavior. After that we classified NSLKDD considered as a training and a dictionary of signatures and choose the best classifier to apply for it. After that, we evaluated the robustness of our hybrid architecture when the unlabeled data are the matrix of formatted log files. Finally, the system updates dynamically the rule-based signature of the training dataset.

The experiment has shown the successful adoption of machine learning to cyber security. The evaluation process employs the accuracy (ACC), the detection rate (DR), false alarm rate (FAR) precision, recall, F1-measure and Receiver Operating Characteristic (ROC) as performance metrics to show the effectiveness of our hybrid model. In the future work, we will reduce the number of involved features in network traffic analysis and use an anomaly detection process only to determine at each probability or deviation from the normal behavior is an unknown behavior a malicious attempt. Studying these deviations, also called outliers is the motivation of our next search project.

References

1. Stats, I.W.: World Internet Users Statistics and 2018 World Population Stats (2018). [En ligne]. https://www.internetworldstats.com/stats.htm
2. Mell, P., Grance, T.: The NIST Definition of Cloud Computing (2011)
3. Ali, M., Khan, S.U., Vasilakos, A.V.: Security in cloud computing: opportunities and challenges. Inf. Sci. **305**, 357–383 (2015)
4. Ramachandra, G., Iftikhar, M., Aslam Khan, F.: A comprehensive survey on security in cloud computing. Procedia Comput. Sci. **110**, 467–472 (2017)
5. Passi, H.: OWASP - Top 10 Vulnerabilities in web applications (updated for 2018 …), 2 January 2018. [En ligne]. https://www.greycampus.com/blog/information-security/owasp-top-vulnerabilities-in-web-applications
6. Shah, J.: Understanding and study of intrusion detection systems for various networks and domains. In: Computer Communication and Informatics (ICCCI) (2017)
7. Indraneel, S., Praveen, V., Vuppala, K.: HTTP flood attack detection in application layer using machine learning metrics and bio inspired bat algorithm. Appl. Comput. Inform. (2017)
8. Moshfeq Salaken, S., Khosravi, A., Nguyen, T., Nahavandi, S.: Extreme learning machine based transfer learning algorithms: a survey. Neurocomputing **267**, 516–524 (2017)
9. Li, Y., Ma, R., Jiao, R.: A hybrid malicious code detection method based on deep learning. Int. J. Secur. Appl. **9**, 205–216 (2015)
10. Diro, A.A., Chilamkurti, N.: Distributed Attack Detection Scheme Using Deep Learning approach for Internet of Things. Future Generation Computer Systems **82**, 761–768 (2017)
11. Lemoudden, M., Amar, M., El Ouahidi, B.: A binary-based MapReduce analysis for cloud logs. Procedia Comput. Sci. **83**, 1213–1218 (2016)
12. Amar, M., Douzi, S., El Ouahidi, B., Lemoudden, M.: A novel approach in detecting intrusions using NSLKDD database and MapReduce programming. Procedia Comput. Sci. **110**, 230–235 (2017)
13. Amar, M., Douzi S., El Ouahidi, B.: Enhancing cloud security using advanced MapReduce k-means on log files. In: ACM International Conference Proceeding Series, pp. 63–67 (2018)
14. Amar, M., Lemoudden, M., El Ouahidi, M.: Log file's centralization to improve cloud security. In: IEEE Xplore, Cloud Computing Technologies and Applications (CloudTech) (2016)
15. Rehman, Z.-U., et al.: Machine learning-assisted signature and heuristic-based detection of malwares in Android devices. Comput. Electr. Eng. **69**, 1–14 (2017)
16. Veeramachaneni, K., Arnaldo, I.: AI^2: training a big data machine to defend. In: Big Data Security on Cloud (BigDataSecurity), IEEE International Conference on High Performance and Smart Computing (HPSC), and IEEE International Conference on Intelligent Data and Security (IDS) (2016)
17. Hervé, B., Tindo, G.: Vers une Nouvelle Architecture de Detection d'Intrusion Reseaux à Base de Reseaux Neuronaux. HAL Archives-ouvertes (2016)
18. Ya, J., Liu, T., Li, Q., Shi, J., Zhang, H., Lv, P.: Mining host behavior patterns from massive network and security logs. Procedia Comput. Sci. **108**, 38–47 (2017)
19. Cathey, R., Ma, L., Goharian, N., Grossman, D.: Misuse Detection for Information Retrieval Systems (2003)
20. Fan, Z., Liu, R.: Investigation of machine learning based network traffic classification. In: Wireless Communication Systems (ISWCS) (2017)
21. Katzir, Z., Elovici, Y.: Quantifying the resilience of machine learning classifiers used for cyber security. Expert Syst. Appl. **92**, 419–429 (2018)

22. Aggarwal, P., Sharma, S.K.: Analysis of KDD dataset attributes - class wise for intrusion detection. Procedia Comput. Sci. **57**, 842–851 (2015)
23. Belavagi, M., Muniyal, B.: Performance evaluation of supervised machine learning algorithms for intrusion detection. Procedia Computer Science **89**, 117–123 (2016)
24. Polamuri, S.: FIve Most Popular Similarity Measures Implementation in Python (2015). [En ligne]. http://dataaspirant.com/2015/04/11/five-most-popular-similarity-measures-implementation-in-python/
25. Mining Similarity Using Euclidean Distance, Pearson Correlation, and Filtering (2010). [En ligne]. http://mines.humanoriented.com/classes/2010/fall/csci568/portfolio_exports/mvoget/similarity/similarity.html
26. SotM 31 - The Honeynet Project (2004). [En ligne]. http://www.honeynet.org/scans/scan31/

Machine-Learned Classifiers for Protocol Selection on a Shared Network

Hamidreza Anvari(✉), Jesse Huard, and Paul Lu

Department of Computing Science, University of Alberta, Edmonton, AB, Canada
{hanvari,jhuard,paullu}@ualberta.ca

Abstract. Knowledge about the state of a data network can be used to achieve high performance. For example, knowledge about the protocols in use by background traffic might influence which protocol to choose for a new foreground data transfer. Unfortunately, global knowledge can be difficult to obtain in a dynamic distributed system like a wide-area network (WAN).

Therefore, we introduce and evaluate a machine-learning (ML) approach to network performance, called *optimization through protocol selection (OPS)*. Using local round-trip time (RTT) time-series data, a classifier predicts the mix of protocols in current use. Then, a decision process selects the best protocol to use for the new foreground transfer, so as to maximize throughput while maintaining fairness. We show that a protocol oracle would choose TCP-BBR for the new foreground traffic if TCP-BBR is already in use in the background, for proper throughput. Similarly, the protocol oracle would choose TCP-CUBIC for the new foreground traffic if only TCP-CUBIC is in use in the background, for fairness.

Empirically, our k-nearest-neighbour (K-NN) classifier, utilizing dynamic time warping (DTW) measure, results in a protocol decision accuracy of 0.80 for $k = 1$. The OPS approach's throughput is 4 times higher than that achieved with a suboptimal protocol choice. Furthermore, the OPS approach has a Jain fairness index of 0.96 to 0.99, as compared to a Jain fairness of 0.60 to 0.62, if a suboptimal protocol is selected.

Keywords: Machine-learned classifier · Protocol selection · Data transfer · Wide-area networks · High-performance network · Fairness · Shared network

1 Introduction

Knowledge about the state of a data network can be used to achieve high performance. For example, knowing which TCP congestion control algorithms (CCA) are being used can be important. Currently, Microsoft Windows uses the Compound CCA, Linux uses the CUBIC CCA (TCP-CUBIC [14], hereafter CUBIC), and Google has been developing the BBR CCA (TCP-BBR [9], hereafter BBR).

E. Renault et al. (Eds.): MLN 2018, LNCS 11407, pp. 98–116, 2019.
https://doi.org/10.1007/978-3-030-19945-6_7

Since, arguably, on any given wide-area network (WAN), there is a mix of protocols in use, knowledge of the background traffic might influence which protocol to choose for a new foreground data transfer.

On the one hand, the mainstream approach in the literature is to tune the parameters of a network protocol, either online (e.g., congestion window [9,14]) or offline (e.g., machine learning [23]). On the other hand, our introduced approach, *optimization through protocol selection (OPS)* is based on choosing an entirely different network protocol in response to the background traffic, which might have a greater impact on performance than tuning protocol parameters. For example, in previous work [4,5], we empirically showed that for bulk-data transfers across a shared WAN, if the background traffic is known to be largely UDP-based, then an appropriate UDP-based protocol like UDT [13] will achieve high throughput for the new foreground traffic. Using a TCP-based protocol with prominent UDP-based traffic resulted in poor overall performance. With OPS, one does not have to invent or tune a new TCP or UDP protocol; one could re-use UDT.

Unfortunately, global knowledge can be difficult to obtain in a dynamic distributed system like a WAN. In a single datacentre or local-area network (LAN), the hardware and software stacks can cooperate to perform flow control. And, network management systems can provide global policy control based on global information. But, it is rare that all paths along a WAN are under a single administrative domain, so many WAN protocols can only react to local network signals in the form of packet loss, packet reordering, or round-trip time (RTT).

Therefore, our main contribution is the introduction of the *optimization through protocol selection (OPS)* mechanism, as well as the design and evaluation of a machine-learning (ML) approach to OPS. The specific ML algorithms and techniques used are simple, but, as a proof-of-concept, showing that OPS can work is notable. Using only RTT time-series data, a classifier predicts the mix of protocols in current use. Then, a decision process selects the best protocol to use for the new foreground transfer, so as to maximize throughput without being too unfair.

Specifically, if BBR is in use in the background, then BBR should also be selected for the new foreground traffic to grab a fair share of the bandwidth. Although the (BBR, BBR) combination of Fig. 1 is not entirely equal, the (BBR, BBR) combination is much more fair than other combinations. In fact, the circa 80 Mb/s imbalance in the (BBR, BBR) combination is mostly due to the foreground traffic having a shorter live time than the background traffic (Sect. 3).

Our empirical data (Fig. 1) also shows that using CUBIC for a foreground data transfer, while BBR is in use in the background, results in lower throughput for the CUBIC stream. Both the (BBR, CUBIC) and (CUBIC, BBR) combinations of Fig. 1 show high unfairness for the CUBIC stream. But, if the background traffic is using CUBIC, then the foreground traffic should also use CUBIC, as shown by the (CUBIC, CUBIC) combination (Fig. 1). And, once again, the apparent imbalance in the (CUBIC, CUBIC) combination is mostly due to the foreground traffic having a shorter live time than the background traffic.

Contrast (CUBIC, CUBIC) with (CUBIC, BBR) where the BBR foreground stream has a shorter live time but still grabs substantially more throughput than the background CUBIC stream (Fig. 1). In fairness, it is important to note that BBR is still in active development at the Internet Engineering Task Force (IETF) and our empirical results should not be taken as a critique. We use CUBIC and BBR for a case study on how the mix of protocols affect performance, and on how OPS can be applied.

For an empirical workload (Sect. 5.3), our k-nearest-neighbour (K-NN) classifier, with $k = 1$, utilizing dynamic time warping (DTW) distance measure, shows modest classification accuracy for the background protocols (between 0.36 and 0.81), but it is sufficient for a protocol decision accuracy of between 0.61 and 0.81. Consequently, the ML approach to OPS achieves throughput close to ideal despite not having a protocol oracle, while avoiding the worst-case performance from a suboptimal protocol choice (Fig. 10).

Using five-fold cross-validation (Fig. 7), we have been able to reach up to 80% average F1-scores in predicting the type of background traffic while deciding between the 6 possible traffic classes. We also experimented with other ML approaches (e.g., naive Bayes, Support Vector Machines, neural networks) with mixed-to-comparable results, but use 1-NN to show the basic efficacy of OPS. Despite violating the feature-independence assumption of some of these classifiers, it was important to at least quantitatively compare multiple algorithms. A parameter-sweep study of the algorithms is left for future work.

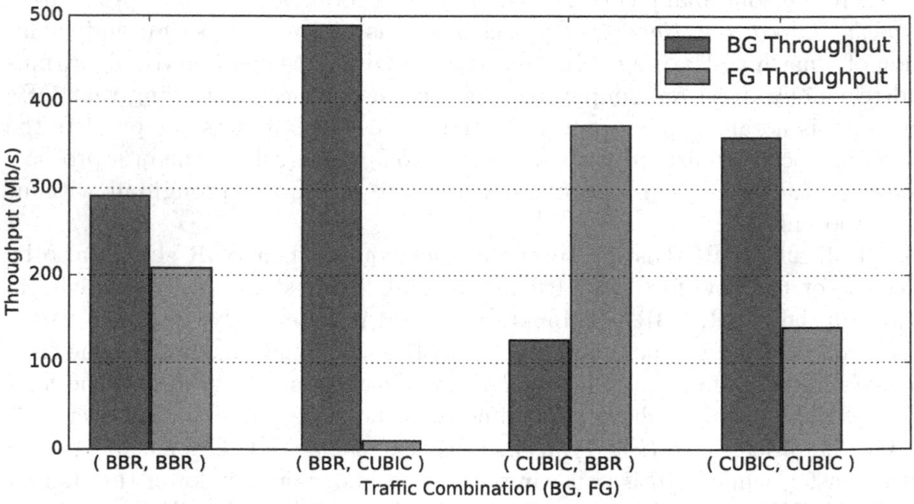

Fig. 1. Variable and unfair throughput in shared networks (*iperf*, FG has less live time)

2 Background and Related Work

2.1 Shared Computer Networks

One primary goal of many network protocols, exemplified by TCP, is to provide reliable data transfer over the network. This reliability is implemented on top of the unreliable physical network, often as part of the *congestion control algorithm (CCA)*. One inherent challenge in designing a CCA, is to offer efficient bandwidth utilization [2]. For TCP, CUBIC [14] and BBR [9] are two state-of-the-art CCAs. CUBIC is the default CCA on most Linux machines, and BBR is a recent algorithm from Google that is an option for Linux.

Networks are not always private. Bandwidth-sharing networks are still the common case for a large number of research and industry users. The workload on shared networks tends to be highly dynamic. As a result, estimating the available bandwidth is one of the challenging tasks in bandwidth-sharing networks.

In networks with shared bandwidth, the protocols should provide a good trade-off between efficiency and being fair to other traffic (i.e., background traffic). Fairness, at a high level of abstraction, is defined as a protocol's tendency towards utilizing an equal share of bandwidth with other concurrent traffic streams. Jain fairness [17] is well-known and serves as a measure of how fair the consumers (i.e., traffic streams) are in sharing system resources (i.e., network bandwidth). The Jain index (Eq. 1, where x_i is bandwidth share of the i-th of n streams) falls between 0 and 1, representing the worst and the perfect fairness, respectively.

$$\mathcal{J}(x_1, x_2, ..., x_n) = \frac{(\sum_{i=1}^{n} x_i)^2}{n \times \sum_{i=1}^{n} x_i^2} \tag{1}$$

While most CCAs include fairness as part of their design goals, fairness problems have been reported for both CUBIC and BBR CCAs [19,20]. We will further discuss TCP in Sect. 3.

2.2 Applying Machine Learning to Networking Problems

Two common challenges in networking problems include, first, probing the network to discover the existing workload and available resources, and, second, efficient utilization of the available resources (i.e., bandwidth).

Probing the Network. Probing or measuring RTT as part of a time-series trace is common. Estimating the available network resources is performed from a local, non-global perspective, viewing the network as a black box. There are a number of studies in this area, mainly investigating the possibility of applying ML techniques to estimate the available bandwidth in high-speed networks [21,25].

Estimating the network workload and the type of background traffic on the network is another important problem. Obtaining such knowledge would allow better and more-efficient adjustments of the network configurations, and choosing appropriate data-transfer tools. The characterization of network background traffic is the topic of our study, discussed in further detail in following sections.

Efficient Network Utilization. ML techniques have been used for designing or optimizing network protocols. Remy used simulation and ML to create a new TCP CCA, via a decentralized partially-observable Markov decision process (dec-POMDP) [24]. Performance-oriented Congestion Control (PCC) is another recent study where an online learning approach is incorporated into the structure of the TCP CCA. Online convex optimization is applied to design a rate-control algorithm for TCP, called Vivace [12].

2.3 Time-Series Classification

Classification Scheme. Classification is a common technique in ML [15]. For this study, we investigated several supervised classification models, namely k-nearest-neighbours (K-NN), a multi-layer perceptron (MLP) neural network, Support Vector Machines (SVM), and naive Bayes (NB) [15]. Even though time-series RTT data violates the assumption of independent features in the feature vector, we still evaluated SVM and NB. Also, although we use cross validation, we have not explored overfitting in detail.

In K-NN classification, the k closest training samples to the input time-series are selected and then the class label for the input is predicted using a majority vote of its K-nearest-neighbours' labels. A multi-layer perceptron (MLP) is a type of feedforward artificial neural network, consisting of one or more fully-connected hidden layers. The activation output of the nodes in a hidden layer are all fed forward as the input to the next layer. For the training phase, the MLP utilizes an algorithm called *backpropagation* [7]. In SVM classification, a hyperplane or a set of hyperplanes, in high-dimensional space, are constructed. The hyperplanes are then used to classify the unseen data-points to one of the pre-defined classes. The hyperplanes are constructed in a way to maximize the marginal distance with the training samples of each class. A naive Bayes classifier is a probabilistic model, assigning a probability for each class conditional to the feature values of the input vector. Notably, the naive Bayes model assumes the features to be independent, while for time-series data the data-point values could be dependant on the previous ones. However, we have included naive Bayes to experiment with its ability to predict the traffic classes given the highly variable nature of the observed data points.

Distance Measure. Time-series classification is a special case where the input data being classified are time-series vectors. While Euclidean distance is widely used as the distance metric in ML algorithms, it is not well-suited for time-series data, since there is noise and distortion over the time axis [18]. Two common approaches for handling time-series data are, (1) quantizing a time-series into discrete feature vectors [11], and, (2) using new distance metrics [3].

We utilize the well-known dynamic time warping (DTW) distance metric, for the time-series data. DTW was originally used for speech recognition. It was then proposed to be used for pattern finding in time-series data, where Euclidean distance results in a poor accuracy due to possible time-shifts in the time-series [6,16].

3 TCP Congestion Control

We briefly review the Transmission Control Protocol (TCP) in terms of its CCA. CUBIC is the default TCP CCA deployed on most Linux-based hosts [14]. BBR is a newer CCA [9]. Some studies show that both CUBIC and BBR manifest unfair bandwidth utilization under various circumstances [19,20]. We have also seen this unfair behaviour in Fig. 1.

To further investigate the interoperability of CUBIC and BBR, we have run a series of experiments, using CUBIC and BBR over a shared network. The experiments were run for a combination of 3 traffic streams, either CUBIC (C) or BBR (B), sharing the network bandwidth (Fig. 2b). Each experiment scenario is labeled by a 3-tuple identifying the three traffic streams on the network. The first two are considered as background streams (BG1 and BG2), running for 120 s. The third stream is considered as foreground traffic (FG), starting with a 30 s offset and running for 60 s. Experimental details are provided in Sect. 4.1.

Both CUBIC and BBR are able to utilize available bandwidth while running in isolation. However, BBR has a significant negative impact on CUBIC in all combinations, regardless of running as a background or foreground stream. CUBIC stream(s) suffer from extreme starvation when running along with BBR streams on a shared networks. In contrast, all-CUBIC and all-BBR scenarios, while not perfect, are considerably fairer in sharing bandwidth compared to heterogeneous combination of the two CCAs. These observations are summarized in Fig. 2c.

4 Research Statement and Methodology

Can ML classify the background traffic in bandwidth-sharing networks? If so, can network performance be improved by selecting an appropriate protocol for the foreground traffic (i.e., OPS)? We apply classification techniques on training data consisting of time-series RTT traces, for six different combinations of CUBIC and BBR CCAs, as summarized in Fig. 2a. Note that the RTT data is local information, gathered by the sender, using probing, and we do not assume any global knowledge or signals from the network, which is a common situation with most shared WANs.

4.1 Network Testbed

We have re-used our controlled network testbed from our previous work [5], consisting of a dumbbell topology with 8 end-nodes, 2 virtual LANs, and 1 virtual

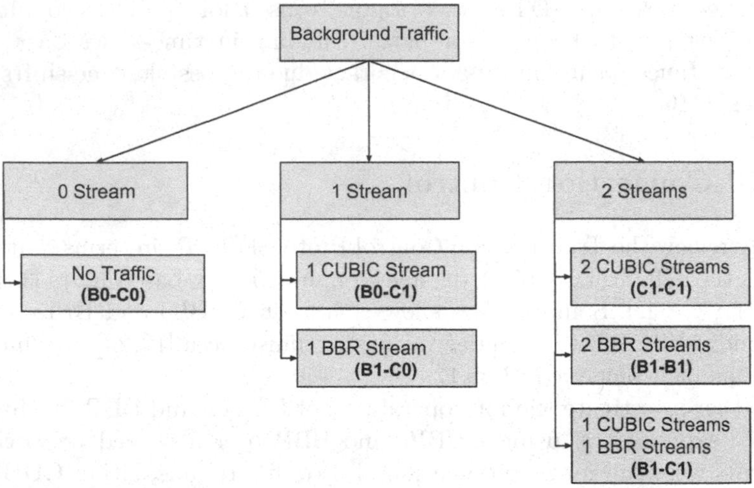

(a) Classes of background traffic to be characterized

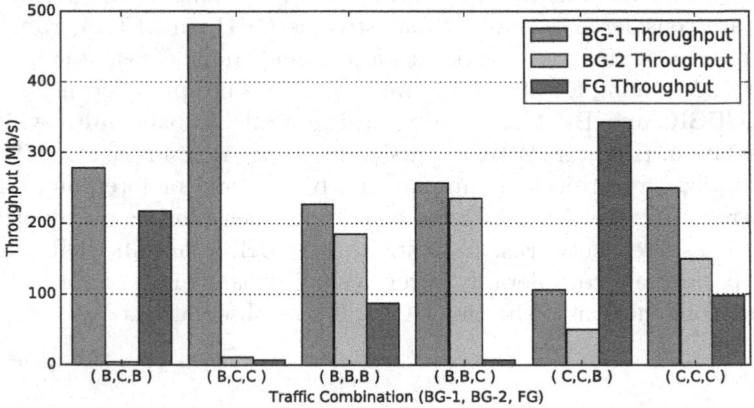

(b) TCP CUBIC and BBR bandwidth Sharing: 3 traffic streams sharing bandwidth (B: BBR, C: CUBIC)

	FG: BBR	FG: CUBIC
BG: BBR	Good Throughput Good Fairness	Poor Throughput Poor Fairness
BG: CUBIC	Good Throughput Poor Fairness	Good Throughput Good Fairness

(c) Interaction of CUBIC and BBR on a shared network

Fig. 2. Interoperability properties of TCP-CUBIC and TCP-BBR

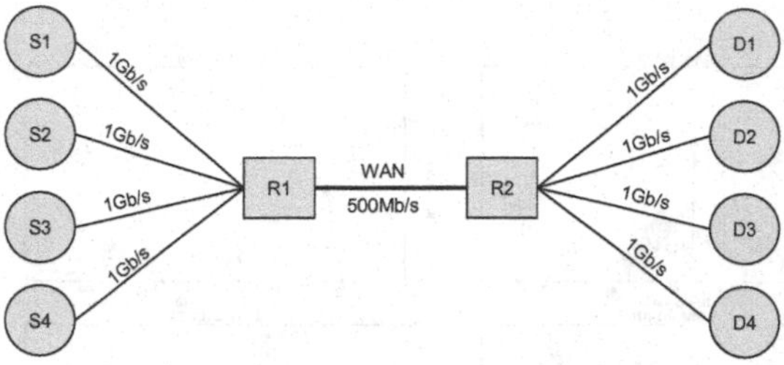

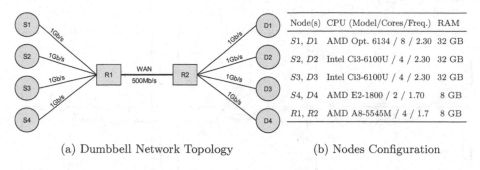

(a) Dumbbell Network Topology

Node(s)	CPU (Model/Cores/Freq.)	RAM
*S*1, *D*1	AMD Opt. 6134 / 8 / 2.30	32 GB
*S*2, *D*2	Intel Ci3-6100U / 4 / 2.30	32 GB
*S*3, *D*3	Intel Ci3-6100U / 4 / 2.30	32 GB
*S*4, *D*4	AMD E2-1800 / 2 / 1.70	8 GB
*R*1, *R*2	AMD A8-5545M / 4 / 1.7	8 GB

(b) Nodes Configuration

Fig. 3. Testbed architecture (Based on our previous study [5])

router which is emulated using two computer nodes (Fig. 3). The Dummynet [8] network emulator is used on nodes R1 and R2 for imposing the desired bandwidth, delay, and router queuing properties. The end-to-end propagation delay (base RTT) for the emulated network is set to 65 ms, as found on many medium-to-large WANs. The bottleneck shared bandwidth is set to 500 Mb/s. Router buffer size is set to 6 MB, roughly about 1.5 BDP[1].

4.2 Data Gathering

For a classifier problem, we need to gather sufficient data for training our model. Our target classes to be decided are the combination of CCAs in background traffic. We designed a controlled experiment to gather RTT data as a time-series, with a 1-second sampling rate, over a one hour period of time. To smooth out the possible noise in the measured RTT value, each probing step consists of sending 10 ping requests to the other end-host, with 1 ms delay in between, and the average value is recorded as the RTT value for that second. We repeated this experiment for the six traffic combinations in Fig. 2a.

For gathering data, we used the aforementioned network testbed to conduct traffic between the two LANs, sharing the same bottleneck bandwidth. We conduct the background traffic between *S*1 and *D*1 pair of nodes. In the cases where we need second background traffic, *S*2 and *D*2 pair of nodes are used respectively. For the foreground traffic we have used *S*3 and *S*4 pair of nodes. Finally, *S*4 and *D*4 nodes are used to probe the RTT over the shared network link. For all data-transfer tasks, we used the *iperf* tool (http://software.es.net/iperf/) for generating TCP traffic of the desired CCA. A sample of the time-series data is presented in Fig. 4.

[1] BDP=Bottleneck-Bandwidth × Delay (RTT). Setting buffer size to less than 1 BDP, so-called shallow buffering, could result in the under-utilization of network bandwidth. In contrast, setting the buffer to large sizes could result in an effect called buffer-bloat where the users experience extremely long delays due to long queuing delays at the router.

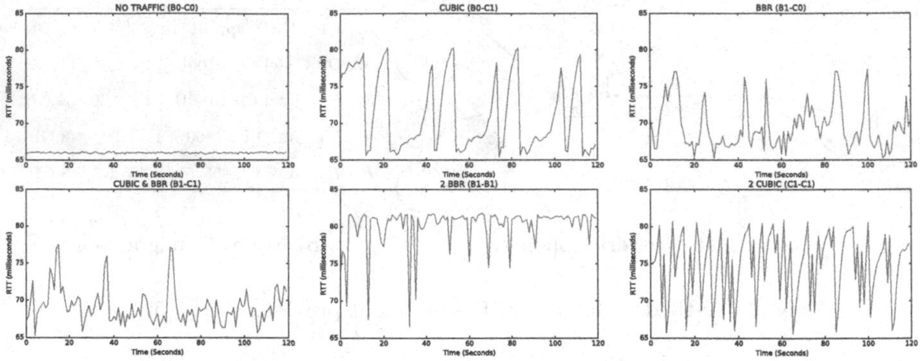

Fig. 4. A sample of RTT time-series data

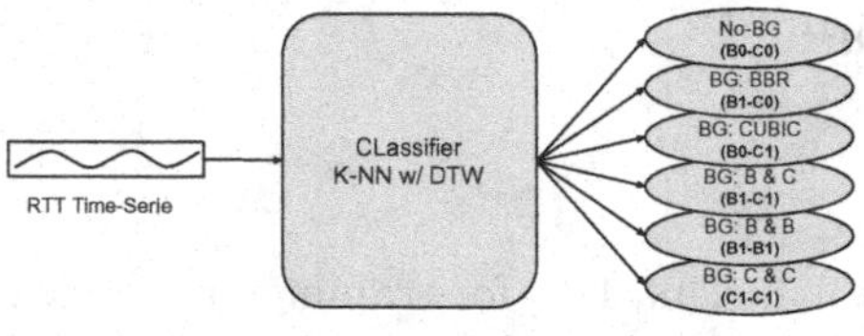

(a) Classifier workflow

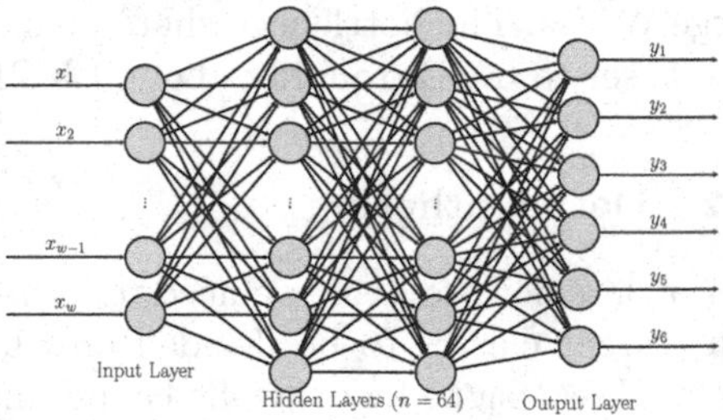

(b) Network Structure for MLP neural network

Fig. 5. Classification models for background traffic classification

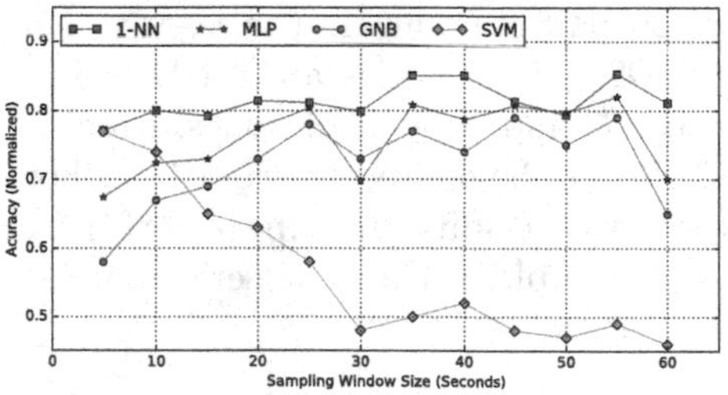

(a) Classification accuracy

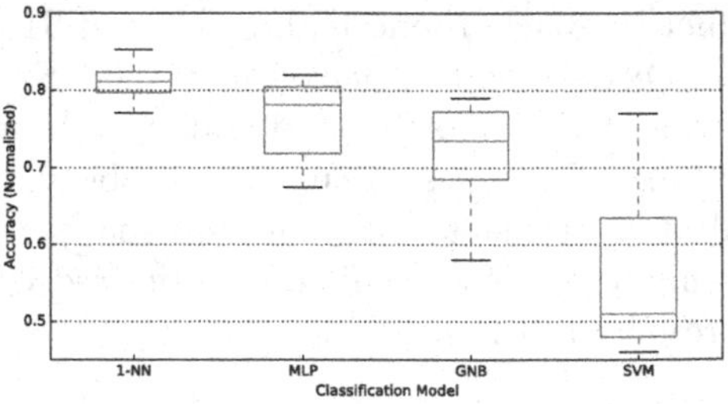

(b) Box Plot for classifiers accuracy

Fig. 6. Performance of classification models for varying time length w of input time-series (Five-Fold Cross-Validation)

4.3 Classification

As briefly reviewed in Sect. 2.3, we implemented K-NN, MLP, SVM, and NB classifiers. As for K-NN classification, the DTW distance metric is used.

Data Preparation. To prepare the gathered data (c.f. Sect. 4.2) for ML, we partition the one-hour RTT time-series into smaller chunks. We developed software to partition the data based on a given parameter w, representing a window size measured in data points (which also corresponds to a window in seconds, given the one second gap between RTT probes). We have experimented with both non-overlapping and overlapping (or sliding) windows. In this paper, we only discuss non-overlapping windows.

Training, Testing and Parameter Tuning. Each classifier takes an RTT time-series of length w as an input vector, and predicts the class label for that time-series as the output. The schematic model of our classifier is depicted in Fig. 5a.

K-NN. For a K-NN classifier, the two main parameters are: k as the number of neighbours involved in a majority vote for decision making, and w the length of the RTT time-series vector to train the network on.

MLP. For the MLP neural network model (Fig. 5b), we designed a network of two fully-connected hidden layers of size 64. The input layer is of size w (i.e., the length of RTT time-series vector) and the output layer consists of six nodes. The output layer generates a one-hot-encoded prediction, consisting of six values where only one value is one (i.e., predicted class) and the rest are zero. The nodes of the two hidden layers deploy Rectified Linear Unit (ReLU) activation, while the output layer uses a Softmax activation function.

SVM and NB. For implementing SVM and naive Bayes classifiers, we have used the default configuration of the ML library we used for implementation (*sci-kit learn*). As for SVM, the Radial Basis Function (RBF) kernel is used. The naive Bayes implementation we used is a Gaussian naive Bayes classifier (GNB).

Selecting the appropriate time-length parameter w is a trade-off between the amount of information provided to the classifier and the RTT probing latency. On the one hand, the larger the w becomes, the more information will be provided to the classifier, providing a better opportunity to recognize the traffic pattern. On the other hand, a larger w means it would take a more time for the probing phase of a real system to gather RTT data to use as input to a classifier. To make an appropriate decision about the parameter value w, we did a parameter-sweep experiment where we calculated the classification accuracy for all the classifiers, varying parameter w from 5 s to 60 s, with a 5 s step (Fig. 6a).

To estimate accuracy and overfitting, we use five-fold cross-validation, where 20% of the data from each traffic class are in a hold-out set, not used for training, and then tested with classifier. The reported results are the average accuracy over the five folds on the cross-validation scheme. Figure 6a represents the average accuracy per window size w (in seconds), calculated for 1-NN, MLP, GNB, and SVM classifiers. We have also included the box plot for each classification model in Fig. 6b.

In our experiment, the 1-NN classifier outperforms the other models in most cases. Also according to the box plot, 1-NN presents the most consistency in accuracy for varying window sizes, with a narrow variation in observed performance. SVM tends to have the best performance after 1-NN for smaller window sizes. However, its accuracy drops significantly as we increase the time-series length. MLP and GNB represent a similar pattern in accuracy, following 1-NN with a slight margin.

For the K-NN model, we have also examined three different configurations with k equal to 1, 5, or 10, where the 1-NN tends to offer a better overall accuracy. Hence, we use K-NN with $k = 1$, where the single closest neighbour to the input would be used for predicting the class label. As for the w parameter, the global peak is for $w = 35 \sim 40$, while there are two local peaks for $w = 10$ and $w = 20$ (Fig. 6a). As the performance gap in the worst case is less than 5% (between 80% for $w = 10$ and 85% for $w = 40$), and considering the aforementioned trade-off, we decided to use the $w = 10$ as the window size for the remaining experiments. It presents a reasonable accuracy relative to the size of training data we used, and we predict that by increasing the size of training data this accuracy would gradually increase.

Notably, the DTW distance metric has a warping parameter for which we have used value 10. We have tested different values for this parameter as well, settling on 10.

In summary, we will be using the classifier K-NN with window size $w = 10$ and $k = 1$. A more-detailed parameter sweep experiment is planned as future work. For reference, the detailed confusion matrix and classification report for five-fold cross-validation of this configuration are provided in Fig. 7.

Infrastructure, Software and Frameworks. In our implementation, we have used the following software and frameworks:

Infrastructure (OS). For all the implementations, data gathering, ML implementations, and evaluation, we have used Linux CentOS release 6.5 Final with kernel version 2.6.32-358.6.2.el6.x86_64.

Network Customization. For implementing the emulated network environment we have used Dummynet network emulator [8]. Our automation scripts for data-gathering and evaluation are implemented in Bash shell scripts.

Machine Learning Environment. All the ML programming and evaluation metrics are implemented in Python language. For K-NN, naive Bayes, and SVM modelling we have used scikit-learn library version 0.19.1 [22]. For implementing MLP neural network, we have used Keras library version 2.2.2 [10] with TensorFlow version 1.10.1 as its backend engine [1].

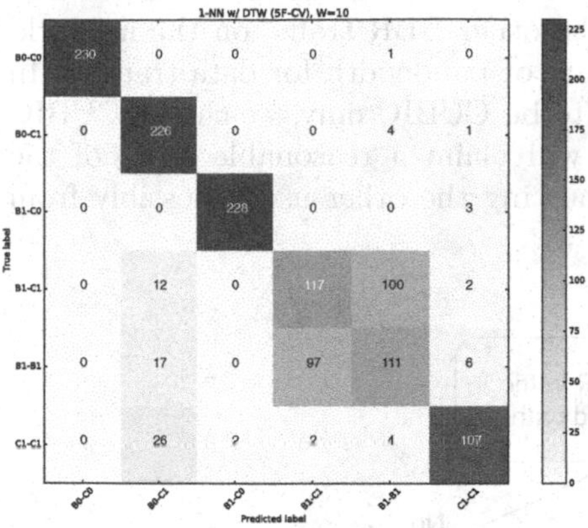

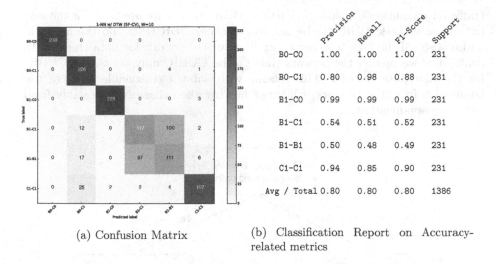

(a) Confusion Matrix

	Precision	Recall	F1-Score	Support
B0-C0	1.00	1.00	1.00	231
B0-C1	0.80	0.98	0.88	231
B1-C0	0.99	0.99	0.99	231
B1-C1	0.54	0.51	0.52	231
B1-B1	0.50	0.48	0.49	231
C1-C1	0.94	0.85	0.90	231
Avg / Total	0.80	0.80	0.80	1386

(b) Classification Report on Accuracy-related metrics

Fig. 7. K-NN (DTW) performance ($w = 10$, $k = 1$)

5 Evaluation and Discussion

We provide and discuss the performance evaluation of the developed classifier for characterizing the background traffic on a shared network environment. We first discuss designing a decision model and a operation cycle to utilize the classifier for data communication on a network. We then investigate the effect of applying this classifier to a real data-transfer scenario. Note that we only present results for the K-NN classification algorithm, where $k = 1$ and $w = 10$.

5.1 Decision Model

Our ultimate goal in designing the classifier model is to utilize it in production use-cases in order to improve on our decision making under different network situations. We need to map each predicted class (Fig. 2a) to an appropriate action. In the case of OPS, the action would be to choose an appropriate network protocol, in particular a CCA, for communicating data over a network. We investigate a *decision model* where one of either CUBIC or BBR is chosen for foreground data transfer over the network.

Earlier in Sect. 3 (Fig. 2c), we summarized the interaction dynamics between CUBIC and BBR. We apply that knowledge to generate a decision model for mapping the classification prediction to an appropriate action. The suggested decision model is presented in Fig. 8. According to this decision model, we consider the presence of BBR on the network as the deciding factor, to either use CUBIC or BBR for sending data on a shared path.

The intuition behind this decision scheme is as follows: BBR significantly hurts CUBIC performance on the network. Hence, if we discover that the existing traffic on the network include at least one BBR stream, we use BBR for the new data transfer task; BBR will be fair to the other BBR traffic on the network, and it also will be able to claim a fair share of bandwidth for data transfer. In contrast, if we predict the network traffic to be CUBIC only, we choose CUBIC for the new data transfer. This scheme will claim a reasonable share of the bandwidth from the network, while not hurting the other users, possibly from the same organization.

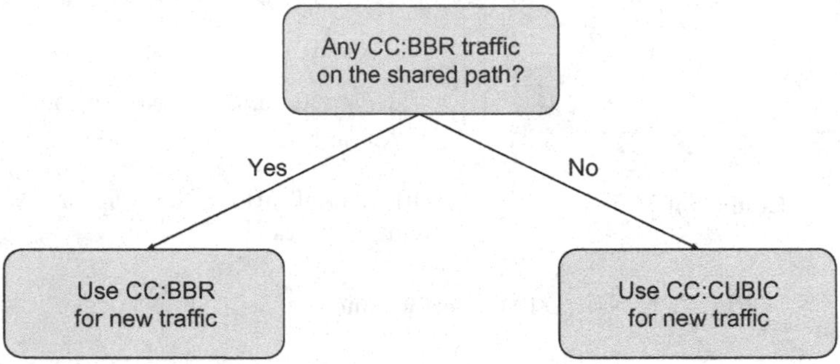

Fig. 8. Binary decision model for choosing a CCA for new data transfer

5.2 Operation Cycle

One intuitive use of the designed classifier and decision model is to feed a RTT time-series, probed from the network, into the classifier; then according to its output, we choose a CCA for sending new data over the network. We call this operation scheme a PDA cycle (Probe-Decision-Action Cycle). To better clarify this concept, we have depicted this operation model in Fig. 9a. While this operation cycle seems reasonable, there is a performance flaw in this cycle: the data-transfer phase could have an impact on the dynamics of RTT over the network. Hence, when we transition to the probing phase immediately after a data-transfer phase, the effect of last data transfer would still be visible in the RTT data; this might mislead the classifier to make a wrong prediction about the type of background traffic.

In an effort to tackle this problem, we add a "Hold" phase to the operation cycle, calling it PDAH Cycle (Probe-Decision-Action-Hold Cycle). This operation mode is depicted in Fig. 9b. Adding this new state to the cycle allows the RTT to go back to its steady state, excluding our own data traffic, and then we probe the RTT to better estimate the type of background traffic on the network. The accuracy of the two modes are presented in Fig. 9c under two scoring schemes: Classification and Decision. Classification accuracy measures a model's ability to predict the exact class label (out of six classes). In contrast, decision accuracy measures the ability to predict the presence of BBR traffic on the network, used for decision making in real use-case scenarios. These accuracy numbers are calculated by running a 5-min network experiments per each operation mode, once for each background traffic of either CUBIC or BBR, while examining the PDA and PDAH cycles to investigate their accuracy in detecting the type of background traffic. The resulting accuracy numbers confirm our hypothesis, adding the hold state considerably improves the prediction accuracy.

As for the time-length of each state, probing takes exactly 10 s (equal to our parameter w), the decision making step (predicting class label) takes only a fraction of a second, the action step (transferring data to the network) is set to take 60 s, and the hold step is also set to 10 s. The last two values are user-provided numbers that could be selected as any arbitrary number. We chose 60 s of real data transfer because it will be a good balance between sending data and a total of 20 s for holding and re-probing the background traffic. In the remainder of this section, we will use these settings for evaluating our classifier in data-transfer scenarios.

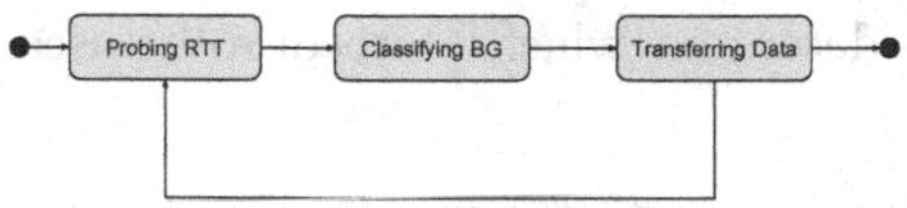

(a) Probe-Decision-Action Cycle (PDA)

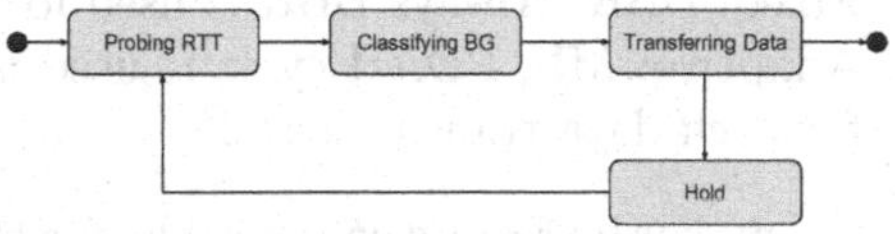

(b) Probe-Decision-Action-Hold Cycle (PDAH)

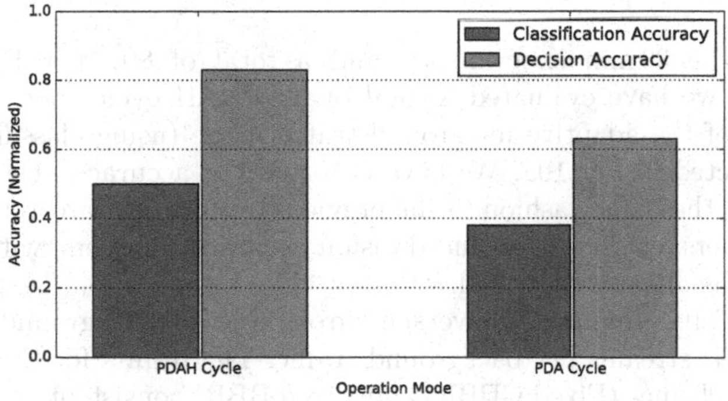

(c) Accuracy of Operation modes. PDA: Probe-Decision-Action Cycle, PDAH: Probe-Decision-Action-Hold Cycle.

Fig. 9. Operation modes for utilizing a classifier in real data-transfer scenarios.

5.3 Data Communication Performance

We present the final performance of our classifier, utilized in a real data-transfer scenario. To summarize, the configuration of our final decision-making model is presented in Table 1.

Table 1. Configuration of decision making model

Component	Used scheme	Scheme parameters
Classifier	K-NN	w (input vector length)=10 k = 1
Distance measure	DTW	max-warping=10
Decision model	Is BBR present?	If yes, use BBR If no, user CUBIC
Operation cycle	PDAH	P = 10 s, D∼0, A = 60 s, H = 10 s

Using this model configuration, we have conducted a series of experiments, running a 30 min long traffic over the network as the background traffic of either CUBIC or BBR scheme. We repeated this test 3 times, each time using one of the following strategies for sending foreground traffic. (A total of 6 experiments were conducted with a combination of 2 background and 3 foreground schemes.):

1. Fixed: CUBIC. Always CUBIC is used for data-transfer phase.
2. Fixed: BBR. Always BBR is used for data-transfer phase.
3. Adaptive: ML. PDAH cycle is used to determine which protocol to be used for each data-transfer cycle.

To make the results comparable, for the first two strategies with a fixed traffic scheme we have included the probing and hold time delays. So the total time length of data-transfer steps and its transitioning times are almost identical in all three cases.

Based on the 30 min experiments, and a total of 80 s per PDAH cycle (10 + 60 + 10), we have evaluated a total of 22 PDAH cycles per experiment. The accuracy of the adaptive foreground traffic mode (using classifier decision model) is depicted in Fig. 10a. We have reported the accuracy of our classifier in Fig. 10a. In the same fashion as in previous results, we are reporting two accuracy numbers, classification and decision accuracy numbers, with the same interpretation as already described.

Finally, Fig. 10 presents the average throughput of background (BG) and foreground (FG) streams per background traffic. The results for the fixed foreground traffic schemes (Fixed:CUBIC and Fixed:BBR) consistent with our early observations in Sect. 3. In Fig. 10b, by introducing BBR traffic in Fixed:BBR (middle bars) the bandwidth share of background CUBIC drops drastically, resulting in a poor fairness with Jain index of 0.62. In contrast, the Adaptive(ML) mode (right-hand-side bars) dynamically probes the network condition and adjusts the network protocol, reclaiming 4 times more bandwidth, comparable to the ideal case of Fixed:CUBIC scenario with a Jain index of 0.99. Similarly, in Fig. 10c, the Fixed:CUBIC mode (left-hand-side bars) suffers from extreme unfair bandwidth sharing in presence of background BBR (Jain index: 0.60). Using the Adaptive(ML) (right-hand-side bars) significantly improves on fairness, again reclaiming 4 times more share of the bandwidth (Jain index: 0.96).

While, according to Fig. 10a, the peak accuracy of our current implementation of PDAH cycles is bounded at 80%, the outcome is quite satisfactory,

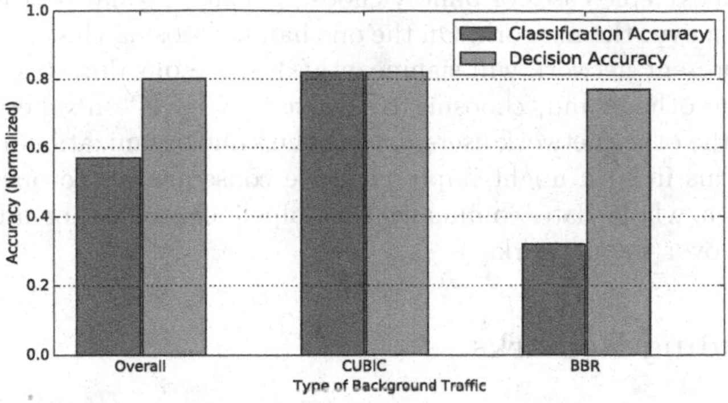

(a) Classifier Accuracy breakdown per type of background traffic. Classification accuracy refers to the accuracy of predicting the exact class label, Decision accuracy refers to the accuracy of predicting the presence of BBR traffic for decision making

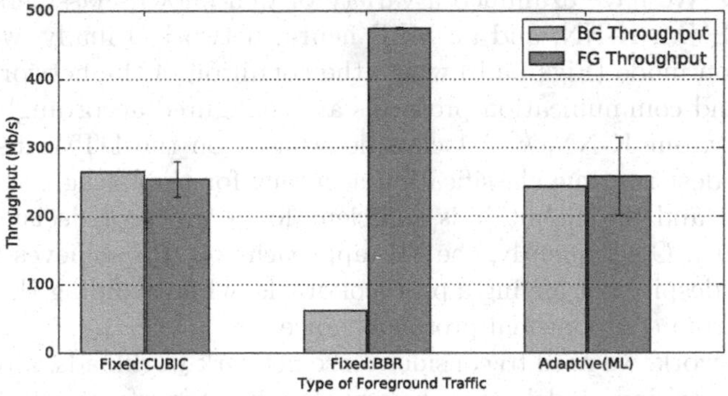

(b) Background Traffic: CUBIC. Average throughput of fixed versus adaptive data-transfer schemes.

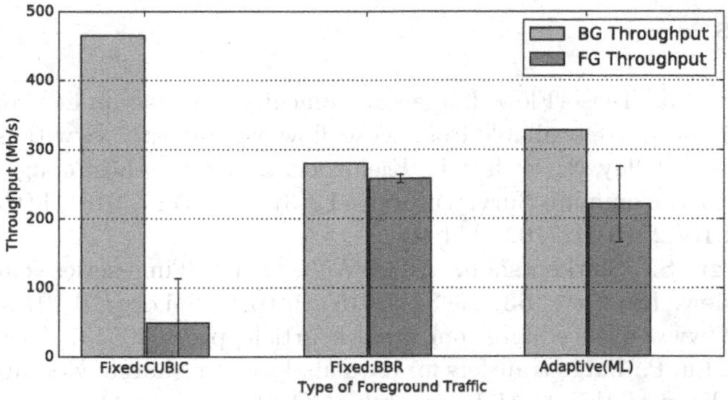

(c) Background Traffic: BBR. Average throughput of fixed versus adaptive data-transfer schemes.

Fig. 10. Performance evaluation of ML-based decision making model comparing the extreme cases of blindly choosing a suboptimal protocol for data communication over the network. On the one hand, choosing the CUBIC scheme over a BBR-present network will significantly degrade our data-transfer performance. On the other hand, choosing BBR over a CUBIC-only network would severely hurt the other network users sharing the same organizational or research bandwidth. This in turn might imply negative consequences to our own workflow, pausing a whole data communication pipeline because of an aggressive traffic stream over the network.

6 Concluding Remarks

Our contribution is the design, implementation, and proof-of-concept evaluation an ML-based approach to optimization through protocol selection, or OPS, for shared-network throughput and fairness. Our approach has a classification model and a decision model, and we use CUBIC and BBR CCAs in an empirical case study. We have examined a variety of well-known classifiers, including naive Bayes, SVM, K-NN, and an MLP neural network. Finally, we created a cyclic operation mode (Figs. 9a,b) where the condition of the network is probed periodically and communication protocols are configured accordingly.

Empirically, our K-NN, $k = 1$ classifier (based on the DTW distance metric) shows modest runtime classification accuracy for the background protocols (between 0.36 and 0.81), but it is sufficient for a protocol decision accuracy of 0.81 (Fig. 10). Consequently, the ML approach to OPS achieves throughput close to ideal, despite not having a protocol oracle, while avoiding the worst-case performance from a suboptimal protocol choice.

For future work, we need to consider more network workloads, alternate classification and decision models (e.g., binary classifiers), perform a full parameter sweep of the algorithms, and further investigate overfitting in the classifiers.

References

1. Abadi, M., et al.: TensorFlow: Large-scale machine learning on heterogeneous systems (2015). software available from tensorflow.org. https://www.tensorflow.org/
2. Afanasyev, A., Tilley, N., Reiher, P., Kleinrock, L.: Host-to-host congestion control for TCP. IEEE Commun. Surv. Tutorials **12**(3), 304–342 (2010). https://doi.org/10.1109/SURV.2010.042710.00114
3. Aghabozorgi, S., Shirkhorshidi, A.S., Wah, T.Y.: Time-series clustering - a decade review. Inf. Syst. **53**, 16–38 (2015). https://doi.org/10.1016/j.is.2015.04.007. http://www.sciencedirect.com/science/article/pii/S0306437915000733
4. Anvari, H., Lu, P.: Large transfers for data analytics on shared wide-area networks. In: Proceedings of the ACM International Conference on Computing Frontiers, CF 2016, pp. 418–423. ACM, New York (2016). https://doi.org/10.1145/2903150.2911718

5. Anvari, H., Lu, P.: The impact of large-data transfers in shared wide-area networks: an empirical study. Procedia Comput. Sci. **108**, 1702–1711 (2017). https://doi.org/10.1016/j.procs.2017.05.211. International Conference on Computational Science, ICCS 2017, 12-14 June 2017, Zurich, Switzerland. http://www.sciencedirect.com/science/article/pii/S1877050917308049
6. Berndt, D.J., Clifford, J.: Using dynamic time warping to find patterns in time series. In: Proceedings of the 3rd International Conference on Knowledge Discovery and Data Mining. AAAIWS 1994, pp. 359–370. AAAI Press (1994). http://dl.acm.org/citation.cfm?id=3000850.3000887
7. Bishop, C.M.: Pattern Recognition and Machine Learning (Information Science and Statistics). Springer, Heidelberg (2006)
8. Carbone, M., Rizzo, L.: Dummynet revisited. SIGCOMM Comput. Commun. Rev. **40**(2), 12–20 (2010). https://doi.org/10.1145/1764873.1764876. http://doi.acm.org/10.1145/1764873.1764876
9. Cardwell, N., Cheng, Y., Gunn, C.S., Yeganeh, S.H., Jacobson, V.: BBR: Congestion-based congestion control. Queue **14**(5), 50:20–50:53 (2016). https://doi.org/10.1145/3012426.3022184. http://doi.acm.org/10.1145/3012426.3022184
10. Chollet, F., et al.: Keras. https://keras.io (2015)
11. Christ, M., Kempa-Liehr, A.W., Feindt, M.: Distributed and parallel time series feature extraction for industrial big data applications. CoRR abs/1610.07717 (2016). http://arxiv.org/abs/1610.07717
12. Dong, M., et al.: PCC vivace: Online-learning congestion control. In: 15th USENIX symposium on networked systems design and implementation (NSDI 2018), pp. 343–356. USENIX Association, Renton (2018). https://www.usenix.org/conference/nsdi18/presentation/dong
13. Gu, Y., Grossman, R.L.: UDT: UDP-based data transfer for high-speed wide area networks. Comput. Netw. **51**(7), 1777–1799 (2007). https://doi.org/10.1016/j.comnet.2006.11.009. Protocols for Fast, Long-Distance Networks. http://www.sciencedirect.com/science/article/pii/S1389128606003057
14. Ha, S., Rhee, I., Xu, L.: Cubic: a new TCP-friendly high-speed TCP variant. SIGOPS Oper. Syst. Rev. **42**(5), 64–74 (2008). http://doi.acm.org/10.1145/1400097.1400105
15. Hastie, T., Tibshirani, R., Friedman, J.: The Elements of Statistical Learning. Springer, New York (2009)
16. Hsu, C.J., Huang, K.S., Yang, C.B., Guo, Y.P.: Flexible dynamic time warping for time series classification. Procedia Comput. Sci. **51**, 2838–2842 (2015). https://doi.org/10.1016/j.procs.2015.05.444. International Conference On Computational Science, ICCS 2015. http://www.sciencedirect.com/science/article/pii/S1877050915012521
17. Jain, R., Chiu, D.M., Hawe, W.R.: A quantitative measure of fairness and discrimination for resource allocation in shared computer system, vol. 38. Eastern Research Laboratory, Digital Equipment Corporation Hudson, MA (1984)
18. Keogh, E., Ratanamahatana, C.A.: Exact indexing of dynamic time warping. Knowl. Inf. Syst. **7**(3), 358–386 (2005). https://doi.org/10.1007/s10115-004-0154-9
19. Kozu, T., Akiyama, Y., Yamaguchi, S.: Improving RTT fairness on cubic TCP. In: 2013 First International Symposium on Computing and Networking, pp. 162–167, December 2013. https://doi.org/10.1109/CANDAR.2013.30
20. Ma, S., Jiang, J., Wang, W., Li, B.: Towards RTT fairness of congestion-based congestion control. CoRR abs/1706.09115 (2017). http://arxiv.org/abs/1706.09115

21. Mirza, M., Sommers, J., Barford, P., Zhu, X.: A machine learning approach to TCP throughput prediction. IEEE/ACM Trans. Netw. **18**(4), 1026–1039 (2010). https://doi.org/10.1109/TNET.2009.2037812
22. Pedregosa, F., et al.: Scikit-learn: machine learning in Python. J. Mach. Learn. Res. **12**, 2825–2830 (2011)
23. Sivaraman, A., Winstein, K., Thaker, P., Balakrishnan, H.: An experimental study of the learnability of congestion control. In: Proceedings of the 2014 ACM Conference on SIGCOMM. pp. 479–490. SIGCOMM 2014. ACM, New York (2014). http://doi.acm.org/10.1145/2619239.2626324
24. Winstein, K., Balakrishnan, H.: TCP ex machina: computer-generated congestion control. In: Proceedings of the ACM SIGCOMM 2013 Conference on SIGCOMM. SIGCOMM 2013, pp. 123–134. ACM, New York (2013). https://doi.org/10.1145/2486001.2486020, http://doi.acm.org/10.1145/2486001.2486020
25. Yin, Q., Kaur, J.: Can machine learning benefit bandwidth estimation at ultra-high speeds? In: Karagiannis, T., Dimitropoulos, X. (eds.) PAM 2016. LNCS, vol. 9631, pp. 397–411. Springer, Cham (2016). https://doi.org/10.1007/978-3-319-30505-9_30

Common Structures in Resource Management as Driver for Reinforcement Learning: A Survey and Research Tracks

Yue Jin(✉), Dimitre Kostadinov, Makram Bouzid, and Armen Aghasaryan

Nokia Bell Labs, Paris-Saclay, Nozay, France
{yue.l.jin,dimitre.kostadinov,makram.bouzid, armen.aghasaryan}@nokia-bell-labs.com

Abstract. In the era of growing digitalization, dynamic resource management becomes one of the critical problems in many application fields where, due to the permanently evolving environment, the trade-off between cost and system performance needs to be continuously adapted. While traditional approaches based on prior system specification or model learning are challenged by the complexity and the dynamicity of these systems, a new paradigm of learning in interaction brings a strong promise - based on the toolset of model-free Reinforcement Learning (RL) and its great success stories in various domains. However, current RL methods still struggle to learn rapidly in incremental, online settings, which is a barrier to deal with many practical problems. To address the slow convergence issue, one approach consists in exploiting the system's structural properties, instead of acting in full model-free mode. In this paper, we review the existing resource management systems and unveil their common structural properties. We propose a meta-model and discuss the tracks on how these properties can enhance general purpose RL algorithms.

Keywords: Resource management · RL · Capacity management · Learning through interactions

1 Introduction

Growing digitalization is rapidly expanding the complexity in computing and cyber-physical systems. These systems in the future will connect to end-users through billions of sensors and IoT devices, and will rely on cloud-enabled telecommunication networks. Moreover, they will operate in permanently evolving environment. The swollen complexity creates a major challenge for system management and control. Traditional approaches to resource management based on system model specification, off-line behavior learning and traffic prediction will be increasingly defied due to the emergence of these complex and dynamically evolving systems.

To deal with this challenge, a new paradigm of continuous learning in interaction brings a strong promise for highly adaptive control mechanisms. Starting with little or no knowledge about system characteristics, the control agents start taking actions and learn on-the-fly about their efficiency through the observed feedback from the environment.

É. Renault et al. (Eds.): MLN 2018, LNCS 11407, pp. 117–132, 2019.
https://doi.org/10.1007/978-3-030-19945-6_8

This approach, referred to as Reinforcement Learning (RL), allows to hide the inherent complexity of the environment and to adapt dynamically to its changing conditions. Taking its origins in the psychology of animal learning, due to its generality, this problem has been studied across multiple fields such as game theory, control theory, operations research or multi-agent systems [1]. The applications are widely spanning across various areas such as robotics [2], manufacturing [3], finance [4], or games [5, 6].

However, current RL methods still struggle to learn rapidly in the incremental, online settings for many practical problems. In [7], many iterations are needed before the RL in tabular form converges. Methods based on deep learning and artificial neural networks rely on batch training with large data sets, extensive off-line self-play [6], or learning asynchronously from multiple simultaneous streams of agent-environment interactions [3]. They don't learn well online. This creates a major barrier for applying RL to practical problems, and more specifically, to resource management problems on which we focus in this paper.

Furthermore, the lack of safeguard in the exploration phase is another obstacle for practical application. In RL, the agent needs to try actions to learn about their efficiency. In other words, actions are tried when their impact on the system is not well understood. In online settings, this can lead to system failure and service degradation.

In [1], the authors point out that we may derive inductive biases for RL such that future learning generalizes well and is thus faster. Indeed, resource management, and more specifically, capacity management systems exhibits many common structures across different application domains such as Cloud computing, telecommunication, or service systems. For instance, the incoming demands usually change stochastically and temporally, and they are processed by servers with stochastic processing time. A higher capacity, i.e. a higher number of servers, reduces the sojourn time of the demands in the system, but increases the cost. Depending on the complexity of the systems, the demands may go through multiple servers before leaving the system. This commonality may lead to common algorithmic structures in RL, e.g. concavity of action-values with respect to states and actions. These algorithmic structures can be exploited to learn faster and safer. In this work, we review resource management in different application domains, reveal its common structures and discuss research tracks on structure exploitation in related literature.

The paper is organized as follows. In Sect. 2, we review resource management problems in different application domains, with a focus on capacity management and its existing approaches. We reveal common system structures and create a meta-model for capacity management in Sect. 3. We discuss existing research tracks for exploiting structures in related literature that may inspire solutions in RL in Sect. 4. We conclude in Sect. 5.

2 Resource Management Systems

Many resource management applications from various fields share common structures. In this section, we survey several such fields, including service systems, production systems, cloud computing and telecommunication networks, and elucidate these structures as well as some of the existing algorithmic approaches. We provide a summary of the papers survey at the end of this section in Table 1.

2.1 Service Systems

Service systems are widely present in our society. They range from telephone call centers to supermarket checkouts and hospital emergence rooms. In these systems, the demand from customers arrives gradually and often changes over time. A group of human agents serve the demand with stochastic processing time. The demand from a customer may have to wait, if all agents are busy. A higher capacity, i.e. a larger number of agents, usually reduces the time the demand spends in the system and provides a higher service level. On the other hand, a higher capacity generates higher cost. Depending on its complexity, the demand may be served in several stages before it leaves the system. The capacity at each stage needs to be adjusted against the varying demand.

The goal of the system is to use the minimum number of agents to provide a good service level. Thus, the staffing problem, i.e. deciding the number of agents on duty over a time horizon, is critical for achieving high system efficiency. Many approaches have been used to solve the staffing problem. [8] develop Delayed-Infinite Server approximation for offered load and use a simple heuristic of "Smallest Staffing Level" to solve the staffing problem. [9] formulate the staffing problem in a dynamic programming setting and show that the optimal policy follows a monotone optimal control. [10] consider a limiting parameter regime and show their Linear Programming based method for the staffing problem is asymptotically optimal. [11] provides a state-of-the-art literature review on staffing and scheduling approaches that account for nonstationary demand.

Other resource management problems in call centers include forecasting incoming calls, deciding how many agents to hire over time, scheduling individual agents to shifts, routing incoming calls to agents and evaluating the performance of call centers. [12] provides a review on the operational problems in telephone call centers and their solutions. Other service systems have similar problems.

2.2 Production Systems

For many production systems, the demand from customers varies stochastically and temporally. The production systems manufacture products to satisfy customer demands. Conventionally the production systems handle the volatility in demand by holding product stocks. Some production systems now have flexible capacity through authorizing overtime production, renting work stations, etc. They can handle the volatility in demand by adjusting capacity as well. A higher capacity reduces the chance of backlogging demand and the need to carry inventory and vice versa. The customer of a production system can be another production system. Repeating this relationship creates a supply chain from several production systems. The capacity in each production system in the supply chain needs to be adjusted for varying end-user demand.

Table 1. Resource management problems in multiple fields.

Paper	Field	Problem	Approach
[8]	Service systems	Staffing problem	Heuristic
[9]	Service systems	Staffing problem	Dynamic programming
[10]	Service systems	Staffing problem	Linear programming
[11]	Service systems	Staffing problem	(Literature review)
[12]	Service systems	Operations problems	(Literature review)
[13]	Production systems	Joint capacity and production problems	Dynamic programming
[14]	Production systems	Joint capacity and inventory management	Search algorithm
[15]	Production systems	Joint capacity and inventory management	Brownian motion approximation
[16]	Production systems	Operations problems	(Literature review)
[17, 18, 25]	Cloud computing	Optimization of resource usage, cost saving	(Survey of resource management problems in cloud computing)
[19]	Cloud computing	Virtual machines allocation	(Survey)
[20]	Cloud computing	Auto scaling	Amazon's manual auto-scaling mechanisms
[21, 22]	Cloud computing	Auto scaling	Predictive auto-scaling
[23, 24]	Cloud computing	Resource management	Using Reinforcement Learning for resource management in Cloud Computing
[26, 27]	Telecommunication networks	Modern mobile networks' management automation	SON paradigm introduced by 3GPP
[28, 29]	Telecommunication networks	Complexity of LTE-A and 5G mobile networks	Automated mechanisms for networks management and cost reduction
[30]	Telecommunication networks	Network Functions Virtualization elasticity	Auto-scaling mechanisms
[31, 32]	Telecommunication networks	5G next gen core (NGC) micro-services elasticity	Auto-scaling mechanisms
[33]	Telecommunication networks	Elasticity of an IMS cloud implementation	Auto-scaling and load-balancing mechanisms

The goal of the system is to minimize the sum of capacity cost, inventory cost and backlog cost while satisfying customer demand. The central problem is to jointly decide production and capacity quantities over time to handle varying demand. Various versions of the problem have been solved by different approaches. [13] formulate the problem in a dynamic programming setting and characterize the optimal policies as order-up-to type. [14] develop quantitative models based on queuing theory that integrate the inventory level decision with the capacity related decisions and build a search algorithm to find optimal solution. [15] employ a Brownian motion approximation for the joint capacity and inventory management problem and derive analytical solutions.

Other problems in the production systems include forecasting demand, deciding the regular capacity, etc. [16] provides a review of operational problems in production and supply chain systems.

2.3 Cloud Computing

Cloud computing has gained in popularity in recent decade due to the flexibility it offers in delivering resources on demand. Here, a set of physical resources, operated by a cloud provider, is shared among cloud users in a leased, usage-based way. One of the main problems in cloud computing is resource management [17–19]. Cloud resources can be classified into four categories: compute, networking, storage and power [18]. In general, resource management is addressed either from the perspective of cloud providers or cloud users. Cloud providers are essentially concerned by efficient utilization of their physical resources and resort to VM (virtual machine) placement and migration algorithms to solve combinatorial problems. In this paper, we focus more on resource management from cloud user perspective where the objective is to cope with the changing load of applications at a minimal cost and under performance constraints; in general, the cost is induced by the leasing price of virtual computing resources.

The most common technique to adjust application resources to incoming workload is scaling. Most public cloud platforms implement manual scaling policies such as threshold-based auto-scaling (e.g. “add a new VM if average CPU usage >80%”), target tracking scaling (e.g. “adjust resources to keep average CPU usage around 50%”), and predictive auto-scaling (e.g. “add N VMs at midnight on New Year”) [20]. Manual auto-scaling policies struggle to adapt to variable traffic. In contrast, other approaches predict the workload and provision resources accordingly [21, 22]. This is done by building a model from past observations, which estimates the needed resources based on workload patterns. However, the model can become invalid when the application context evolves, e.g. user requests become more complex, or the volume of data to be processed increases. Some works explore RL approaches to manage capacity in Cloud [7]. Apply Reinforcement Learning to determine the number of servers and application placement dynamically for a web application [23]. Apply Reinforcement Learning to reconfigure individual virtual machines over CPU, I/O, memory and storage for an E-Commerce application. In [24], the authors apply Reinforcement Learning to determine the number of servers for a user-defined application to generate Fibonacci numbers. These works show the feasibility of managing Cloud resources with Reinforcement Learning and highlight the existing technical challenges.

Other problems in the cloud computing include resource allocation, resource scheduling, resource brokering, etc. [25] provides a survey of resource management problems in cloud computing.

2.4 Telecommunication Networks

Resources management is one of the major activities in the operation of telecommunication networks. For example, the paradigm of Self-Organizing Networks (SON) has been introduced by 3GPP [26, 27] to automate the management of modern mobile networks such as LTE-A [28] and 5G [29]. It fosters a self-organizing behavior which reacts dynamically to variations in network performance indicators, amplified due to the increasing number of wireless devices and services. Thus, the final goal is to optimize the global network performance and to ensure the requested quality of service for end-users while reducing the underlying costs for the operator.

Among various problems addressed in SON, the auto-scaling aspects introduced by Network Functions Virtualization [30] and 5G next gen core (NGC) micro-services [31, 32] are of particular interest to our study. Here, virtual resources or micro-services can be scaled up/down to adjust dynamically the network capacity. For example, the cloud implementation of IMS [33] can be deployed as a Virtual Network Function where all the components are scalable using stateless load-balancing. Similarly, micro-service instances of NGC such as session or subscriber management, can be added/removed on demand to adjust the capacity of their functions.

3 Structural Properties of Capacity Management

In this section, we summarize the common attributes of capacity management problems in different application domains as covered above and present the resulting structural properties.

3.1 Common Attributes (Meta-Model)

- W: demand processed by the system. The demand is usually stochastic and often non-stationary. It can be considered as a scalar quantity (homogeneous demand) or composed of multiple types of demands, $W = \langle W_1, W_2, .., W_m \rangle$.
- D: capacity of the system. Depending on the complexity of the system, it may process the demand in one or more stages, each of which has its own resource pool. The vector $D = \langle D_1, D_2, .., D_n \rangle$ of the number of servers (resources) at each resource pool forms the capacity of the system.
- K_D: cost of the capacity. It's an increasing function in the capacity.
- ρ: utilization of the capacity. This represents the percentage of time on average the servers are busy in each resource pool, $\rho = \langle \rho_1, \rho_2, \cdots, \rho_n \rangle$.
- L: service level received by the demand. If there are multiple types of demands, the service level is on each type, $L = \langle L_1, L_2, .., L_m \rangle$. A higher capacity provides a better service to the demand, for instance reducing the time the demand spends in the system.

- K_L: benefit from the service level. For instance, it can be the revenue received from the demand by providing a service level L. It's an increasing function in the service level.

Figure 1 provides a graphical presentation of the common attributes of a capacity management system, including a set of interconnected resource pools which represent the different processing stages that a demand may go through and receive a certain service level. The average capacity utilization of each resource pool is also shown by the horizontal dashed lines.

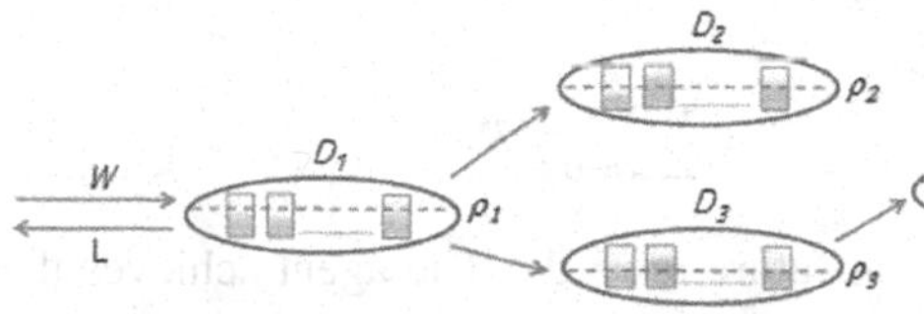

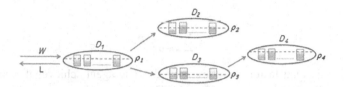

Fig. 1. General attribute of a capacity management system

The high-level architecture of a capacity management system can take many forms. Figure 1 provides an instantiation of it. A concrete example of the instantiation could be a distributed virtualized application, with a scalable gateway as an application front-end (Node 1 on Fig. 1, with capacity D_1). The application gateway can initially contact an authentication server (Node 2 on Fig. 1, with capacity D_2) for user authentication and generation for usage token. The gateway will then allow the user to interact with a scalable service (Node 3 on Fig. 1, with capacity D_3), using the generated token, that is in turn using another scalable service (Node 4 on Fig. 1, with capacity D_4) to respond to user query.

The capacity management problem is usually defined on discrete time intervals, which we refer to as periods. Decisions are made in each period on the capacity given the situation at the beginning of the period. The objective is to achieve a balance between good service and low capacity cost over the long run.

3.2 Reinforcement Learning Representation and Algorithm

When the distribution of the demand and the service level for given demand and capacity are unknown, one can use RL to solve the capacity management problem. RL uses Markov Decision Process (MDP) as the underlying model for an environment with unknown information and continuously learns to control it in interactions. In this section we briefly introduce the basic elements in RL and instantiate them with the common attributes of the capacity management systems.

In RL, an agent observes a state S of the system and takes an action A. The state of the system transits to S' and the agent receives a reward R associated with the transition (S, A, S'). This interaction is shown in Fig. 2. It can keep iterating over finite or infinite time horizon. The objective is to maximize the return G, the long-term discounted cumulative rewards.

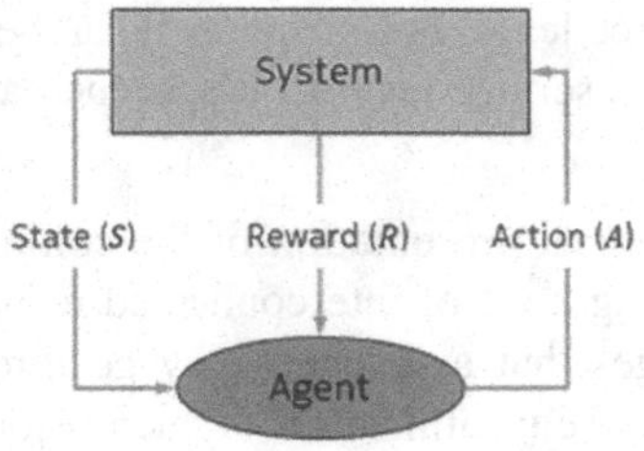

Fig. 2. Reinforcement Learning concepts.

$$G = \sum_{t=0}^{\infty} \gamma^t R_t \tag{1}$$

where γ is the discount factor on future rewards. The agent achieves the objective by selecting the action to take under each state. The mapping from a state to an action is referred to as a policy π, $A = \pi(S)$.

The agent uses the notion of action-values to reason about finding the optimal policy. The action-values $Q(S, A)$ represents the expected cumulative rewards with first taking action A in state S and following a certain policy thereafter. The agent empirically estimates the action-values through iterations. For the transition (S, A, S') with reward R, it updates the action-value

$$Q(S,A) \leftarrow Q(S,A) + \alpha[R + \gamma \max_{A'} Q(S', A') - Q(S,A)] \tag{2}$$

where $\alpha(\ll 1)$ is the learning rate. This method is commonly referred to as Q-learning. It has been shown that if all state-action pairs are updated infinite times, the action-values converge to their values under the optimal policy. At convergence, the optimal policy π^* is formed by choosing the action that gives the highest action-value, $\pi^*(S) = \text{argmax}_A Q(S, A)$.

For capacity management, the state S can be the demand that has arrived in the last period, the occupancy (the demand that occupies the system) at the beginning of the current period, the capacity utilization, etc. The action A can be the capacity to use in the current period, i.e. $A = D$. The cost of capacity is $K_D(D)$. The demand during the current period receives a service level L under the capacity D. The benefit $K_L(L)$ can be the revenue from serving the demand with service level L. The reward R can be specified as the revenue minus the capacity cost.

$$R = K_L(L) - K_D(D) \tag{3}$$

The return G can then be the long-term profit of operating the system to serve demands.

3.3 Common Structural Properties

In capacity management across application domains, there are various common structural properties. Some of them are dependencies between attributes at a single processing stage. Other are dependencies between attributes at different stages. In this section, we provide instances for both kinds of dependencies.

3.3.1 In-Stage Functional Dependencies

One instance of the dependencies at a single stage is the dependency between service level and utilization. At a given demand, a higher capacity increases the service level and reduces the utilization. The service level is thus decreasing in the utilization. Moreover, the rate the service level decreases is often increasing in the utilization, as shown in Fig. 3. This is characterization of the functional dependency between the service level and utilization. These and other possible functional dependencies can be used to enhance the performance of the general purpose RL algorithms by making them more specific to capacity management.

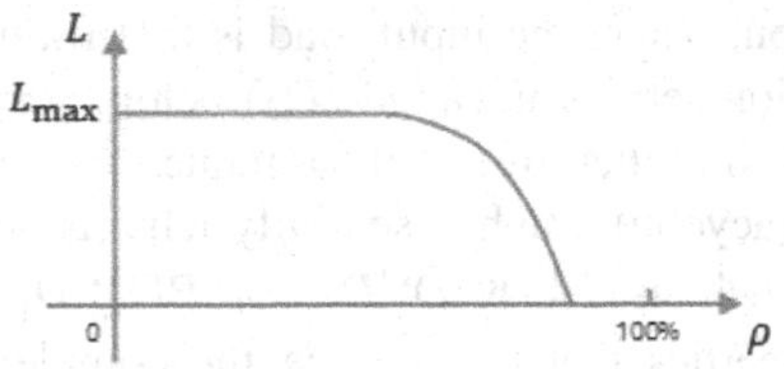

Fig. 3. System saturation effect: service level vs utilization

3.3.2 Cross-Stage (Causal) Dependencies

Recall that $D = \langle D_1, D_2, .., D_n \rangle$ is a vector of decision variables in a capacity management system, i.e. each D_i is the capacity decided for a given resource pool. To express the topological dependencies between different processing stages in the system, we will rely on the theory of causality. Namely, we assume the optimal solution to the capacity allocation problem can be expressed through Structural Equation Modeling (SEM) [34] as a solution to a set of equations

$$\begin{cases} D_i^* = f_i(\bar{D}_{-i}^*, W, U_i), i = 1, \ldots, n \\ W = f_{n+1}(U_{n+1}) \end{cases} \tag{4}$$

where $\bar{D}_{-i}^*$ is the set of optimal decision variables excluding D_i^*, W is the demand (workload), and U_i are independent exogenous random variables. The latter represent noise components in different parts of the system: in the processing stages and the demand. For each of the remaining (endogenous) variables an SEM provides an equation expressing an autonomous mechanism through which the nature assigns their values [34]. Note however that the values of variables D_i^* are not assigned exclusively by some natural processes as in the cases covered in the literature, but they correspond to the output of an ideal optimization algorithm, given an instantiation of the exogenous variables. An optimal solution to a resource allocation problem will depend on the inherent dynamics of the system as well as the optimization criteria defined by the reward function. By extension, this can still be considered as a natural process.

In case of a linear model, we have the following parametrized form of SEM

$$\begin{cases} D^* = \mathbf{A}D^* + \mathbf{B}W + U \\ W = U_{n+1} \end{cases} \tag{5}$$

where $\mathbf{A}$ and $\mathbf{B}$ are the matrices of respective linear coefficients. Both the non-parametric (4) and linear models (5) have the graphical representation shown on Fig. 4 (left). So far it does not reflect any structural property of the system and suggests an (almost) full-mesh dependency graph among the endogenous variables which stipulates that any optimal decision variable D_i^* depends on all the others and on the demand. For a distributed system this graph can be much simpler when some coefficients in (5) are equal to zero. Furthermore, one can reasonably expect that the simplified graph would follow the topology of the system, i.e. its structural decomposition into different processing stages, the service workflow, etc. Figure 4 (right) illustrates an example of a distributed web application, where the input load is first received at front-end servers (D_1), then an authentication service is called (D_2) before forwarding the request to a pool of workers (D_3), which in their turn make a request to database resources (D_4). In this case, the simplified (acyclic) graph essentially tells us about a set of conditional independence relations such as $D_2^* \perp\!\!\!\perp D_3^* | D_1^*$, or $P\left(D_4^* | D_1^*, D_2^* D_3^*, W\right) = P(D_4^* | D_3^*)$. Those are important properties that can reduce the complexity of the optimization algorithm, although their specific exploitation modalities are out of the scope of the current paper. Furthermore, these are statistically testable assumptions; i.e. they can be verified through experiments or from the observed data.

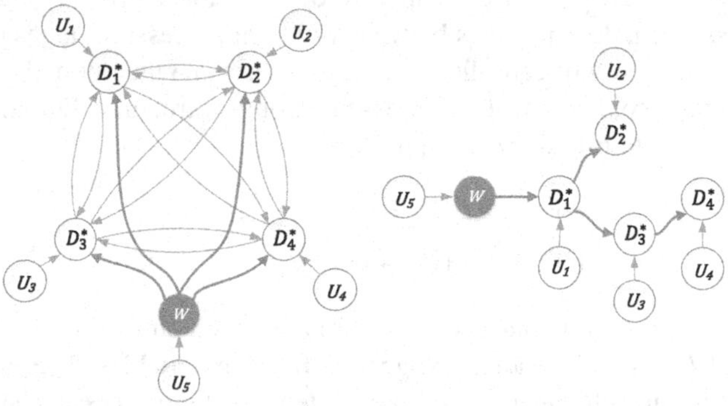

Fig. 4. Graphical representation of a Structural Equation Model: (left) the internal system structure is not exploited, (right) the system topology defines an acyclic causal graph.

4 Research Tracks on System Structure Exploitation

The standard RL methods are developed for general problems. They don't exploit structural properties specific to concrete problems they are solving. However, concrete problems do have structural properties beyond those from the underlying MDP model assumed by RL. Exploiting these structures may be the key to make RL faster. As pointed out in [1], we may need to derive inductive biases that generalize well in future learnings to accelerate and safeguard RL.

In this paper, we don't provide solutions to exploit common structural properties yet, but we review research tracks that may inspire these solutions. Some of the tracks derive analytically algorithmic structures in action values and policies from the common structural properties. They exploit the algorithmic structures for developing computationally efficient solutions for problems with full information. Others learn algorithm structures from data of related problems and generalize these structures to the problem on hand for faster learning. These may give ideas on how to derive and exploit algorithmic structures in RL and learn faster in problems with unknown information. We provide a summary of the papers survey at the end of this section in Table 2.

Table 2. System structure exploitation in multiple research tracks

Paper	Field	Problem	Structural property
[9, 35]	Service systems	Staffing problem	Monotonicity
[36, 37]	Wireless communication	Rate and power control	Supermodularity and monotonicity
[38]		Partially observable MDP	(Introductory tutorial)
[39]	Causal inference	Causality learning	(Survey of observational causality learning)
[40]	Causal inference	Causality learning	Causality learning with small interventions
[41]	Causal inference	Causality learning	PC algorithm for observational causality learning
[42]	Causal inference	Causality	(Books introducing the theory of causal inference)
[43]	Transfer learning	Transfer learning	(Blog on transfer learning)
[44]	Transfer learning	Reusing knowledge acquired when performing previous tasks	Reusing lower layers of trained neural networks
[45]	Transfer learning	Reusing knowledge acquired when performing previous tasks	Learning domain invariant representations
[46]	Transfer learning	Reusing knowledge acquired when performing previous tasks	Using transfer learning for Reinforcement Learning

4.1 Operations Research

In Operations Research, the problems are formulated as advanced analytical models based on common system properties, and the algorithmic structures are analytically derived based on mathematical principles. These structures are then exploited for great gains in computational efficiency.

In [9], the authors address the staffing problem for a queuing system over finite horizon, considering transient queueing effects. They establish analytically the sub-modularity of the objective function with respect to the staffing level and initial queue size. They show the optimal policy follows a monotone control, i.e. the optimal policy should prescribe increased staffing level for higher initial queue lengths. The monotonicity is one of the key assumptions in developing a computationally efficient heuristic algorithm to solve the staffing problem in [35]. It results in a significant reduction in searching the state space.

In [36], Q-learning based stochastic control algorithms are introduced for rate and power control in wireless MIMO transmission systems. They exploit the supermodularity and monotonic structure properties derived in [37]. Finally, an introductory tutorial on structural results in partially observed Markov decision processes (POMDPs) is given in [38]. They use lattice programming methods to characterize the structure of the optimal policy of a POMDP without brute force computations.

4.2 Causal Inference

Elucidation of cause-effect relationships is a central driver in science and philosophy. A comprehensive theory of causality elaborated in [34] relies on graphical models where the directed edges represent stable and autonomous causal mechanisms.

The causal relationships can be inferred through interventions or passive observations [39]. The first consists in intervening on some variables by altering the process which generates them, i.e. enforcing their values, and observing their effect on other variables. A natural problem in this case is to minimize the number of observations, necessary to discover all causal relationships [40].

Passive observational methods are used to discover a causal graph when interventions or random experiments cannot be applied. These methods can be classified into two main categories [41]. While the 'search and score' algorithms explore the space of possible graphs by evaluating at what extend each of them conforms to observations, the constraint-based approaches are based on statistical tests for conditional independence. A well-known algorithm in this category is the PC algorithm [42]. It starts with a complete, undirected graph and deletes recursively edges, based on statistically inferred conditional independence relationships.

Causal Inference can be used to identify the factors in states S and actions A that effectively cause the cumulative reward G. Based on the identification, we can strip off irrelevant factors and achieve a simplification in states and actions. This shrinks the space of states and actions and provides a boost to the learning speed in RL. For instance, in the game of blackjack, the win or loss is determined only by the values of the cards and not their suits. We can strip the information of the suits off the state and have a more compact state representation.

4.3 Transfer Learning

Transfer learning (TL) is a design methodology which consists in transferring knowledge from one domain or system (source) to another (target). It is motivated by the fact that traditional machine learning techniques rely on huge amounts of labelled data to train a model, while in many systems, such as critical systems (e.g. aircraft, or autonomous cars) or large networks, the labelled data are not available or are difficult to obtain. The main idea of TL is to use the knowledge acquired when learning to perform one task to improve the learning process of a related, but a different task. The extensive use of Deep Learning in recent years has contributed to the development of the TL approaches [43]. Examples include using pre-trained neural networks whose lower layers capture low level features that can be reused for new problems [44] or learning domain invariant representations from unlabeled data [45].

Transfer Learning has also been applied in the domain of RL. The existing algorithms are organized around five comparison criteria [46]: (i) task difference assumption: what differences are allowed between the source and the target systems, e.g. different states, or different actions, (ii) source task selection: how the relevant source systems are identified, (iii) transferred knowledge: what should be transferred, e.g. actions to use in some states, action-value functions, or a policy, (iv) task mapping: how to overcome differences between the source and the target tasks, e.g. mapping the actions of the source system to actions of the target system, and (v) allowed learners: by which type of learners the transferred knowledge can be used, e.g. if action-value function is transferred, then the target system should use temporal difference learning method. Those criteria correspond to questions that need to be answered when transferring knowledge from one system to another.

We have argued in this paper that the resource management systems in various application fields often share common structures. Some of these systems are more available for experimentations and machine learning than others. We can then expect that, due to their common structural properties, the resource management knowledge and models learned in one system can become beneficial in the others, which is similar to the paradigm of TL. So, the common structural properties we have identified can help answering some of the above-mentioned questions in terms of TL: how to identify similar systems where the knowledge can be transferred from, and to identify the relevant information to be transferred. These structural properties also constitute a pivotal model for building mappings between the differences in source and target systems.

5 Conclusion

Future digitalized systems will have high complexity and operate in a constantly evolving environment. RL brings a strong promise for autonomously managing resources in these highly complex, dynamic and often not very well understood systems. However, current RL methods still struggle to learn rapidly in the incremental, online settings of many practical problems. The reason may be that current RL methods are developed for general problems. They don't exploit the structures of the concrete problems they are solving. If we derive inductive biases for RL (e.g. algorithmic structures) from common system structures or learn them from related domains or problems, future learning may be able to generalize these biases well and be faster. In addition, such knowledge could be used to make learning safer. Indeed, RL relies on exploration where actions are randomly taken which might be harmful for the system. Having some common knowledge/rules such as "if the QoS is poor, do not explore by reducing resources in the system" could prevent such behavior. In this work, we have considered a category of resource management problems and focused on making RL faster. We reviewed this space in various application domains with a focus on capacity management. We then identified common structures in capacity management. Even though we don't provide solutions to derive and exploit inductive biases for RL, we surveyed research tracks that derive and exploit such inductive biases in different fields and which may give inspirations for further research.

Incorporating inductive biases gives leverage to RL processes and makes it faster and safer. However, this relies on the assumption that the biases are present in the problems under consideration. The assumption may not always hold. The incorporated biases can then be obstacles rather than boosts. How can we test whether the assumption holds? How can we stay adaptive in case the assumption is violated? These remains open questions for future research.

References

1. Sutton, R.S., Barto, A.G.: RL: An Introduction, 2nd edn. The MIT Press, Cambridge, London (2017)
2. Clark, J. This Preschool is for Robots. Bloomberg (2015)
3. Gu, S., Holly, E., et al.: Deep reinforcement learning for robotic manipulation with asynchronous off-policy updates. In: IEEE International Conference on Robotics and Automation (ICRA), Singapore (2017)
4. Pit.ai. https://www.pit.ai/
5. Mnih, V., Kavukcuoglu, K., Silver, D., et al.: Human-level control through deep Reinforcement Learning. Nature **518**, 529–533 (2015)
6. Silver, D., Hassabis, D.: AlphaGo: mastering the ancient game of Go with Machine Learning. Google Research Blog (2016)
7. Jin, Y., Bouzid, M., Kostadinov, D., Aghasaryan, A.: Model-free resource management of cloud-based applications using RL. In: International Workshop on Network Intelligence (NI/ICIN2018), Paris, France (2018)
8. Liu, Y., Watt, W.: Stabilizing customer abandonment in many-server queues with time-varying arrivals. Oper. Res. **60**(6), 1551–1564 (2012)
9. Fu, M.C., Marcus, S.I., Wang, I.: Monotone optimal policies for a transient queueing staffing problem. Oper. Res. **48**(2), 327–331 (2000)
10. Bassamboo, A., Harrison, J.M., Zeevi, A.: Design and control of a large call center: asymptotic analysis of an LP-based method. Oper. Res. **54**(3), 419–435 (2006)
11. Defraeye, M., Van Nieuwenhuyse, I.: Staffing and scheduling under nonstationary demand for service: a literature review. Omega **58**, 4–25 (2016)
12. Gans, N., Koole, G., Mandelbaum, A.: Telephone call centers: tutorial, review, and research prospects. Manuf. Serv. Oper. Manage. **5**(2), 79–141 (2003)

13. Tan, T., Alp, O.: An integrated approach to inventory and flexible capacity management subject to fixed costs and non-stationary stochastic demand. OR Spectrum **31**(2), 337–360 (2009)
14. Buyukkaramikli, N.C., van Ooijen, H.P., Bertrand, J.W.: Integrating inventory control and capacity management at a maintenance service provider. Ann. Oper. Res. **231**(1), 185–206 (2015)
15. Bradley, J.R., Glynn, P.W.: Managing capacity and inventory jointly in manufacturing systems. Manage. Sci. **48**(2), 273–288 (2002)
16. Snyder, L.V., Atan, Z., Peng, P., Rong, Y., Schmitt, A.J., Sinsoysal, B.: OR/MS models for supply chain disruptions: a review. IIE Trans. **48**(2), 89–109 (2015)
17. Parikh, S., Patel, N., Prajapati, H.: Resource management in cloud computing: classification and taxonomy. CoRR (2017)
18. Jennings, B., Stadler, R.: Resource management in clouds: survey and research challenges. J. Netw. Syst. Manage. **23**, 567–619 (2015)
19. Mann, Z.A.: Allocation of virtual machines in cloud data centers - a survey of problem models and optimization algorithms. ACM Comput. Surv. **48**(1), 11 (2015)
20. Amazon: AWS Auto Scaling. https://aws.amazon.com/autoscaling/
21. Jacobson, D., Yuan, D., Joshi, N.: Scryer: Netflix's Predictive Auto Scaling Engine. Netflix Technology Blog (2013)
22. Roy, N., Dubey, A., Gokhale, A.: Efficient autoscaling in the cloud using predictive models for workload forecasting. In: IEEE CLOUD 2011, Washington, pp. 500–507 (2011)
23. Li, H., Venugopal, S.: Using RL for controlling an elastic web application hosting platform. In: International Conference on Automatic Computing, pp. 205–208 (2011)
24. Rao, J., Bu, X., Xu, C.-Z., Wang, K.: A distributed self-learning approach for elastic provisioning of virtualized cloud resources. In: 19th Annual IEEE International Symposium on Modelling, Analysis, and Simulation of Computer and Telecommunication Systems, pp. 45–54 (2011)
25. Manvi, S.S., Shyam, G.K.: Resource management for Infrastructure as a Service (IaaS) in cloud computing: a survey. J. Netw. Comput. Appl. **41**, 424–440 (2014)
26. SON: Self-Organizing Networks. https://www.3gpp.org/technologies/keywords-acronyms/105-son
27. Hämäläinen, S., Sanneck, H., Sartori, C.: LTE self-organising networks (SON): Network Management Automation for Operational Efficiency. Wiley, Chichester (2012)
28. Sesia, S., Toufik, I., Baker, M.: LTE - The UMTS Long Term Evolution: From Theory to Practice, 2nd edn. Wiley, Chichester (2011)
29. Rodriguez, J.: Fundamentals of 5G Mobile Networks. Wiley, Chichester (2015)
30. Network Functions Virtualisation – Update White Paper. ETSI (2013)
31. Evolution of the cloud-native mobile core, Nokia White Paper (2017)
32. Evolving Mobile Core to Being Cloud Native. Cisco White Paper (2017)
33. Project Clearwater - IMS in the Cloud. http://www.projectclearwater.org/
34. Pearl, J.: Causality: Models, Reasoning and Inference, 2nd edn. Cambridge University Press, New York (2009)
35. Yoo, J.: Queueing models for staffing service operations. Ph.D. dissertation. University of Maryland, College Park, MD (1996)
36. Djonin, D.V., Krishnamurthy, V.: Q-learning algorithms for constrained markov decision processes with randomized monotone policies: application to MIMO transmission control. IEEE Trans. Signal Process. **55**(5), 2170–2181 (2007)
37. Djonin, D.V., Krishnamurthy, V.: MIMO transmission control in fading channels—a constrained markov decision process formulation with monotone randomized policies. IEEE Trans. Signal Process. **55**(10), 5069–5083 (2007)

38. Krishnamurthy, V.: Structural Results for Partially Observed Markov Decision Processes (2015). arXiv:1512.03873. https://arxiv.org/abs/1512.03873
39. Rosenbaum, P.: Design of Observational Studies. Springer, New York (2010). https://doi.org/10.1007/978-1-4419-1213-8
40. Shanmugam, K., Kocaoglu, M., Dimakis, A., Vishwanath, S.: Learning causal graphs with small interventions. In: NIPS 2015, Cambridge, MA, USA, pp. 3195–3203 (2015)
41. Le, T., Hoang, T., Li, J., Liu, L., Liu, H.: A fast PC algorithm for high dimensional causal discovery with multi-core PCs. In: IEEE/ACM Transactions on Computational Biology and Bioinformatics (2015). https://doi.org/10.1109/tcbb.2016.2591526
42. Spirtes, P., Glymour, C., Scheines, R.: Causation, Prediction, and Search, 2nd edn. MIT Press, Cambridge (2000)
43. Ruder, S.: Transfer Learning - Machine Learning's Next Frontier. Blog post (2017). http://ruder.io/transfer-learning/
44. Bingel, J., Søgaard, A.: Identifying beneficial task relations for multi-task learning in deep neural networks. In: EACL, pp. 164–169 (2017)
45. Glorot, X., Bordes, A., Bengio, Y.: Domain adaptation for large-scale sentiment classification: a deep learning approach. In: 28th International Conference on Machine Learning, pp. 513–520 (2011)
46. Taylor, M., Stone, P.: Transfer learning for reinforcement learning domains: a survey. J. Mach. Learn. Res. **10**, 1633–1685 (2009)

Inverse Kinematics Using Arduino and Unity for People with Motor Skill Limitations

Nicolás Viveros Barrera[1], Octavio José Salcedo Parra[1,2], and Lewys Correa Sánchez[2(✉)]

[1] Department of Systems and Industrial Engineering, Faculty of Engineering, Universidad Nacional de Colombia, Bogotá D.C., Colombia
{nviverosb,ojsalcedop}@unal.edu.co
[2] Faculty of Engineering, Intelligent Internet Research Group, Universidad Distrital "Francisco José de Caldas", Bogotá D.C., Colombia
osalcedo@udistrital.edu.co,
lcorreas@correo.udistrital.edu.co

Abstract. In this document, the creation process of an application is detailed that can use various sensors connected and managed by an Arduino UNO board to capture the movement from the extremities in people with limited movement. It is noteworthy to mention that the results of said application are not discussed here but only the creation process is described. The first part is dedicated to describing the hardware required to use the app's programmed technology. A brief overview of the Arduino platform is also given followed by a description of the sensors used for calculation and capture of people's movement. The selection of the sensors is justified, and their operation is presented. The second part focuses on the construction of the application starting by offering a synthesized view of the Unity platform up to the development process. Additionally, the basic concepts to generate the 3D models are explained with the purpose of allowing anyone that reads this document to replicate the project in a simple manner.

Keywords: Arduino · Models · Unity

1 Introduction

Inverse kinematics (IK) is the calculation process of a system of articulations with the objective of capture their movement and use it for different purposes such as animation and videogames [1]. When movement is registered, it can be replicated in a virtual environment giving realistic mobility to virtual entities. Instead of calculating the position or rotation of objects in the workspace over time, the idea is to mimic the position or rotation of said object in the real world and assign this behavior to virtual objects.

This article focuses on the elaboration of an inverse kinematics system that uses sensors made from Arduino boards which capture the movement of people by attaching sensors to their bodies. This can be applied to the field of medicine and will also include a virtual reality app for mobile phones developed in Unity. It will enable to virtually experience the movements made by the user in real time.

É. Renault et al. (Eds.): MLN 2018, LNCS 11407, pp. 133–145, 2019.
https://doi.org/10.1007/978-3-030-19945-6_9

Inverse kinematics can be obtained in various methods since its most basic form only the movement of the individual must be captured through cameras that follow the extremities of the target and create a mathematical model applied to a virtual character as seen in Kinect [2]. However, it is intended to use an accelerometer, a gyroscope and a magnetometer connected to an Arduino board that calculates the changes in the model rotation and sends them to the app for their interpretation in real time so that there is an interaction with the data.

2 Related Works

There are various ways to establish an inverse kinematics system, so a summary is given of some projects involving different technologies for different purposes. The *capture system for movement in real time* and the *free software Arduino-Unity-IMU* projects are crucial for this project.

The first project will serve as a guideline to work with the Arduino platform and the chosen sensor. The same sensors are used except for the motion sensor (or IMU) which is different in this case. The research carried out in the inverse kinematics model is taken as a basis since it has the same tools and the same goal as this project which consists on using virtual reality technologies. The second project provides source code to use the new sensor with nine degrees of freedom that is going to be used. Even if it is a simple project, it is useful since it avoids programming code from scratch by providing another alternative with less hardware than other schemes.

A. *Style-based inverse kinematics*

The style-based inverse kinematic project [3] presents a kinematics system inspired in learning human poses. It can not only memorize various human poses but also generate them through a cloud of possible poses that consider their range and make the system choose one. This range represents the positions that the machine did not learn directly but that it generated from a database of pre-registered positions.

It can be clarified that the inverse kinematics system is not implemented with any hardware technology but rather describes the mathematical and computational model developed. At the beginning of the document, the mathematical entities (vectors) are described to elaborate the 3D human model. Then, the learning process is described for recognition of new poses and addition to the database. The poses are synthesized, and the algorithm is optimized. Finally, the possible uses of motion recognition and capture for image and video are briefly mentioned. The algorithm is capable of seeing a person's movement in a series of photos or videos and replicate it in a 3D model.

B. *Inverse kinematics for robotic arms*

[4] The project describes the process to develop an inverse kinematics model that can be implemented in Unity in order to control a robotic arm coherently. The article gives a brief overview of a simple mathematical model then proceeds to describe an implementation of the model in the form of a code that allows the user to play with each articulation. The movement is meant to be realistic and coherent and is programmed in C# which is the same language used for the scripting of the entire app.

The limitations of the system's movement are established so that once it becomes realistic it can be restricted. This assures that the generated movement range is not unrealistic and does not leave the workspace of a real robot.

3 Design and Functionality

A brief overview is given on the functionalities of the end product and the way in which the user can interact with the product starting by functional and non-functional requirements up to specifying the use case scenarios which serve as foundation.

The path followed by the user when using this technology starts with the sensors which must be fed with some sort of supply source that is connected to the Arduino board. Then, the sensors must be placed in the person's extremities, right in the middle of two articulations, i.e., if this technology is used with one person that has difficulty moving one arm then one sensor must be placed in the bicep, another one on the radial muscle and another one in the palm of the hand. This will adequately capture the movement of the targeted extremity.

Once the sensors are added to the user's extremity, then the Arduino board must be paired with the phone containing the app to then choose the type of activity that is going to be performed. It is worth clarifying that the virtual reality is used in the application only during the chosen activity, not during the calibration process of the sensors nor in the app menus.

After choosing the activity, the sensors must be calibrated for each activity and each time that a new activity is chosen. Finally, the user can perform the activity as many times as he wishes.

A. *Functional requirements*

- The application must allow the user to pair his phone via Bluetooth with the Arduino board.
- The app must allow the user to choose which extremity (out of the main four) he wishes to use with the sensors since he cannot use more than one extremity at a time due to the limited number of available sensors and the transmission speed of Bluetooth technology.
- The application must include a different activity for each extremity so only four activities are available.
- The application must be able to give scores to the players only during its execution since no record of the user's progress is kept. Such task is in charge of external agents.
- All medical related matters are avoided at all cost or using the solution as an alternative to therapy. This project explores the possibility of using this technology not its results in the medical field.
- The application must enable the calibration of sensors via Bluetooth.
- The user cannot perform activities involving more than one extremity at a time.
- The user must carry out the activity corresponding to each extremity without the need to interact with the phone (only within a virtual reality visor) until the activity is concluded.

B. *Non-functional requirements*

- The application must present a friendly environment to the user during the duration of the activity, while facilitating the movement of the prosthesis or extremity.
- The application must be capable of responding to different screen sizes, so it is not limited to a single resolution.
- The application must work only on Android-based phones.
- The communication between the sensors and the phone must not be slow, since it would completely block the use of sensors for motion capture.
- Since the Arduino board and the phone work at different clock frequencies, the project must offer a synchronization method for hardware resources to avoid loss of data.
- The number of sensors required by the project must not exceed three units per sensor.
- Only a Bluetooth module with the Arduino board will be used so the motion capture modules will be connected with cables due to the connectivity and the supply of the circuit.
- The calibration of the sensors cannot exceed one minute.
- The calibration process of the sensors must deliver accurately enough results, so that the external noise does not interfere in the use of the application.

C. *Use case scenarios*

1. *Turning on the motion capture unit*
   The user shall connect the Arduino board with a supply source of 5 V which in turn will feed all the required circuits. When turning on the unit, the user shall press a button to turn on the Bluetooth module.
2. *Pairing the phone with the board*
   The user must pair the Arduino board from the phone to then open the app and start the calibration process of the sensors.
3. *Sensor calibration*
   The user must place the motion capture sensors in the extremity that he chooses to perform the activity. This will initiate the calibration process of the sensors to use the information in an inverse kinematics model.
4. *Choosing the target extremity*
   After the calibration, the user shall be able to choose one of the four extremities, on which the motion capture modules are attached.
5. *Performing the activity*
   Once the extremity has been chosen, the user can carry out the corresponding activity without the need to interact with the phone (only by having the virtual reality visor) until the activity is finished (Fig. 1).

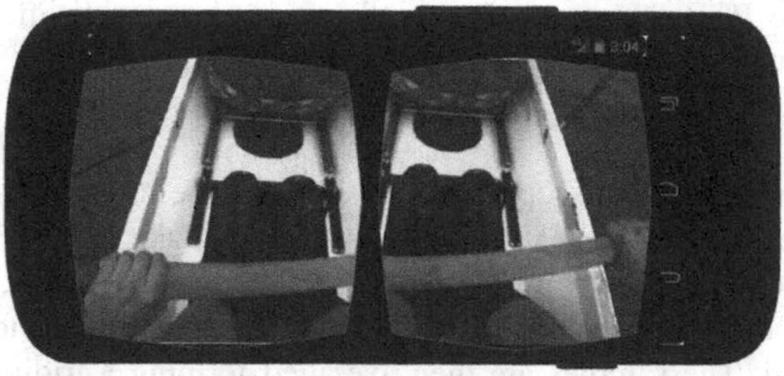

Fig. 1. Example of an activity for lower extremities. Source: Authors

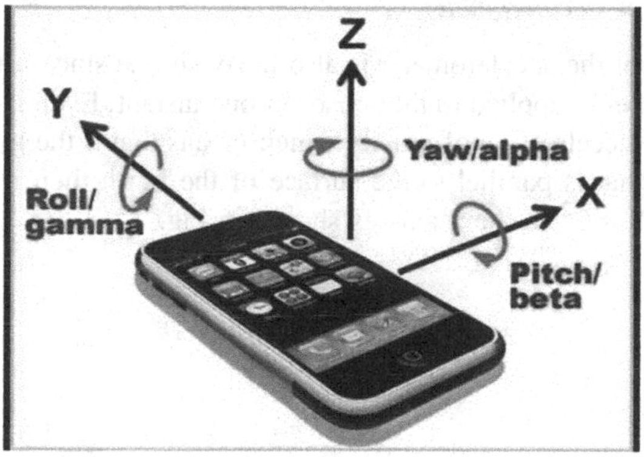

Fig. 2. Gyroscope of a telephone. Source: [6]

4 Implementation

To implement this technology, the article *Position and orientation of a probe with use of the IMU MPU9250 and a ATmega328 microcontroller* [5] is used as a basis. This project has a similar implementation to the one intended in this project with the exception that the former has medical purposes.

A. *Software*

1. *Sensor calibration*

As stated in the previous subsection, the sensors used do not take perfect measurements and it necessary to calibrate them so that the collected data is reliable.

The calibration of the first two sensors (accelerometer and gyroscope) is simple since it is easy to identify the sources of noise and distortion affecting these sensors. Nonetheless, the calibration of the magnetometer is more complicated because it has two sources of noise: Hard iron and soft iron. Noise is introduced by the hard iron because the current circulating through a circuit creates a magnetic field that interferes with the measurements taken by the sensor. The hard iron issue is easy to correct although it requires the capture of many samples in each axis. The soft iron is harder to calibrate since it consists on the interference related to the Earth's magnetic field, in elements of the PCB that are not normally under a magnetic field. This process requires a large number of samples and calculations.

2. *Calibration of the gyroscope*

The calibration of the gyroscope is fairly simple since it can measure the angular speed of the object which means that if the object remains still over a surface, its readings should be equal to zero in all axes.

To calibrate the gyroscope, the sensor is left still in a surface and 1000 measurements are captured. The samples are then averaged forming a tridimensional vector. This average is a constant value that is subtracted to each real-time measurement of the gyroscope.

3. *Calibration of the accelerometer*

The calibration of the accelerometer is also fairly simple since it measures the total acceleration that is being applied to the sensor in one instant. Even if the device is still, it must measure the acceleration of gravity which means that if the gyroscope is placed over a flat surface that is parallel to the surface of the Earth then it must measure an acceleration of $-9.8\ \text{m/s}^2$ in the z axis as shown in Fig. 3.

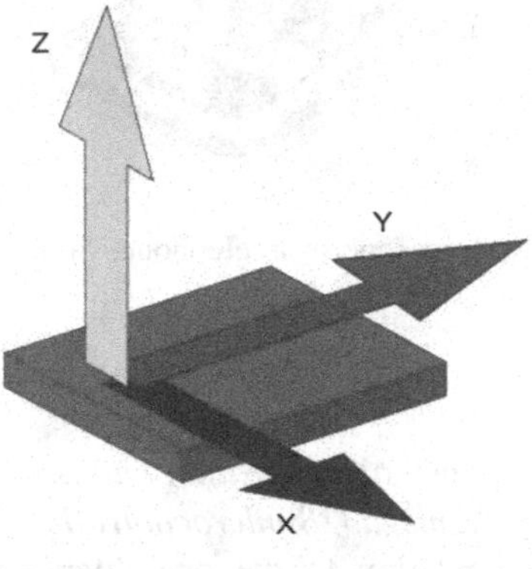

Fig. 3. Axes of an accelerometer. Source: Google images

To calibrate an accelerometer, the sensor must be placed on flat surface to deliver a measurement of $-9.8\ \text{m/s}^2$. 1000 samples are averaged in said position and a value of the sensor's noise or error is estimated. Subtracting this value to each measurement of the accelerometer is an active filtering strategy that makes the data more reliable.

4. *Calibration of the magnetometer*

To understand the adjustments required by the magnetometer, it is necessary to first understand what is being measured since it can lead to establishing the steps needed to calibrate the sensor.

The magnetometer measures the magnetic field in each Cartesian axis so the result of each measurements is a three-dimension intensity vector. To calibrate the hard iron of this sensor, the following equations must be established:

$$A = CS - CT$$

$$B = CS + CT$$

CT is the magnetic field of the Earth and CS is the magnetic field produced by the sensor's current flow. Afterwards, after adding both equations, the Earth's magnetic field cancels out and the result is twice the magnetic field of the sensor.

$$A + B = CS - CT + CS + CT = 2CS$$

$$A + B = 2CS$$

To determine A and B, 1000 measurements of the sensor must be captured in a specific orientation and then rotate the sensor by 180° so that the following 1000 measurements register the Earth's magnetic field as negative. This process is repeated for each Cartesian axis.

Figure 4 shows an example of the concept used to calibrate the magnetometer in the x axis, first pointing into one direction, then rotating the device by 180° and then measuring in the new opposite direction.

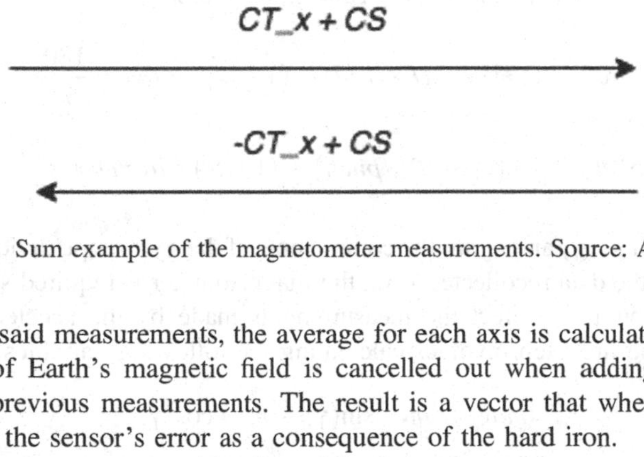

Fig. 4. Sum example of the magnetometer measurements. Source: Authors

By adding said measurements, the average for each axis is calculated. Finally, the measurement of Earth's magnetic field is cancelled out when adding the averaged values of the previous measurements. The result is a vector that when divided by 2 corresponds to the sensor's error as a consequence of the hard iron.

To calculate the error caused by the soft iron, more measures must be captured for each axis. Nonetheless, due to the extreme difficulty of calculating this noise, it was decided to use third party software [7] to calibrate the sensor and deliver the noise vector for soft iron. The reader can check the link in [8] for information on using this third-party software.

B. *Calculation of the sensor rotation in degrees*

Before starting this section, the names of the target angles are given for convention purposes. Figure 2 illustrates the rotations that need to be determined. The names of said angles are in English and they will be used in this article to keep the same conventions. The angles of Roll (γ) and Pitch (β) come from the rotation of the device in each horizontal axis, i.e. the axes which are parallel to the Earth's horizon. The angle of Yaw (α) refers to the angle resulting from the rotation of the object in its vertical axis (or perpendicular to the horizon).

With the purpose of obtaining the rotation of the MPU9250 module, the data from the gyroscope and accelerometer are used to find γ and β. In contrast, since the accelerometer measures the linear acceleration of an object, measuring the angle α it very hard with only two sensors so the data from the magnetometer was used.

[5] To calculate γ and β, the data from the accelerometer and gyroscope were used by finding the measurements in each axis of the accelerometer and keeping the vertical axis as reference point. The calculations can be made to find the gravity measured in each axis which leads to finding:

$$aRoll = atan2(a_x, a_z) \tag{1}$$

$$aPitch = atan2(a_y, a_z) \tag{2}$$

Equations 1 and 2 use the atan2 function from C# that calculates the angle between two points. Said points are the components x (ax) and z (az) of the accelerometer vector and the components y (ay) and z (az) of the same vector.

$$roll = \propto *(g_x * dT + roll) + (1 - \propto) * aRoll * \frac{180}{\pi} \tag{3}$$

$$pitch = \propto *(g_y * dT + pitch) + (1 - \propto) * aPitch * \frac{180}{\pi} \tag{4}$$

The values of $\mathbf{g_x}$, $\mathbf{g_y}$ and $\mathbf{g_z}$ are the components of the gyroscope divided by 131 [5]. To calculate α, the data recollected from the magnetometer is required since a rotation in that angle would not affect the measurements made by the accelerometer in the vertical axis. The first step involves calculating the following variables.

$$mag_x = m_z * \sin(\gamma) - m_y * \cos(\gamma) \tag{5}$$

$$mag_y = M_z + M_y + M_x \tag{6}$$

$$\propto = \text{atan}\left(\frac{mag_x}{mag_y}\right) * rad2deg \tag{7}$$

Where $M_x = m_x * cos(\beta)$, $M_y = m_y * sin(\beta) * sin(\gamma)$ and $M_z = m_z * sin(\beta) * cos(\gamma))$. The variables m_x, m_y and m_x correspond to the values of the magnetometer in each tridimensional axis after being normalized. *rad2deg* is the multiplication constant to convert from radians to degrees.

It is worth mentioning that in order to rotate an object in Unity with the data captured in real time by this filter, the data must be converted into a quaternion with the *Quaternion Euler* function included in Unity.

This way of rotating objects introduces a problem known as the Cardan blockage in which the two rotation axes are aligned when one of them is exactly rotated by 90°. This leads to losing degrees of freedom.

C. *Calculation of the rotation with quaternions*

Since the calculation of the rotation in degrees did not consider the errors generated after several iterations and that the Unity platform handles a different coordination system for rotation (not in degrees), it is more convenient for accuracy purposes to calculate quaternions since it eludes coordinate changes and the degree issue.

With the purpose of calculating the rotation of the object, AHRS (Attitude and Heading Reference System) open source algorithms were used. In particular, the Open-Source-AHRS-with-x-IMU [9] which provides an accurate calculation of the rotation angles in quaternions keeping in mind the error generated over time.

D. *Programming and communication of the Arduino board with the phone*

In previous sections, the wiring of the Arduino board was discussed. However, the programming task and the communication with the phone are a different matter.

1. *Calibration of the magnetometer*

In order to program the board, the open source algorithm MPU9250BasicAHRS was modified to work with the Bluetooth module. The original code can be found in the references as well the final version used in the current project [10].

To send data via Bluetooth, a serial port was created from the Arduino board which sent information at high speeds to the phone with basic messages printed in the console.

2. *Communication with the phone*

Since the Unity platform only offers basic support for smartphone hardware, it is required to use a Plugin that has been natively programmed for Android so that it can access hardware elements within Unity-based programming. Unity does not have native support for some hardware pieces such as the camera, the sensor or the Bluetooth module.

To accomplish such task, an owner plugin [11] was bought in the Unity Assets store because the open source alternatives were very limited and poorly documented. The plugin used can create a channel between both devices so that they can exchange messages by simulating a terminal, even at high speeds. This translates into the capacity to create a communication channel were a fast exchange process is allowed in one direction, from the Arduino board to the phone. This enables the real time tracking of the 9DOF sensor data.

Although the purpose of this project is to create an Android application that can track the mobility of a person's extremities, there are obstacles within the project such as programming the Plugin that do not concern its scope so their elaboration is discarded.

E. *Implementation in Unity*

The implementation was simple and divided into two simple scenes. The first scene is a menu to select the Bluetooth device corresponding to the HC-06 module connected to the Arduino board. Once the device has been chosen, a terminal is enabled where the user can see the data sent by the sensors as they are printed on the screen. From this point on, the player can move to the next scene where the 3D model is found with a moving arm that can be controlled by the MPU9250 sensor. The scene of the game is fairly simple. The player can only move the arm and his movement is reflected on the 3D model.

5 Results

As mentioned in the requirements, this project does not concern any medical point of view and will not be tested on real patients due to the difficulty of doing so and the scope. However, the results are compared with similar projects that seek to establish inverse kinematics or virtual reality activities using motion sensors.

The project meets the basic requirements since it can capture the rotation of a user and reflect it in a virtual reality 3D model. The interface is simple yet it is user-friendly and easy-to-use. Figure 5 shows a screenshot taken from the final application.

Fig. 5. Final application. Source: Authors

With the purpose of measuring the results qualitatively, the app was used in a testing subject called José Villanueva who suffered a stroke that paralyzed the right section of his body. To carry out the tests, it was asked that he flexed his arm for one minute a total of three times giving one-minute breaks between each attempt. In the first table, the results of each activity are detailed (Table 1).

Table 1. Data capture before the application

Attempt	Flexions per minute (before)
1	5.3
2	5.1
3	4.8

Then, after giving a break to the subject, he was asked to use the app and play a match of the game. This was repeated three time giving two-minute breaks between each match since they last longer. In the end, the first flexing exercise was repeated three times with one-minute breaks. The results are shown in Table 2.

Table 2. Data capture after the application

Attempt	Flexions per minute (before)
1	5.5
2	5
3	5

Over the previous data, the motion was improved by 1.93% in the testing subject.

6 Comparison of Results

A brief comparison is made of similar projects which will be used to assess the efficiency and relevance of the project.

Oscillatory Motion Tracking with x-IMU

This project calculated not only the rotation of the sensors but also their position and it delivers more efficient calculations. However, other types of 9DOF sensors are used which are way better calibrated but at a substantially higher cost.

Furthermore, only cyclic movement was calculated which is convenient for projects such as this one since it is the type of movement carried out in the activity. A table of results is presented where the transmission and clock speeds are compared between both projects.

As previously stated, the project *Oscillatory Motion Tracking with x-IMU* has the advantage of using better calibrated results and, even at a lower clock speed, they calculation of quaternions is more efficient.

Table 3 shows a comparison between the hardware devices used in the selected projects. The *x-IMU* project has an improvement of 586% in the data sending speeds and the capture of the cyclic movement is more precise than the capture rotational movement carried out in this project.

Table 3. Project comparison

Criterion	x-IMU	Project
Transmission speed	267	43
Type of movement	Cyclic	Rotational
Clock speed	65 MHz	100 MHz

Finally, the sensor used in the current project has a clock speed 1.53 times higher than the one used in the *x-IMU* project which offers a faster operation in the calculation of data inside the sensor.

7 Conclusions

It is concluded that this type of implementations is viable, as long as the adequate equipment is available. Furthermore, this method has the advantage of being cheaper than other projects involving motion capture and inverse kinematics. Table 4 compares the estimated costs of the elaboration of this project with other motion capture methods.

Table 4. Price comparison

Method	Price (USD)	Number of terminals
Current	140	4
Xbox Kinect	180*	4
OptiTrack	289 [12]	4
Rokoko Smart Suite	2495 [13]	4

Although the method is feasible in terms of implementation, it suffers from complications regarding its portable nature and working with motion which are the source of supply and the transmission of information. Hence, motion capture models such as the one offered by Kinect are very practical in that area.

The model implemented has the potential of delivering very good results when tested in real subjects. Nonetheless, the current implementation is very limited in terms its effectiveness.

This price is calculated considering that a virtual reality visor different from the phone is required which must be connected to a PC that supports virtual reality. However, since the prices of PCs vary constantly they are not included which means that the implementation costs are probably higher.

References

1. Inverse Kinematics - Roblox Developer. http://wiki.roblox.com/index.php?title=Inversekinematics
2. Mukherjee, S., Paramkusam, D., Dwivedy, S.K.: Inverse kinematics of a NAO humanoid robot using kinect to track and imitate human motion. In: 2015 International Conference on Robotics, Automation, Control and Embedded Systems (RACE). IEEE (2015)
3. Grochow, K., Martin, S.L., Hertzmann, A., Popovic, Z.: Style-based inverse kinematics. University of Washington (2004)
4. Inverse Kinematics For Robotic Arms - Alan Zucconi. https://www.alanzucconi.com/2017/04/10/robotic-arms/

5. Treffers, C., Van Wietmarschen, L.: Position and orientation determination of a probe with use of the IMU MPU9250 and a ATmega328 microcontroller. Electrical Engineering, Mathematics and Computer Science, TUDelft (2016)
6. Reference image for a gyroscope. https://www.google.com.co/search?q=phone+gyro&source=lnms&tbm=isch&sa=X&ved=0ahUKEwi75eSV1tjaAhXkc98KHck1DwQAUICigB&biw=1920&bih=870#imgrc=EtdQu11uKVo3WM
7. MagMaster Software download. https://github.com/YuriMat/MagMaster/archive/master.zip
8. Advanced hard and soft iron magnetometer calibration for dummies. https://diydrones.com/profiles/blogs/advanced-hard-and-soft-iron-magnetometer-calibration-for-dummies
9. Open-Source-AHRS-With-x-IMU. https://github.com/xioTechnologies/Open-Source-AHRS-With-x-IMU
10. VRImu Arduino. https://github.com/nviverosb/VRImu
11. Arduino Android & Microcontrollers/Bluetooth. https://assetstore.unity.com/packages/tools/input-management/android-microcontrollers-bluetooth-16467
12. Motion Capture Suits. https://optitrack.com/products/motion-capture-suits/
13. Rokoko Smart Suit. https://www.rokoko.com/en/shop

DELMU: A Deep Learning Approach to Maximising the Utility of Virtualised Millimetre-Wave Backhauls

Rui Li[1](✉), Chaoyun Zhang[1], Pan Cao[2], Paul Patras[1], and John S. Thompson[3]

[1] School of Informatics, University of Edinburgh, Edinburgh, UK
rui.li@ed.ac.uk
[2] School of Engineering and Technology, University of Hertfordshire, Hatfield, UK
[3] School of Engineering, University of Edinburgh, Edinburgh, UK

Abstract. Advances in network programmability enable operators to 'slice' the physical infrastructure into independent logical networks. By this approach, each network slice aims to accommodate the demands of increasingly diverse services. However, precise allocation of resources to slices across future 5G millimetre-wave backhaul networks, to optimise the total network utility, is challenging. This is because the performance of different services often depends on conflicting requirements, including bandwidth, sensitivity to delay, or the monetary value of the traffic incurred. In this paper, we put forward a general rate utility framework for slicing mm-wave backhaul links, encompassing all known types of service utilities, i.e. logarithmic, sigmoid, polynomial, and linear. We then introduce DELMU, a deep learning solution that tackles the complexity of optimising non-convex objective functions built upon arbitrary combinations of such utilities. Specifically, by employing a stack of convolutional blocks, DELMU can learn correlations between traffic demands and achievable optimal rate assignments. We further regulate the inferences made by the neural network through a simple 'sanity check' routine, which guarantees both flow rate admissibility within the network's capacity region and minimum service levels. The proposed method can be trained within minutes, following which it computes rate allocations that match those obtained with state-of-the-art global optimisation algorithms, yet orders of magnitude faster. This confirms the applicability of DELMU to highly dynamic traffic regimes and we demonstrate up to 62% network utility gains over a baseline greedy approach.

1 Introduction

The 5th generation mobile networks (5G) embrace a new wave of applications with distinct performance requirements [1]. For example, ultra-high definition video streaming and immersive applications (AR/VR) typically demand very high data throughput. Autonomous vehicles and remote medical care are stringently delay-sensitive, belonging to a new class of Ultra-Reliable Low-Latency Communications (URLCC) services [2]. In contrast, Internet of Things (IoT)

É. Renault et al. (Eds.): MLN 2018, LNCS 11407, pp. 146–165, 2019.
https://doi.org/10.1007/978-3-030-19945-6_10

applications, including smart metering and precision agriculture, can be satisfied with a best-effort service. In order to simultaneously meet such diverse performance requirements, while enabling new verticals, mobile network architectures are adopting a *virtually sliced* paradigm [3]. The core idea of slicing is to partition physical network infrastructure into a number of logically isolated networks, i.e. slices. Each slice corresponds to a specific service type, which may potentially belong to a certain tenant operator.

At the same time, cellular and Wi-Fi base stations (BSs) are deployed massively, in order to increase network capacity and signal coverage. Millimetre wave (mm-wave) technology is becoming a tangible backhauling solution to connect these BSs to the Internet in a wireless fashion at multi-Gbps speeds [4]. In particular, advances in narrow beam-forming and multiple-input multiple-output (MIMO) communications mitigate the severe signal attenuation characteristic to mm-wave frequencies and respectively multiply achievable link capacities [5].

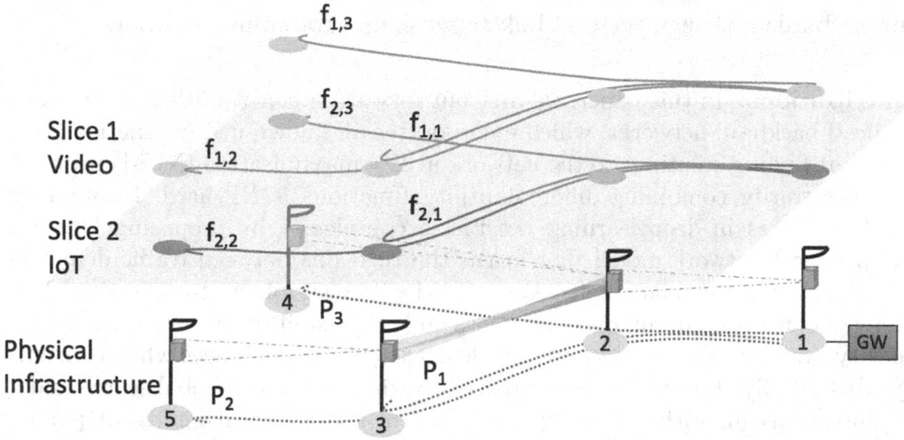

Fig. 1. Example of sliced backhaul over physical lamppost based mm-wave infrastructure. Slice 1 accommodates video streaming flows with sigmoid utilities and Slice 2 carries traffic from IoT applications, which have logarithmic utility.

Partitioning sliced mm-wave backhauls, and in general backhauls that employ any other communications technology, among traffic with different requirements, as in the example shown in Fig. 1, is essential for mobile infrastructure providers (MIPs). By and large, MIPs aim to extract as much value as possible from network resources, yet achieving this in sliced backhauls is not straightforward. In this example, five BSs are inter-connected via mm-wave directional links, forming a shared backhaul. The notion of rate utility is widely used to quantify the worth of an allocation of resources to multiple flows. The question is: *what type of utility is suitable to such multi-service scenarios?* Logarithmic utility as proposed in [6] has been adopted for elastic services and remains suitable for best-effort IoT traffic. On the other hand, applications such as video streaming typically throttle below a threshold, whilst an increase in service level is mostly imperceptible by users when the allocated rate grows beyond that threshold. Hence, the utility of such traffic can be modelled as a step-like sigmoid [7]. Had there been real-time applications to accommodate, their utility is typically formulated through polynomial functions [8,9]. Further, in the case of traffic for which the MIP allocates resources solely based on monetary considerations, a linear utility function can be employed. However, as the application scenarios diversify, a single type of utility cannot capture the distinct features of different service types. Therefore, we argue that a mixed utility must be considered. Unfortunately, combining all these utility functions may lead to non-concave expressions and computing in a timely manner the optimal rate allocation that maximises their value becomes a challenging task. Global search metaheuristics explore the feasible solution space intelligently to find global maxima [10], yet often involve unacceptably long computational times. Thus they fail to meet 5G specific delay requirements in highly dynamic environments, where application demands change frequently. Greedy approaches can be used to overcome the runtime burden, though these will likely settle on sub-optimal solutions.

Contributions: In this paper, we first put forward a general utility framework for sliced backhaul networks, which incorporates all known utility functions. We show that finding solutions to the network utility maximisation (NUM) problem when arbitrarily combining different utility functions is NP-hard. Inspired by recent advances in deep learning, we tackle complexity by proposing DELMU, a deep neural network model that learns the relations between traffic demands and optimal flow rate allocations. Augmented with a simple post-processing algorithm that ensures minimum service levels and admissibility within the network's capacity, we show that DELMU makes close-to-optimal inferences while consuming substantially shorter time as compared to state-of-the-art global search and a baseline greedy algorithm. In view of the current technological trends, we particularly focus on backhauls that operate in mm-wave bands. However, our utility framework and deep learning approach can be applied to other systems that operate in microwave or sub-gigahertz bands.

The remainder of the paper is structured as follows. In Sect. 2 we discuss the system model and in Sect. 3 we formulate the general NUM problem in the context of sliced mm-wave backhauls. We present the proposed deep learning approach to solving NUM in Sect. 4 and show its performance in Sect. 5. We review relevant related work in Sect. 6. Finally, in Sect. 7 we conclude the paper.

2 System Model

We consider a backhaul network deployment with $\mathcal{B}$ base stations (BSs) interconnected via mm-wave links.[1] Each BS is equipped with a pair of transceivers,

[1] Although we primarily focus on mm-wave backhauls, due to their potential to support high-speed and low latency communications, the optimisation framework and deep learning solution we present next are generally applicable to other technology.

hence is able to transmit and receive simultaneously, while keeping the footprint small to suit dense deployment. To meet carrier-grade requirements and ensure precise TX/RX beam coordination, the network operates with a time division multiple access (TDMA) scheme. We assume carefully planned deployments where BSs have a certain elevation, e.g. on lampposts, hence interference is minimal and blockage events occur rarely.

We focus on settings where the backhaul network is managed by a single MIP and is partitioned into I logical slices to decouple different services (e.g. as specified in [3]). $\mathcal{F}$ user flows traverse the network and are grouped by traffic type i corresponding to a specific slice, i.e. $\mathcal{F} = \cup_{i\in\{1,\ldots,I\}}\mathcal{F}_i$. The MIP's goal is to adjust the flow rates according to corresponding demands, in order to maximise the overall utility of the backhaul network. Flow demands are defined by upper and lower bounds. Lower bounds guarantee minimum flow rates, so as to ensure service availability, whilst upper bounds eliminate network resources wastage. We assume a controller (e.g. 'network slice broker' [11]) has complete network knowledge, periodically collects measurements of flow demands from BSs, solves NUM instances, and distributes the flow rate configurations corresponding to the solutions obtained.

Link Capacity: To combat the severe path loss experienced at mm-wave frequencies and boost capacity, BSs employ multiple input multiple output (MIMO) antenna arrays. We consider K array elements deployed at each base station for TX/RX. In backhaul settings, the stations' locations are fixed and the channel coherence time is typically long; hence it is reasonable to assume full knowledge of the channel state information is available at both transmitter and receiver sides. Given the channel matrix $\mathbf{H_{m,n}}$ from BS m to BS n, the received signal at BS n can be computed as

$$\mathbf{y_n} = \mathbf{H_{m,n}}\mathbf{x_m} + \mathbf{n_{m,n}}, \tag{1}$$

where $\mathbf{x_m}$ is an K-dimensional signal transmitted by BS m, and $\mathbf{y_n}$ are the received symbols at BS n. The singular value decomposition (SVD) of $\mathbf{H_{m,n}}$ is:

$$\mathbf{H_{m,n}} = \mathbf{U_n}\mathbf{\Sigma}\mathbf{V_m^H}, \tag{2}$$

where $\mathbf{U_n}$ and $\mathbf{V_m}$ are $K \times K$ unitary matrices, i.e. $\mathbf{U_n}\mathbf{U_n^H} = \mathbf{I}$ and $\mathbf{V_m}\mathbf{V_m^H} = \mathbf{I}$, and $\mathbf{\Sigma}$ is an $K \times K$ non-negative diagonal matrix containing the singular values of $\mathbf{H_{m,n}}$. The k-th diagonal entries of $\mathbf{\Sigma}$, i.e. σ_k, represents the k-th channel gain, and is also the k-th non-negative square root of the eigenvalues of matrix $\mathbf{H_{m,n}}\mathbf{H_{m,n}^H}$.

The parallel channel decomposition can be implemented efficiently for mm-wave systems as follows [12]. The transmitter precoding performs a linear transformation on the input vector $\mathbf{\tilde{x}_m}$, i.e. $\mathbf{x_m} = \mathbf{V_m}\mathbf{\tilde{x}_m}$, and the received signal $\mathbf{y_n}$ is linearly decoded by $\mathbf{U_n^H}$, i.e. $\mathbf{\tilde{y}_n} = \mathbf{U_n^H}\mathbf{y_n}$. Therefore, the link capacity $c_{m,n}$ between base station m and n can be computed as:

$$c_{m,n} := \max_{\substack{\mathbf{Q_m}: \\ Tr(\mathbf{Q_m})\leq P_{\max}}} B\log_2\det(\mathbf{I} + \mathbf{H_{m,n}}\mathbf{Q_m}\mathbf{H_{m,n}^H}), \tag{3}$$

where $\mathbf{Q_m} = \mathbf{V_m V_m^H}$ is the transmission covariance matrix, B is the channel bandwidth, and $P_{\max}$ is the maximum transmit power. Without loss of generality, we assume that all BSs have the same maximum transmit power budget.

For a channel known at the transmitter, the optimal capacity can be achieved by the well-known channel diagonalisation and the water-filling power allocation method [13]. For all BS m, by employing the optimal transmit pre-coding matrix $\mathbf{V_m} = \mathbf{X_{m,n}}$, where $\mathbf{X_{m,n}}$ denotes the eigenvector matrix of $\mathbf{H_{m,n}^H} * \mathbf{H_{m,n}}$, the MIMO channel capacity maximisation can be reformulated as:

$$c_{m,n} := \max B \sum_{k=1}^{K} \left(\log \left(1 + \frac{\lambda_k p_m^k}{\epsilon^2} \right) \right), \tag{4}$$

$$\text{s.t.} 0 \leq \sum_{k=1}^{K} p_m^k \leq P_{\max}, \tag{5}$$

$$p_m^k \geq 0, \forall k, m \tag{6}$$

where $\lambda_k = \sigma_k^2$, and ϵ^2 denotes the noise power. If the power allocated on the k-th sub-channel is p_m^k at BS m, then (5) specifies the total transmit power constraint. The optimal water-filling power allocation yields $p_m^k = \max\{0, \mu - \epsilon^2/\lambda_k\}$, where $\mu > 0$ is the water-filling level such that $\sum_{k=1}^{K} p_m^k = P_{\max}, \forall m$ [13].

3 Problem Formulation

Our objective is to find the optimal end-to-end flow rates that maximise the utility of sliced multi-service mm-wave backhaul networks. We first introduce a general network utility framework, based on which we formulate the NUM problem, showing that in general settings this is NP-hard.

3.1 Utility Framework

Recall that network utility refers to the value obtained from exploiting the network, which can be monetary, resource utilisation, or level of user satisfaction. For any flow f we consider four possible types of utility functions of flow rate r, depending on which slice $\mathcal{F}_i$ that flow belongs to. The utilities considered are parameterised by α_i and β_i, whose values have practical implications, such as the amount billed by the MIP for a service. Given an allocated rate r, we distinguish the following types of services that can be mapped onto slices, whose utilities we incorporate in our framework:

1. Services for which the MIP aims to maximise solely the attainable **revenue**. Denoting $\mathcal{F}_1$ the set of flows in this class, their utility is formulated as a linear function [14]:

$$U_{\text{lnr}}(r) = \alpha_1 r + \beta_1, \quad \forall f \in \mathcal{F}_1. \tag{7}$$

   We note that $U_{\text{lnr}}(r)$ is both concave and convex.

2. Flows $f \in \mathcal{F}_2$ generated by applications that require certain level of **quality of service**, e.g. video streaming, and whose corresponding utility is thus formulated as a sigmoid function [7]:

$$U_{\mathrm{sig}}(r) = \frac{1}{1 + e^{-\alpha_2(r-\beta_2)}}, \quad \forall f \in \mathcal{F}_2. \tag{8}$$

   Observe that $U_{\mathrm{sig}}(r)$ is convex in $[0, \beta_2)$ and concave in (β_2, ∞), therefore non-concave over the entire domain.
3. **Delay sensitive** flows, $f \in \mathcal{F}_3$, whose utility is modelled as a polynomial function [8]:

$$U_{\mathrm{ply}}(r) = \alpha_3(r^{\beta_3}), \quad \forall f \in \mathcal{F}_3, \tag{9}$$

   where β_3 is in the range $(0, 1]$, for which the above expression is concave.
4. **Best-effort** traffic, $f \in \mathcal{F}_4$, that does not belong in any of the previous classes, and whose utility is commonly expressed through a logarithmic function [6]:

$$U_{\mathrm{log}}(r) = \log(\alpha_4 r + \beta_4), \quad \forall f \in \mathcal{F}_4. \tag{10}$$

   It is easy to verify that $U_{\mathrm{log}}(r)$ is also concave.

Our general utility framework encompasses all the four types of traffic discussed above (which may be parametrised differently for distinct tenants), therefore we express the overall utility of the sliced backhaul network as

$$\begin{aligned} \mathcal{U} := \sum_{f \in \mathcal{F}} U(r) = \sum_{f_1 \in \mathcal{F}_1} U_{\mathrm{lnr}}(r_1) + \sum_{f_2 \in \mathcal{F}_2} U_{\mathrm{sig}}(r_2) \\ + \sum_{f_3 \in \mathcal{F}_3} U_{\mathrm{ply}}(r_3) + \sum_{f_4 \in \mathcal{F}_4} U_{\mathrm{log}}(r_4). \end{aligned} \tag{11}$$

Arbitrary combinations of both concave and non-concave utility functions may result in non-concave expressions $\mathcal{U}$, as exemplified in Fig. 2. In this figure, we show the total utility when combining 4 flows with different utility functions, two of them sigmoidal and two polynomial, each with different parameters. We assume the rates of each type of flow increase in tandem. Observe that even in a simple setting like this one, the network utility is highly non-concave and finding the optimal allocation that maximises it is non-trivial. We next formalise this problem with practical mm-wave capacity constraints, following which we discuss its complexity.

3.2 Network Utility Maximisation

Consider a set of flows that follow predefined paths, $P_j, j \in \{1, 2, ..., J\}$, to/from the local gateway, where the number of possible routes in the network is J. We denote $f_{i,j}$ a flow on slice i that traverses path P_j, which is allocated a rate

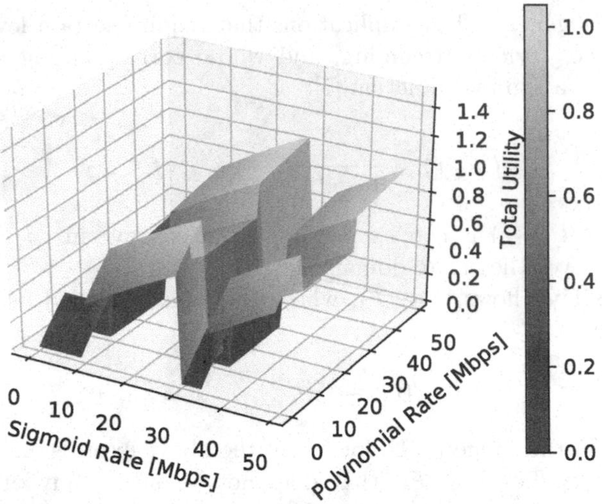

Fig. 2. Total utility when combining four flows with different utility functions; namely, two have sigmoid utility parametrised by $\alpha_2 = 0.08$, $\beta_2 = 15$ in the [10–30] Mbps range, and respectively $\alpha_2 = 0.08$, $\beta_2 = 40$ in the [35–50] Mbps range; the other two flows have polynomial utility with $\alpha_3 = 0.03651$, $\beta_3 = 0.9$ between [0–10] Mbps, and $\alpha_3 = 0.03$, $\beta_3 = 0.6$ in [30–50] Mbps. Rates increased in tandem for each type of flow.

$r_{i,j}$. By contract, $r_{i,j}$ shall satisfy $\delta_{i,j} \leq r_{i,j} \leq d_{i,j}$, where $\delta_{i,j}$ is the minimum rate that guarantees service availability, and $d_{i,j}$ is the upper bound beyond which the service quality cannot be improved. $d_{i,j}$ is no less than $\delta_{i,j}$ by default. Furthermore, each path P_j consists of a number of mm-wave links, and the link between BSs m and n is subject to a link capacity $c_{m,n}$. We use $\tau^{s}_{j,m} \in \{0,1\}$, $s \in \{Tx, Rx\}$, to indicate whether node m transmits or receives data of flows traversing path P_j. The total network utility in (11) can be rewritten as:

$$\sum_{f \in \mathcal{F}} U(r) = \sum_{i=0}^{I} \sum_{j=0}^{J} U_i(r_{i,j}). \tag{12}$$

Finding the flow rate allocation vector $\mathbf{r_{i,j}}, \forall i, j$, that maximises this utility requires to periodically solve the following optimisation problem:

$$\max \sum_{i=0}^{I} \sum_{j=0}^{J} U_i(r_{i,j}) \tag{13}$$

$$\text{s.t. } \delta_{i,j} \leq r_{i,j} \leq d_{i,j}, \forall i, j; \tag{14}$$

$$\sum_{i=0}^{I} \sum_{j=0}^{J} \tau^{s}_{j,m} \frac{r_{i,j}}{c_{m,n}} \leq 1, \{m,n\} \in P_j, s \in \{Tx, Rx\}. \tag{15}$$

In the formulation above, (13) is the overall objective function and (14) specifies the demand constraints. Each BS can transmit and receive to/from one and only one BS simultaneously, and the total time allocated at a single node for all flow Tx/Rx should not exceed 1, which is captured by (15). Here $r_{i,j}/c_{m,n}$ denotes the time fraction allocated to flow $f_{i,j}$ on link $l_{m,n}$.

3.3 Complexity

In what follows we briefly show that the network utility optimisation problem formulated above, where the objective function is a linear combination of linear, sigmoid, polynomial, and logarithmic functions, is NP-hard. By Udell and Boyd [15] any continuous function can be approximated arbitrarily well by a suitably large linear combination of sigmoidal functions [15]. Thus $\sum U(r)$ can be regarded as a sum of sigmoids and a larger number of other sigmoidal functions. Following the approach in [15], we can reduce an integer program

$$\begin{aligned} &\text{find } \mathbf{r} \\ &\text{s.t. } A\mathbf{r} \le Z;\ \mathbf{r} \in \{0,1\}^n, \end{aligned}$$

to an instance of a sigmoidal program

$$\begin{aligned} &\max \sum_i g(r_i) = \sum_i r_i(r_i - 1) \\ &\text{s.t. } A\mathbf{r} \le Z;\ 0 \le r_i \le 1. \end{aligned}$$

Here $g(r_i)$ enforces a penalty on non-integral solutions, i.e. the solution to the sigmoidal program is 0 if and only if there exists an integral solution to $A\mathbf{r} = Z$. Since the integer program above is known to be NP-hard [16], the reduced sigmoid program is also NP-hard, and therefore the NUM problem we cast in (13)–(15) is also NP-hard.

4 DELMU: A Deep Learning Approach to NUM

To tackle the complexity of the optimisation problem formulated in the previous section and compute solutions in a timely manner, we propose DELMU, a deep learning approach specifically designed for sliced mm-wave backhauls and also applicable to other technologies. In essence, our proposal learns correlations between traffic demands and allocated flow rates, to make inferences about optimal rate assignments. We show that, with sufficient training data, our deep neural network finds solutions close to those obtained by global search, while requiring substantially less runtime.

4.1 Convolutional Neural Network

We propose to use a Convolutional Neural Network (CNN) to imitate the behaviour of global search. We train the CNN by minimising the difference between ground-truth flow rates allocations (obtained with global search) and those inferred by the neural network. In general CNNs preform weight sharing across different feature channels [17]. This significantly reduces the number of model parameters as compared to traditional neural networks, while preserving remarkable performance. At the same time, our approach aims to work well with a limited amount of training data, which makes CNNs particularly suitable for our problem. Therefore, we design a 12-layer CNN to infer the optimal flow rate and illustrate its structure in Fig. 3. The choice is motivated by recent results that confirm neural network architectures with 10 hidden layers, like ours, can be trained relatively fast and perform excellent hierarchical feature extraction [18].

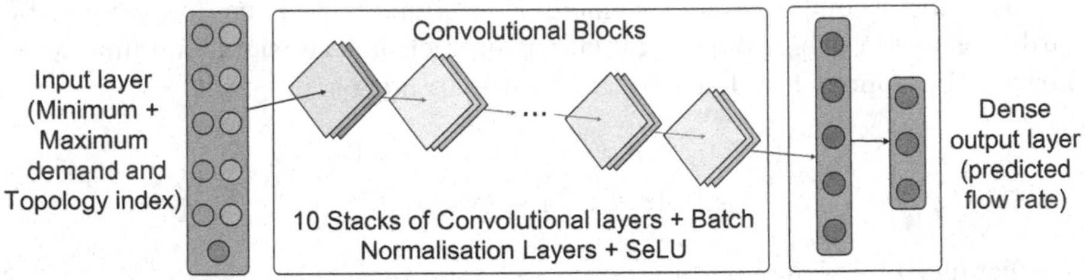

Fig. 3. Proposed Convolutional Neural Network with 10 hidden layers, which takes traffic demand and topology index as input, and infers the optimal flow rate allocations.

The minimum and maximum traffic demand, and topology information are concatenated into a single vector, which will be subsequently fed to a sequence of convolutional blocks. Each block consists of a one-dimensional convolutional layer and a Scaled Exponential Linear Unit (SELU) [19], which takes the following form:

$$\text{SELU}(x) = \omega \begin{cases} x & x > 0 \\ \eta e^x - \eta & x \leqslant 0. \end{cases} \tag{16}$$

Here $\omega = 1.0507$ and $\eta = 1.6733$ by default. Employing SELU functions aims at improving the model representability, while enabling self-normalisation without requiring external techniques (e.g. batch normalisation). This enhances the robustness of the model and eventually yields faster convergence. Features of traffic demands are hierarchically extracted by convolutional blocks, and they are sent to fully-connected layers for inference. We train the CNN using a stochastic gradient descent (SGD) based method named Adam [20], by minimising the following mean square error:

$$L_e = \frac{1}{Q \times I \times J} \sum_{q=0}^{Q} \sum_{i=0}^{I} \sum_{j=0}^{J} (r_{q,i,j} - r'_{q,i,j})^2. \tag{17}$$

Q denotes the number of training data points, $r_{q,i,j}$ denotes the allocated rate allocated to flow j on slice i, with demand instance q, as suggested by global search. $r'_{q,i,j}$ is the corresponding rate inferred by the neural network. We train the CNN with 500 epochs, with an initial learning rate of 0.0001.

4.2 Post-processing Algorithm

The output of the CNN on its own occasionally violates the constraints (14) and (15), because the model is only fed with traffic demands without embedding of constraints. We address this issue by designing a post-processing algorithm that adjusts the CNN solutions to fall within feasible domains, while maintaining minimum utility degradation and very short computation times. The idea is to first decrease recursively with a large step-length the rate of flows that breach the constraints, then increase repeatedly with a smaller step-length the rate of flows that can achieve the largest utility gains.

Algorithm 1 CNN Post-Processing Algorithm

1: Compute the time between each pair of nodes $t^s_{m,n}$
2: Compute the utility of each flow $u_{i,j} = U_i(r_{i,j})$
3: **while** Any $t^s_{m,n} > 1$ **do**
4:     Find the link $l_{m,n}$ with the maximum $t^s_{m,n}$
5:     deStepLen $= \min\{10, r_{i,j} - \delta_{i,j}\}$
6:     **for** Flows satisfying $\tau^s_{j,m} == 1$ or $\tau^s_{j,n} == 1$ for $l_{m,n}$ **do**
7:         Potential utility decrease $u'_{i,j} = U_i(r_{i,j} - \text{deStepLen})$
8:     **end for**
9:     Find the $f_{i,j}$ with the minimum non-zero $\Delta u_{i,j} = u_{i,j} - u'_{i,j}$
10:     Decrease rate of $f_{i,j}$, i.e. $r_{i,j} = r_{i,j} - \text{deStepLen}$
11:     Update $t^s_{m,n}$ and $u_{i,j}$
12: **end while**
13: **while** Any flow rate can be increased **do**
14:     inStepLen $= \min\{1, d_{i,j} - r_{i,j}\}$
15:     Potential utility increase $u''_{i,j} = U_i(r_{i,j} + \text{inStepLen}), \forall f_{i,j}$
16:     Find the $f_{i,j}$ with the maximum $\Delta u_{i,j} = u''_{i,j} - u_{i,j}$
17:     Increase rate of $f_{i,j}$, i.e. $r_{i,j} = r_{i,j} + \text{inStepLen}$
18:     Update $t^s_{m,n}$ and $u_{i,j}$
19: **end while**

Algorithm 1 shows the pseudo-code of this procedure. The routine starts by computing the total time on each link for all traversing flows, i.e. $t^s_{m,n} = \sum_i \sum_j \tau^s_{j,m} r_{i,j} / c_{m,n}$ (line 1) and the utility of each individual flow based on the rate allocation returned by CNN (line 2). Then it searches recursively for a flow to decrease (lines 3–12). At each step, Algorithm 1 selects the link with the highest total time (line 4) and reduces the rate of the flow traversing the link with minimum possible utility loss (lines 5–10). Then the total link time and the flow utilities are updated (line 11). The process (lines 4–11) is repeated until the time for all links comply with the time constraints. Next, we increase iteratively a flow that yields the maximum potential utility gain, while ensuring that all constraints are satisfied (lines 13–19). This is done by tentatively increasing each flow, with a step-length that complies with the demand constraint (line 14), computing the corresponding utility increment (line 15), then finding the flow with maximum possible utility increase (line 16), and confirming the rate increment for that flow (line 17). Before the next round of increasing the rates, Algorithm 1 recomputes the total time on all links to verify that further rate increases are possible, and updates the utility of each flow (line 18).

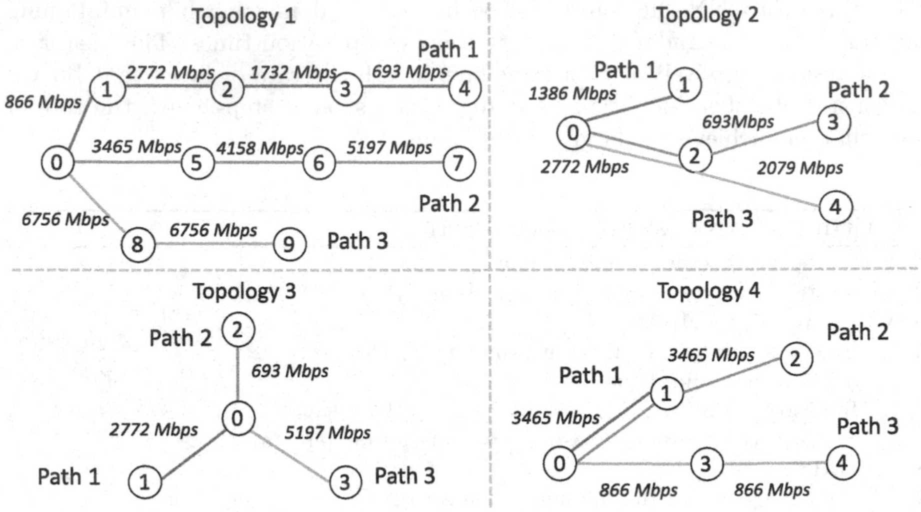

Fig. 4. The four network topologies used for evaluation. Circles represent the BSs, flow paths are shown with lines of different colour, and link capacities are labelled. (Color figure online)

5 Performance Evaluation

We evaluate the proposed DELMU solution, which encompasses the CNN structure and the post processing algorithm, on different backhaul topologies under a range of conditions. Specifically, we use four different topologies as illustrated in Fig. 4, where the number of BSs varies from 4 to 10, and link rates range from 693 Mbps to 6.8 Gbps. Each path carries up to four types of flows, i.e. with individual sigmoid, linear, polynomial, and logarithmic utilities. For each topology, we generate randomly 10,000 combinations of flow demands in the range [0–750] Mbps in increments of 50 Mbps. The corresponding minimum service rates are generated uniformly at random in the range [0–100] Mbps as integer values, and are capped by the maximum flow demand. The parameters shown in Table 1 are used to model utility.

To train and subsequently test the neural network, we run a global search (GS) algorithm, the optimality of which is proven in [10], on each of the 10,000 network settings described above. We use 80% of the results obtained to construct a synthetic dataset that we use in the training process, which effectively seeks to minimise the mean square error expression defined in (17), by means

Table 1. α_i and β_i parameters for the utility functions used in the evaluation.

Utility type	Linear	Sigmoid	Polynomial	Logarithmic
α_i	0.00133	0.08000	0.03651	0.00229
β_i	0	350	0.5	1

of SGD. We use the remaining 20% of cases for as ground truth for testing the accuracy of the optimal rate allocation inferences that DELMU makes. More precisely, we compare the performance of DELMU in terms of total network utility and computational time, against the solutions obtained with GS and those computed with a baseline greedy approach that we devise. We discuss both benchmarks in more detail in the following subsection.

To compute solutions with the GS and greedy algorithms, and make inferences with the proposed CNN, we use a workstation with an Intel Xeon E3-1271 CPU @ 3.60 GHz and 16 GB of RAM. The CNN is trained on a NVIDIA TITAN X GPU using the open-source Python libraries TensorFlow [21] and TensorLayer [22]. We implement the greedy solution in Python and employ the GS solver of MATLAB®.

5.1 Benchmarks

The GS method works by starting from multiple points within the feasible space and searching for local optima in their vicinity, then concluding on the global optimum from the set of local optima obtained [10]. With default settings, which we employ in our evaluation, the GS generates 1,000 starting points using the scatter search algorithm [23], then eliminates those starting points that are not promising (judging by the corresponding value of the objective function and constraints). It then repeatedly executes a constrained nonlinear optimisation solver, i.e. `fmincon`, to search for local maxima around the remaining start points. Eventually the largest of all local maxima is taken as the global maximum, if one exists. We let the local optimisation routine work with the default Interior Point algorithm, which satisfies bounds at all iterations and can recover from non-numeric results. We note that simpler approximations such as semidefinite programming are constrained to convex optimisation problems, thus inappropriate for our task.

We also engineer a baseline greedy algorithm for the purpose of evaluation, with the goal of finding reasonably good solutions *fast*. The greedy approach starts by setting all flow rates to the minimum demand and then recursively chooses a flow to increase its rate, with the aim of achieving maximum utility gain at the current step, as long as the constraints (14)–(15) are respected. A solution is found when there are no remaining flows whose rates can be further increased. For fair comparison, the greedy approach takes exactly the same flow demands and the corresponding minimum service rates as used by GS and DELMU. A step size of 1 Mbps is employed.

5.2 Total Utility

We first examine the overall utility performance of the proposed DELMU, in comparison with that of the greedy and the GS solutions. Figure 5 illustrates the distributions of the total network utility for the 12 flows traversing the network, over the 2,000 instances tested. We observe that, among the 4 topologies used, the distribution of the total utility obtained by DELMU is almost the same as that of the optimal solution obtained with GS, as confirmed by the similar median values, the distance between the first and third quartiles, as well as the whiskers (minima and maxima). Specifically, the median values of the total utility attained by GS in Topologies 1–4 are 5.23, 4.07, 4.66, and 4.75, while those achieved by the proposed DELMU are 5.09, 3.88, 4.56, and 4.64. In sharp contrast to the DELMU's close-to-optimal performance, the greedy solution attains the medians of 3.30, 3.32, 2.81, and 3.16 utility units in the 4 topologies considered. Among these, for the case of Topology 3, DELMU obtains a 62% total utility gain over the greedy approach. It is also worth remarking that, although a greedy approach can perform within well-defined bounds from the optimum when working on submodular objective functions [24], this is clearly suboptimal in the case of general utility functions as addressed herein.

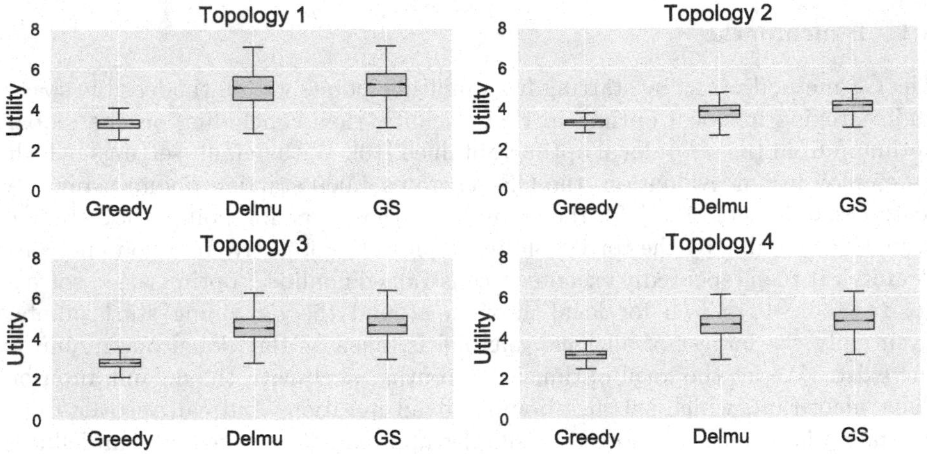

Fig. 5. Distribution of the total utility attained by the proposed DELMU, and the benchmark GS and greedy algorithms, for the four topologies shown in Fig. 4. Numerical results.

5.3 Decomposing Performance Gains

To understand how DELMU achieves close-to-optimal utility, and why the benchmark greedy solution performs more poorly, we examine one single instance for each topology, and dissect the utility values into the components corresponding to each type of traffic (i.e. slice). Figure 6 illustrates the sum of utilities for each type of traffic, attained with the greedy, CNN, and GS approaches. We note that the greedy solution tends to allocate more resources to traffic with logarithmic utility (in all topologies) and respectively polynomial utility (in Topologies 2, 3, and 4). In contrast, the CNN allocates higher rates to traffic subject to sigmoid utility in all the scenarios studied, which results in higher overall utility. This is because the greedy approach gives more resources to the flows that yield utility gains in the first steps of the algorithm's execution and fails to capture the inflection point of the traffic with sigmoid utility, which can contribute to a higher overall utility, under limited resource constraints. Furthermore, the allocations of rates to different traffic types by DELMU show close resemblance to the GS behaviour, which confirms the fact that DELMU achieves overall close to optimal utility allocations, at a lower computational cost, as we will see next.

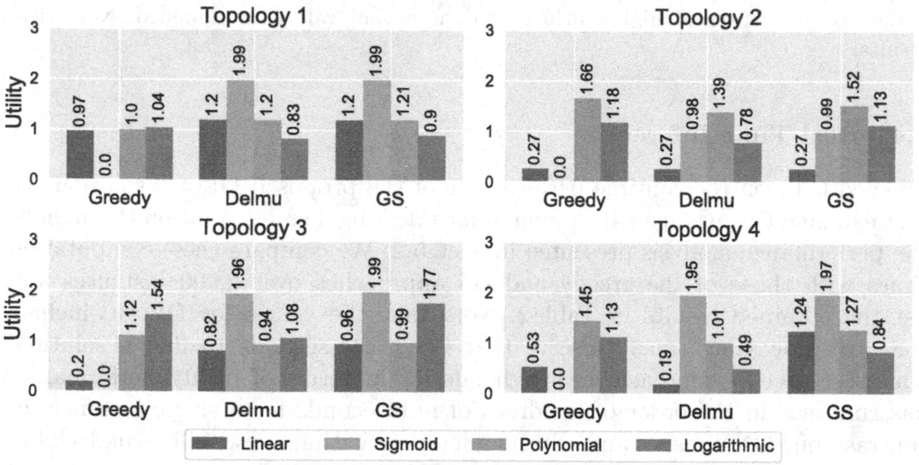

Fig. 6. An example instance of the utility corresponding to each traffic type in each topology. Bars represents the sum utility of flows in the same slice. Numerical results.

We delve deeper into the utility attained by each flow on each slice, along different paths, and in Fig. 7 compare the performance of our approach and the benchmarks considered in the case of Topology 1. Flows corresponding to slices that have linear, sigmoid, polynomial, and respectively logarithmic utility are indexed from 1 to 4. Again, observe that the greedy approach assigns zero utility to traffic subject to sigmoid utility, in stark contrast with the GS method. While DELMU obtains the highest gains from traffic with linear and sigmoid utility on paths 2 and 3, greedy dedicates most of the network resources to traffic with logarithmic and exponential utility, without obtaining significantly more utility from these types of flows. DELMU achieves accurate inference, as the performance is nearly the same with that of GS for all flows.

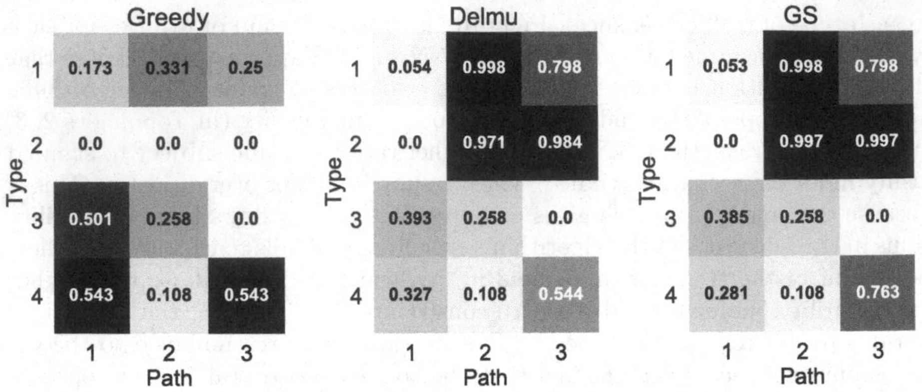

Fig. 7. Utility of all data flows (on different slices and over different paths) attained by greedy, DELMU, and GS in one demand instance in Topology 1. In each subfigure, darker shades represent higher utility and the actual values are labelled. Numerical results.

5.4 Real-Time Inference

To shed light on the runtime performance of the proposed DELMU solution, we first examine the average time required for inferring a single solution throughout the performance analysis presented in Sect. 5.2. We compare these computation times with those of the greedy and GS approaches over 2,000 instances and list the obtained results in Table 2. Note that the values for DELMU include the post-processing time. Observe that GS takes seconds to find a solution, while the greedy approach, although inferior in terms of utility performance, has runtimes in the order of hundreds of milliseconds for a single instance. In contrast, our CNN makes and adjusts inferences within a few milliseconds. That is, as compared to the greedy algorithm, CNN generally requires two orders of magnitude smaller computation time. On the other hand, the GS algorithm, although working optimality, has three orders of magnitude higher runtimes as compared to DELMU. Lastly, note that the CNN inference itself requires ~1.5 ms per instance, and hence the post-processing dominates the overall execution time in the first two topologies. We conclude that the proposed DELMU is suitable for highly dynamic backhauls.

Table 2. Average computation time required to obtain a single solution to the NUM problem in Topologies 1–4 using GS, greedy, and the proposed CNN mechanism.

Topology index	1	2	3	4
GS	8.4339 s	4.6075 s	3.4492 s	4.8311 s
Greedy	0.1500 s	0.1590 s	0.1178 s	0.1345 s
DELMU	**0.0036 s**	**0.0035 s**	**0.0025 s**	**0.0026 s**

We complete this analysis by investigating the ability of the proposed DELMU solution to handle network dynamics in sliced mm-wave backhaul settings, including changes in traffic demand due to e.g. on/off behaviour of user applications and variations in capacity triggered e.g. by occasional blockage on the mm-wave links. We consider Topology 3 in Fig. 4, transporting a mix of flows with linear, polynomial, and logarithmic utility and different lifetimes, considering a 10 Mbps minimum level of service, in all cases when a flow is active. Precisely, in Fig. 8 we examine the time evolution of the throughput DELMU allocates to flows on each slice, according to a sequence of events. In particular, flows subject to sigmoid utility start with 0 Mbps demands, whilst all flows of the other types on all path have each an initial demand of 200 Mbps. After 100 ms, a flow with sigmoid utility on path 2 (i.e. $f_{2,2}$) becomes active, adding a 400 Mbps demand to the network. At time 200 ms, partial link blockage occurs on the link between BS 0 and BS 1, causing the corresponding capacity $c_{0,1}$ to drop from 2,772 Mbps to 693 Mbps. $f_{2,2}$ finishes 100 ms later.

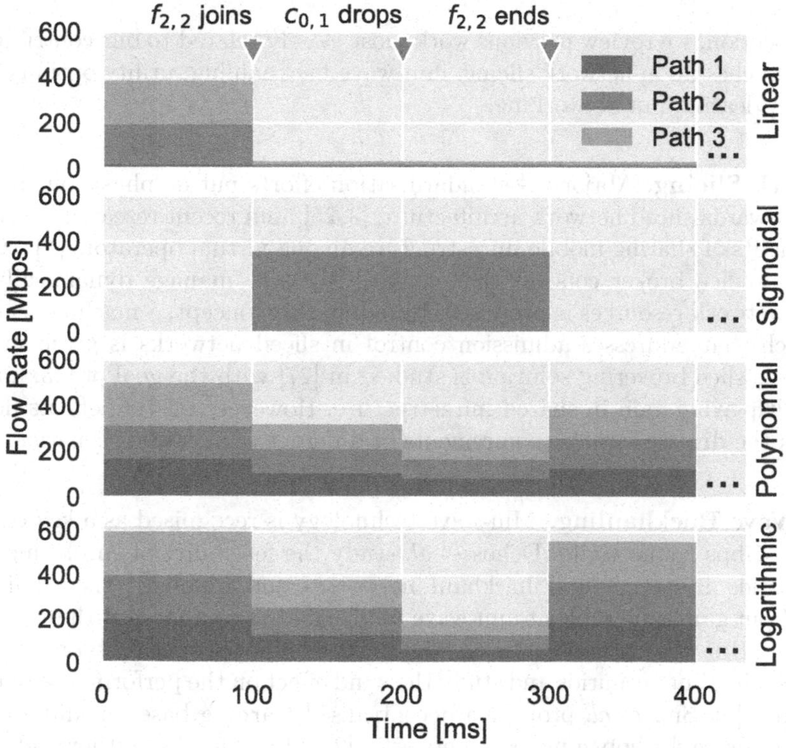

Fig. 8. Rate allocations performed by DELMU for flows of different slices and paths over time in Topology 3 (see Fig. 4), as a sequence of demand and capacity changes occur as labeled at the top of the figure. Numerical results.

Observe in the figure that the DELMU performs a correct allocation as soon as a change occurs and, given the millisecond scale inference times, the transition is almost instantaneous even at the 100 ms granularity. For instance, when $f_{2,2}$ joins, the allocation of network resources is immediately rearranged, so that the request of $f_{2,2}$ is mostly satisfied, whereas the rest of flows receive reduced rates. In this case, all the flows with linear utility are reduced to close to the minimum level of service, i.e. each to 11 Mbps rate. The drop in $c_{0,1}$ capacity at 200 ms leads to a significant degradation of the rates assigned to flows with polynomial and logarithmic utility, while the linear and sigmoid flows remain unaffected. Eventually, at 300 ms, when flow $f_{2,2}$ finishes, the rate of the flows with polynomial and logarithmic utility are increased, yet remain below the values assigned initially, due to the inferior $c_{0,1}$ capacity. Hence, the proposed DELMU is suitable for highly dynamic backhaul environments, as it makes close to optimal inferences fast and is able to adapt to sudden changes.

6 Related Work

In this section, we review previous work most closely related to our contribution, which touches upon network slicing, mm-wave backhauling, utility optimisation, and deep learning in networking.

Network Slicing. Major 5G standardisation efforts put emphasis on the evolution towards sliced network architectures [3,25], and recent research highlights the benefits of sharing mobile infrastructure among virtual operators [11,26,27]. In [11], a slice broker concept that enables MIPs to manage dynamically the shared network resources is proposed. Based on this concept, a machine learning approach that addresses admission control in sliced networks is given in [26]. An online slice brokering solution is studied in [27] with the goal of maximising the multiplexing gain in shared infrastructure. However, existing efforts do not address the diverse service requirements of different application scenarios.

Mm-Wave Backhauling. Mm-wave technology is recognised as a key enabler of multi-Gbps connectivity. Dehos *et al.* study the feasibility of employing mm-wave bands in access and backhaul networks, and highlight the significant throughput gain achievable at mm-wave frequencies as compared with microwave bands [4]. Hur *et al.* propose a beam alignment scheme specifically targeting mm-wave backhauling scenarios and study the wind effect on the performance of backhaul links [5]. Sim *et al.* propose a decentralised learning based medium access protocol for multi-hop mm-wave networks [28]. In [29], the authors advocate a max-min fair flow rate and airtime allocation scheme for mm-wave backhaul networks. These efforts however do not consider network utility and disregard sliced multi-service settings.

Network Utility Maximisation (NUM). With growing popularity of inelastic traffic, optimising a mix of both concave and non-concave utilities has been studied [8,30,31]. Fazel *et al.* propose a sum-of-square method to solve non-concave NUM problems that tackle primarily polynomial utility [8]. Hande *et al.* study the sufficient conditions for the standard price-based (sub-gradient based dual) approach to converge to global optima with zero duality gap, which relies on capacity provisioning [30]. Chen *et al.* consider NUM with mixed elastics and inelastic traffic, and develop a heuristic method to approximate the optimal [31]. Recent work investigates convex relaxation of polynomial NUM and employs distributed heuristics to approximate the global optimal [9], and Udell and Boyd define a general class of non-convex problems as sigmoidal programming and propose an approximation algorithm [15]. The limitation of these heuristics lies within their convergence times that is in the order of seconds, which can hardly meet the latency requirements of 5G networks. In contrast, our deep learning approach infers close to optimal rate allocations within milliseconds.

Deep Learning in Networking. With the increase in computational power and data sets availability, a range of deep learning applications in the computer and communications networking domain are emerging [32]. A fully-connected neural network is used in [33] to find optimal routes in wired/wireless heterogeneous networks. Zhang *et al.* employ a dedicated CNNs to infer fine-grained mobile traffic consumption from coarse traffic aggregates [34], improving measurement resolution by up to 100× while maintaining hight accuracy. CNNs have also employed been in [35], where the authors incorporate a 3D-CNN structure into a spatio-temporal neural network, to perform long-term mobile traffic forecasting. To the best of our knowledge, our work is the first that uses deep learning to solve utility optimisation problems in sliced backhauls.

7 Conclusions

In this paper we tackled utility optimisation in sliced mm-wave networks by proposing DELMU, a deep learning approach that learns correlations between traffic demands and optimal rate allocations. We specifically deal with scenarios where traffic is subject to conflicting requirements and maximise non-concave utility functions that reconcile all services, while overcoming the inherent complexity of the problems posed. We demonstrated that the proposed convolutional neural network attains up to 62% utility gains over a greedy approach, infers close to optimal allocation solutions within orders of magnitude shorter runtimes as compared to global search, and responds quickly to network dynamics.

Acknowledgement. We gratefully acknowledge the support of NVIDIA Corporation with the donation of the Titan Xp GPU used for this research.

References

1. NGMN: 5G White Paper. Next generation mobile networks (2015)
2. Schulz, P., et al.: Latency critical IoT applications in 5G: perspective on the design of radio interface and network architecture. IEEE Commun. Mag. **55**(2), 70–78 (2017)
3. 3GPP: Technical Specification Group Services and System Aspects; System Architecture for the 5G System. 3GPP TS 23.501, December 2017
4. Dehos, C., González, J.L., De Domenico, A., Ktenas, D., Dussopt, L.: Millimeter-wave access and backhauling: the solution to the exponential data traffic increase in 5G mobile communications systems? IEEE Commun. Mag. **52**(9), 88–95 (2014)
5. Hur, S., et al.: Millimeter wave beamforming for wireless backhaul and access in small cell networks. IEEE Trans. Commun. **61**(10), 4391–4403 (2013)
6. Kelly, F.: Charging and rate control for elastic traffic. Trans. Emerg. Telecommun. Technol. **8**(1), 33–37 (1997)
7. Yin, X., Jindal, A., Sekar, V., Sinopoli, B.: A control-theoretic approach for dynamic adaptive video streaming over HTTP. In: Proceedings of ACM SIGCOMM, pp. 325–338 (2015)
8. Fazel, M., Chiang, M.: Network utility maximization with nonconcave utilities using sum-of-squares method. In: Proceedings of IEEE CDC-ECC 2005, no. 1, pp. 1867–1874 (2005)
9. Wang, J., Ashour, M., Lagoa, C., Aybat, N., Che, H., Duan, Z.: Non-concave network utility maximization in connectionless networks: a fully distributed traffic allocation algorithm. In: Proceedings of IEEE ACC, pp. 3980–3985 (2017)
10. Ugray, Z., Lasdon, L., Plummer, J., Glover, F., Kelly, J., Martí, R.: Scatter search and local NLP solvers: a multistart framework for global optimization. INFORMS J. Comput. **19**(3), 328–340 (2007)
11. Samdanis, K., Costa-Perez, X., Sciancalepore, V.: From network sharing to multi-tenancy: the 5G network slice broker. IEEE Commun. Mag. **54**(7), 32–39 (2016)
12. Ayach, O.E., Rajagopal, S., Abu-Surra, S., Pi, Z., Heath, R.W.: Spatially sparse precoding in millimeter wave MIMO systems. IEEE Trans. Wirel. Commun. **13**(3), 1499–1513 (2014)
13. Raleigh, G.G., Cioffi, J.M.: Spatio-temporal coding for wireless communication. IEEE Trans. Commun. **46**(3), 357–366 (1998)
14. Ahuja, R.K., Magnanti, T.L., Orlin, J.B.: Network Flows: Theory, Algorithms, and Applications. Prentice-Hall Inc., Upper Saddle River (1993)
15. Udell, M., Boyd, S.: Maximizing a Sum of Sigmoids. Optimization and Engineering, pp. 1–25 (2013)
16. Papadimitriou, C.H., Steiglitz, K.: Combinatorial Optimization: Algorithms and Complexity. Courier Corporation, North Chelmsford (1998)
17. Goodfellow, I., Bengio, Y., Courville, A.: Deep Learning. MIT Press, New York (2016)
18. Srivastava, R.K., Greff, K., Schmidhuber, J.: Training very deep networks. In: Advances in Neural Information Processing Systems, pp. 2377–2385 (2015)
19. Klambauer, G., Unterthiner, T., Mayr, A., Hochreiter, S.: Self-normalizing neural networks. In: Proceedings of NIPS (2017)
20. Kingma, D., Ba, J.: Adam: a method for stochastic optimization. In: Proceedings of ICLR (2015)
21. Abadi, M., et al.: TensorFlow: a system for large-scale machine learning. In: Proceedings of OSDI, vol. 16, pp. 265–283 (2016)

22. Dong, H.: TensorLayer: a versatile library for efficient deep learning development. In: Proceedings of ACM Multimedia Conference (2017)
23. Glover, F.: A template for scatter search and path relinking. In: Hao, J.-K., Lutton, E., Ronald, E., Schoenauer, M., Snyers, D. (eds.) AE 1997. LNCS, vol. 1363, pp. 1–51. Springer, Heidelberg (1998). https://doi.org/10.1007/BFb0026589
24. Son, K., Eunsung, O., Krishnamachari, B.: Energy-efficient design of heterogeneous cellular networks from deployment to operation. Comput. Netw. **78**, 95–106 (2015)
25. 3GPP: Technical Specification Group Services and System Aspects; Study on Architecture for Next Generation System. 3GPP TS 23.799, December 2016
26. Sciancalepore, V., Samdanis, K., Costa-Perez, X., Bega, D., Gramaglia, M., Banchs, A.: Mobile traffic forecasting for maximizing 5G network slicing resource utilization. In: Proceedings of IEEE INFOCOM (2017)
27. Sciancalepore, V., Zanzi, L., Costa-Perez, X., Capone, A.: ONETS: online network slice broker from theory to practice. arXiv preprint arXiv:1801.03484 (2018)
28. Sim, G.H., Li, R., Cano, C., Malone, D., Patras, P., Widmer, J.: Learning from experience: efficient decentralized scheduling for 60GHz mesh networks. In: Proceedings of IEEE WoWMoM (2016)
29. Li, R., Patras, P.: WiHaul: max-min fair wireless backhauling over multi-hop millimetre-wave links. In: Proceedings of ACM Workshop HotWireless, pp. 56–60 (2016)
30. Hande, P., Zhang, S., Chiang, M.: Distributed rate allocation for inelastic flows. IEEE/ACM Trans. Netw. **15**(6), 1240–1253 (2007)
31. Chen, L., Wang, B., Chen, L., Zhang, X., Dacheng, Y.: Utility-based resource allocation for mixed traffic in wireless networks. In: Proceedings of IEEE INFOCOM Workshops, pp. 91–96 (2011)
32. Zhang, C., Patras, P., Haddadi, H.: Deep learning in mobile and wireless networking: a Survey. arXiv preprint arXiv:1803.04311 (2018)
33. Kato, N., et al.: The deep learning vision for heterogeneous network traffic control: proposal, challenges, and future perspective. IEEE Wirel. Commun. **24**(3), 146–153 (2017)
34. Zhang, C., Ouyang, X., Patras, P.: ZipNet-GAN: inferring fine-grained mobile traffic patterns via a generative adversarial neural network. In: Proceedings of ACM CoNEXT, pp. 363–375
35. Zhang, C., Patras, P.: Long-term mobile traffic forecasting using deep spatio-temporal neural networks. In: Proceedings of ACM MobiHoc, pp. 231–240 (2018)

Malware Detection System Based on an In-Depth Analysis of the Portable Executable Headers

Mohamed Belaoued[1(✉)], Bouchra Guelib[2], Yasmine Bounaas[2], Abdelouahid Derhab[3], and Mahmoud Boufaida[1]

[1] LIRE Laboratory, Software Technologies and Information Systems Department, University of Constantine 2, Constantine, Algeria
belaoued.mohamed@gmail.com,boufaida_mahmoud@yahoo.fr

[2] Software Technologies and Information Systems Department, University of Constantine 2, Constantine, Algeria
guelibbouchra@gmail.com,Y_asmine@windowslive.com

[3] Center of Excellence in Information Assurance (COEIA), King Saud University, Riyadh, Saudi Arabia
abderhab@ksu.edu.sa

Abstract. Malware still pose a major threat for cyberspace security. Therefore, effective and fast detection of this threat has become an important issue in the security field. In this paper, we propose a fast and highly accurate detection system of Portable Executable (PE) malware. The proposed system relies on analyzing the fields of the PE-headers using a basic way and a more in-depth way in order to generate a set of standard attributes (SAT), and meaningful attributes (MAT) respectively. The decision phase is conducted by leveraging several machine learning classifiers, which are trained using the best K attributes according to two different feature selection methods. The experimental results are very promising, as our system outperforms two state-of-the-art solutions with respect to detection accuracy. It achieves an accuracy of 99.1% and 100% using 10-folds cross validation and train-test split validation, respectively. In both validation approaches, we only use less than 1% out of the initial set of 1329 extracted attributes. Also, our system is able to analyze a file in 0.257 s.

Keywords: Malware detection · Machine learning · Portable Executable

1 Introduction

Today's IT infrastructures are facing increasing cyber attacks, aiming at deliberately violating one or more security properties. Malware, such as worms, viruses and Trojans, still account for the majority of cyber-attacks. The number of released malware is growing exponentially every year, reaching 780 million malware samples in 2017 [9]. Malware are computer programs, which are designed

É. Renault et al. (Eds.): MLN 2018, LNCS 11407, pp. 166–180, 2019.
https://doi.org/10.1007/978-3-030-19945-6_11

to perform unauthorized actions without the user's consent, such as: stealing or damaging data, disrupt normal operations of the IT system, etc., causing considerable damages to people, organizations, and critical infrastructures [4]. Therefore, it is essential to implement systems capable of rapidly detecting and eliminating these threats.

Malware detection techniques can be broadly classified in three categories, which are signature-based, behavior-based, and heuristic-based [2]. Signature-based techniques are widely used by antivirus tools, as they can accurately detect known malware. However, they are unable to detect previously unseen as well as polymorphic ones, which have the ability to change their signatures. Behavior-based malware detection techniques [7] overcome the limitations of signature-based ones. However, they are time-consuming when it comes to analyze the malware. Moreover, the accuracy of the heuristic-based techniques is not always sufficient as many false positives and false negatives are recorded [13]. The need for new detection methods is dictated by malware's high polymorphism, and their constant proliferation. To deal with the above issues, one approach, which is adopted in our work, is to combine heuristic-based analysis with data mining and machine learning techniques. The latter have shown higher detection accuracy since their first use [12]. However, the main challenge of such an approach is choosing the adequate nature as well as the proper number of attributes in order to achieve a high detection accuracy in the shortest possible time.

In this paper, we propose a malware detection system based on a static and an in-depth analysis of the information stored within the headers' fields of the Portable Executable (PE) file [11]. Our choice is motivated by the fact that the PE headers have a standard structure, and thus contain a predefined set of fields. Therefore, extracting attributes from those fields remains less time-consuming compared to other types of attributes such as: API calls or machine instructions (Opcode). Also, they have shown very good performances in correctly distinguishing between malicious and benign programs. However, most of works, which considered PE-header information, combined them with other features, such as imports, exports, and resources, in order to increase the accuracy. Moreover, some approaches, which only considered PE-header fields, obtained poor results. We argue that PE-header fields alone have a great potential in discriminating between malware and benign programs. For this purpose, we conduct two types of analysis. The first one, named basic analysis, in which the values of all fields are stored in what we call a standard attributes (SAT). The second, considers the meaning of each field, as well as its legitimate value, in order to generate what we call meaningful attributes (MAT). After extracting attributes, it is necessary to remove irrelevant ones, and group the rest of them in smaller subsets based on their relevance. To this end, we use two well-known, and widely used filter-based feature selection techniques (FST), namely the chi-square test [10] and Mutual Information [1]. Finally, the decision process is made by training several classification algorithms separately, and then the best performing ones are combined using a voting ensemble technique [6].

Experimental results demonstrate encouraging results. The proposed system outperforms two state-of-the-art solution with respect to detection accuracy. It achieves an accuracy of 99.1% and 100% using 10-folds cross validation and train-test split validation, respectively. This result is obtained by using only less than 1% out of the initial set of 1329 extracted attributes. The process to analyze a file, starting from feature extraction to detection, takes 0.257 s.

This paper is organized as follows: Sect. 2 presents related work. Section 3 provides a detailed description of our proposed detection system. In Sect. 4, we present the obtained experimental results. Finally, Sect. 5 concludes our work and underlines our future perspectives.

2 Related Work

In this section, we present recent methods that consider PE-header information as the main features for malware detection. Therefore, we do not consider the methods that use the full PE structural information (i.e., imports, exports, resources, etc.) or combined header information with other features. To date, and to the best of our knowledge, there are only two methods that employed such attributes, and are discussed below.

In their work published in 2015, Belaoued and Mazouzi [3] proposed a real-time malware detection system based on the analysis of the information stored in the PE Optional-Header fields (OH). This system considered the OH fields as raw features and did not conduct any in-depth analysis of the latter fields. Indeed, the proposed attributes were generated by the concatenation of the fields name and its value for the whole samples of the dataset. Once the features are generated, the frequency of appearance of each feature in malicious and legitimate files is considered. They used a feature selection method based on chi-square test and Phi (ϕ) coefficient, which allowed to divide the features into groups according to their relevance. The decision phase, consists in using several classification algorithms separately. The proposed malware detection system achieved an accuracy equal to 97.25 % under the Train-Test split validation, and an average detection time of 0.077 s/file using 14 features.

The work of Kumar et al. [8], proposed a new concept based on the meaning of PE-Header attributes. In their work, authors introduced what they called derived features, which are composed of raw and integrated features. The raw attributes are similar to those used by Belaoued and Mazouzi [3] but they are limited to 54 features, what we consider too restrictive. As for the derived attributes, they are obtained by an in-depth analysis of the standard values of the PE-headers' fields. The idea was to compare the observed values to the expected values according to the standard PE specifications [11]. For instance, the authors proposed an attribute called TimeDateStamp, which is an integer indicating the number of seconds that elapsed since December 31, 1969 at 4:00 pm. The latter is set to true if it has a valid value in that interval, otherwise it is set to false. Using 68 features, the proposed system achieved an accuracy of 98.4% under the 10-folds cross validation, and 89.23% on a previously unseen samples, which raises concerns about the relevance of the used attributes.

In this work, we aim to improve the accuracy of the aforementioned methods by taking advantage of their strengths. Indeed, for the raw attributes which we called standard attributes (SAT), we will use those proposed by Belaoued and Mazouzi [3], since they take into consideration all the possible values that a field can take. However, we will expand the analysis the to PE-file header (FH) as well. Moreover, we propose our own set of meaningful attributes (MAT), and improving those that have already been used by Kumar et al. [8].

3 Description of the Proposed Detection System

In this section, we describe the architecture of the proposed system for PE-malware detection, and as shown in Fig. 1, it is composed of three main modules namely: the *Attribute extractor and pre-processing* module, the *Attribute selector* module, and finally the *Decision* module.

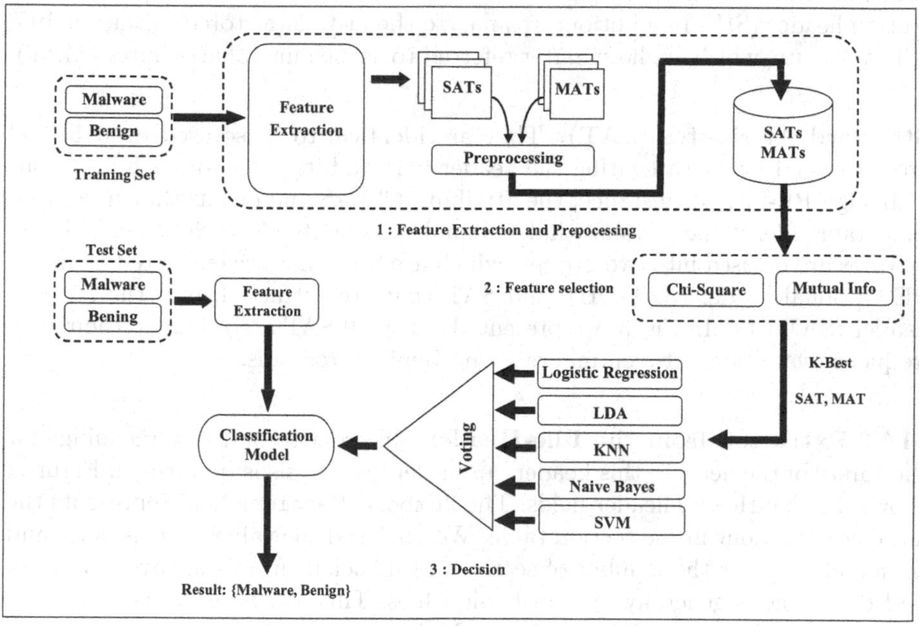

Fig. 1. The architecture of the proposed system

3.1 Feature Extraction Module

The feature extraction module aims at providing a set of attributes that allow the distinction between malicious and legitimate programs. In our work, the features used are those stored in the PE-headers' fields. Indeed, and as shown in Fig. 2, the PE file format is composed of several parts including headers that contain rich meta-data about the file.

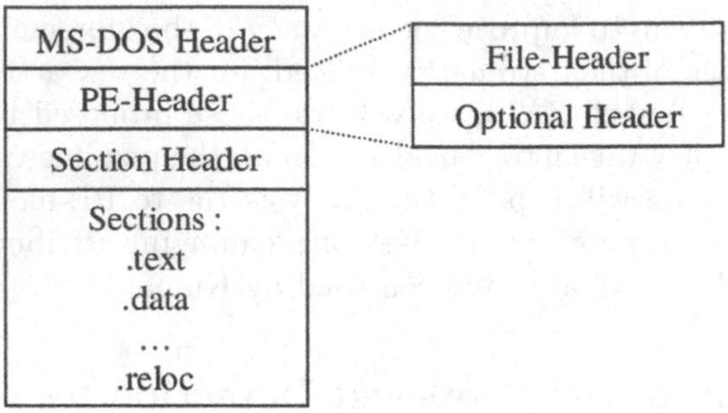

Fig. 2. PE file format overview [5]

In this work, we seek new meanings for the information contained in these fields. Thus, and in addition to the standard attributes (SAT), we conduct an in-depth analysis of the PE Optional Header (OH), the File header (FH), and the section header (SH). In addition, we analyze the metadata stored in the FileInfo (FI) attribute, which are hereinafter referred to as meaningful attributes (MAT).

Standard Attributes (SAT). They are identical to those used in [3–5] and are generated by concatenating the header field and its value for each malicious or benign PE-file. For instance, the attribute CheckSum0 means that in a given executable file, it has a value that is equal to 0 at its CheckSum field. Those features are devised into two groups, which are SAT that are extracted from the PE-Optional header (SAT-OH), and SAT that are extracted from the PE-File header (SAT-FH). In Fig. 3, we present the Top-10 SAT according to their high frequency variation between malware and benign programs.

MAT Extracted from the File-Header. In order to find new meanings for the values of the fields in this header, an in-depth analysis is performed. Figure 4 shows the File-Header header fields: The `NumberOfSections` field represents the number of sections in the section table. We analyzed the values of this field, and we found out that the number of sections in malicious files is always ≤ 3 or ≥ 9 and this value is generally ≥ 10 in benign files. This allows us to generate two MATs under the name "NumberSectionMalware" and "NumberSectionBenin". If the value of those fields is in the defined intervals, they take the value "True". Otherwise they remains "False". Note that this field has been considered as raw feature in [8]. The remaining FH fields follow the standard values of the PE format in both malicious and legitimate files, and thus we do not consider them.

MAT Extracted from the Optional Header. The majority of the works that employed SAT are based on the values of the OH fields, and were able to achieve a high accuracy. Indeed, the OH contains more information than the FH.

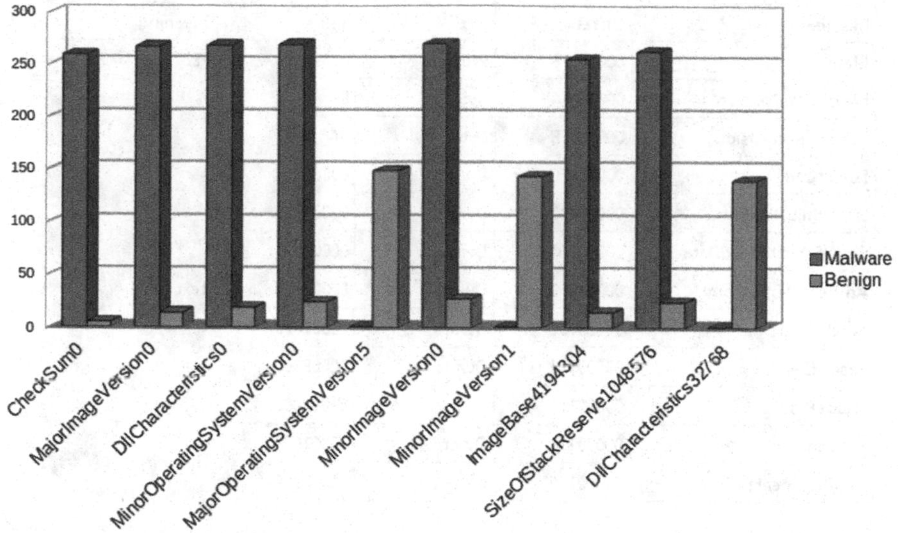

Fig. 3. Top 10 SAT with the highest frequency variation between malware and benign files

Member	Offset	Size	Value	Meaning
Machine	000000E4	Word	014C	Intel 386
NumberOfSections	000000E6	Word	0003	
TimeDateStamp	000000E8	Dword	48025297	
PointerToSymbolTa...	000000EC	Dword	00000000	
NumberOfSymbols	000000F0	Dword	00000000	
SizeOfOptionalHea...	000000F4	Word	00E0	
Characteristics	000000F6	Word	010F	Click here

Fig. 4. File Header fields overview

The structure of the OH is presented in Fig. 5. The OH fields should respect the standard PE format values, such as:

- The value of `ImageBase` field must be a multiple of 64K.
- The value of `FileAlignment` field must be a power of 2 (between 512 and 65536).
- The value of `SectionAlignment` field must be greater than the `FileAlignment` one.
- The value of `SizeOfImage` field must be a multiple of the `FileAlignment`.

According to our analysis, and differently from Kumar et al. [8], the values of the aforementioned fields respect the standard PE specifications in both malware and benign files, and thus they are ignored and no MAT is generated from them.

Member	Offset	Size	Value	Meaning
Magic	000000F8	Word	010B	PE32
MajorLinkerVersion	000000FA	Byte	07	
MinorLinkerVersion	000000FB	Byte	0A	
SizeOfCode	000000FC	Dword	0000C600	
SizeOfInitializedData	00000100	Dword	00021E00	
SizeOfUninitializedData	00000104	Dword	00000000	
AddressOfEntryPoint	00000108	Dword	0000BE85	.text
BaseOfCode	0000010C	Dword	00001000	
BaseOfData	00000110	Dword	0000E000	
ImageBase	00000114	Dword	01000000	
SectionAlignment	00000118	Dword	00001000	
FileAlignment	0000011C	Dword	00000200	

Fig. 5. Optional Header fields overview

MAT Extracted from the Section Header. The section table is located between the PE header and the raw data for the image sections. The section table is essentially a directory containing information about each section of the image. The header of the sections is partially different from what we have seen so far because it does not have a constant size. Indeed, there are as many section headers as there are sections in the file [11]. Important information are stored in this table, which motivates us to study this table. Figure 6 shows the section table in a PE file.

Name	Virtual Size	Virtual Address	Raw Size	Raw Address	Reloc Address	Linenumbers	Relocations N...	Linenumbers ...	Characteristics
Byte[8]	Dword	Dword	Dword	Dword	Dword	Dword	Word	Word	Dword
.text	0000C42C	00001000	0000C600	00000400	00000000	00000000	0000	0000	60000020
.data	00009868	0000E000	00000C00	0000CA00	00000000	00000000	0000	0000	C0000040
.rsrc	000211E8	00018000	00021200	0000D600	00000000	00000000	0000	0000	40000040

Fig. 6. Sections table fields overview

- `Name`: This is the name of the section, and it has 8 characters and is filled with #0 if the name has less. An analysis consists in studying the values of this field. The PE format proposes standard section names which are: ".data", ".rdata", ".reloc", ".rsrc", ".text". We used a Boolean attribute called `StandardSectionName`, and if a section has a name different from the standard names, the attribute takes as value "False". Otherwise, it remains

"True". In the work of Kumar et al. [8], it is considered as an integer, which counts the number of sections that have illegitimate names.

- `SizeOfRawData`: This is the actual size occupied by the section in the file. The value of this field must be a multiple of 512, an attribute is used under the name `SizeOfRawData` which takes the value "True" if this condition is respected.Otherwise, it remains "False". This field has not been investigated in [8].

MAT Based on the FileInfo. The FileInfo, as shown in Fig. 7, is a set of strings, which contain metadata about the PE file such as FileVersion, ProductVersion, ProductName, CompanyName, etc. Benign programs incorporate rich metadata while malware authors avoid putting metadata.

Property	Value
CompanyName	Microsoft Corporation
FileDescription	Assistant Accessibilité Microsoft
FileVersion	5.1.2600.5512 (xpsp.080413-2105)
InternalName	ACCWIZ
LegalCopyright	© Microsoft Corporation. Tous droits réservés.
OriginalFilename	ACCWIZ.EXE
ProductName	Système d'exploitation Microsoft® Windows®

Fig. 7. Example of a file info metadata

Our analysis confirms that all benign files have this field, unlike malicious files, which rarely have it. Therefore, an attribute named `FileInfoFound` is used to test the presence of this field. This attribute takes a value "True" if the field is found. Otherwise, it remains "False". Moreover, we also noticed that even if the FileInfo is present in malicious files, its fields contain null values. Therefore, an additional attribute called `FullDescriptionFound` is used, and its value is set to "True" if the FileInfo fields (e.x. CompanyName, FileDescription, FileVersionLegalCopyright, ProductName, productVersion) contains values. In the work of Kumar et al. [8] they limited the analysis to the presence or absence of the FileInfo metadata.

The in-depth analysis of the PE-headers fields allowed us to generate six (06) different MATs, which are summarized in Table 1.

Table 1. Summary of the proposed MAT

Feature	Type	Expected value
`NumberOfSectionMalware`	Boolean	≤3 or ≥9
`NumberOfSectionBenign`	Boolean	≥10
`StandardSectionName`	Boolean	.data, .rdata, .reloc, .rsrc, .text
`SizeOfRawData`	Boolean	Multiple of 512
`FileInfoFound`	Boolean	-
`FullDescriptionFound`	Boolean	-

3.2 Feature Selection Module

After extracting attributes from the PE headers (i.e., 1329 attributes), it is necessary to reduce their number, in order to keep only the most relevant ones. For this purpose, we use a feature selection method based on chi-square test and mutual information. For both methods, we select the K best attributes that will provide the highest accuracy and the lowest detection time.

Chi-Square-Based Feature Selection. The chi-square test of independence [10] is a statistical method proposed by Pearson in 1900, and it is used to determine whether there is a relationship (correlation) between two categorical variables. In our case the two variables in question are the attribute and the PE file category. The chi-square score (D^2) is calculated using the following formula:

$$D^2 = \sum \frac{(O_{r,c} - E_{r,c})^2}{E_{r,c}} \tag{1}$$

where:

O (Observed) denotes the observed frequencies of each attribute.
E (Expected) denotes the expected values.

Mutual-Info-Based Feature Selection. Mutual information (MI) is a statistical method that focuses primarily on measuring the amount of information shared between two randomly selected variables. It calculates the amount of information shared between them and gives a non-negative value. A nil value of MI indicates that the two variables observed are statistically independent. The mutual info is calculated using the following formula:

$$I(X,Y) = \sum_{x \in X} \sum_{y \in Y} p(x,y) \log \frac{p(x,y)}{p(x)p(y)} \tag{2}$$

where:

X: Benign Class
Y: Malware Class
x: Attributes of Benign class
y: Attributes of Malware class

3.3 Decision Module

The decision phase consists of categorizing a file as malicious or benign. The proposed decision module is composed of several classifiers, which are trained and tested separately. Moreover, the best performing ones are grouped together using the voting ensemble technique in order to improve the overall accuracy. The used classifiers are the following:

- Logistic regression (LR)
- Naive Bayes (NB)
- K-Nearest Neighbors (KNN)
- Decision Tress (DT)
- Linear Discriminant Analysis (LDA)
- Support Vector Machines (SVM)

In our work, we used the ensemble voting classifier, which is a combination set of different classifiers that are trained and evaluated continuously. The final decision for a prediction is made by majority vote, as presented in Fig. 8.

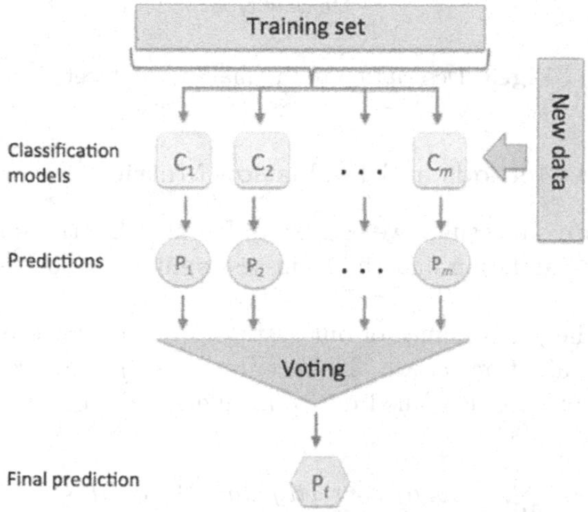

Fig. 8. Architecture of voting classifier (source: rasbt.github.io)

4 Experimental Results

4.1 Dataset

In order to evaluate the performance of our malware detection system, we used the dataset that has been employed by [3,4]. Which is composed of 214 benign files and 338 malicious ones, all in MS Windows PE format. The distribution of malicious files by category is presented in Fig. 9.

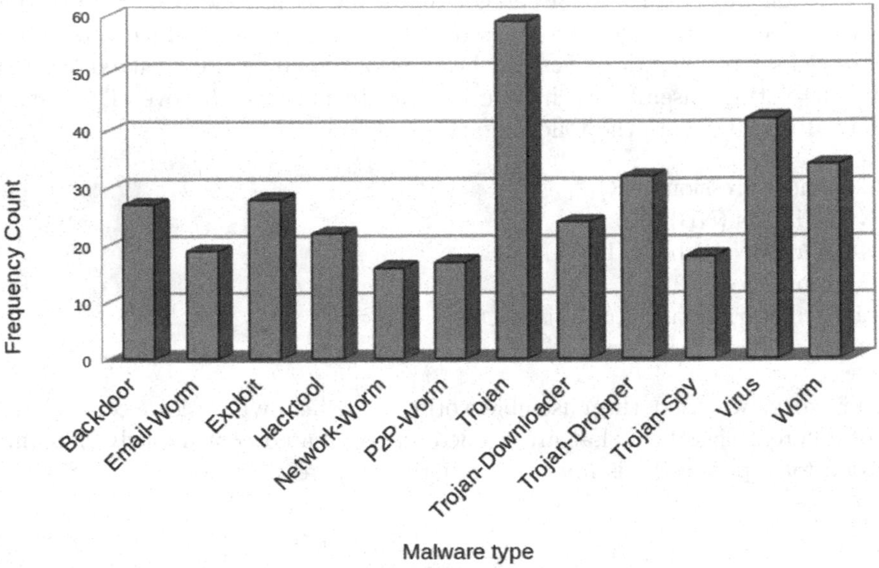

Fig. 9. Description of the malware dataset.

4.2 Validation Methods and Evaluation Metrics

In order to validate our results, we use two different validation methods, namely the 10-folds cross-validation and the train-test split with 80% training set and 20% test set.
In order to test the performance of our system, we consider the accuracy (AC) as an the main evaluation metric. The accuracy (AC) measure the percentage of files that are correctly classified out of all files, and can expressed using the following formula:

$$AC = \frac{nb.\ of\ correctly\ classified\ files}{total\ number\ of\ files} \tag{3}$$

In addition to accuracy, we evaluate the performance of our system in terms of detection time (DT).

4.3 Results

Experimental results are obtained by using both validation methods (10-fold cross validation and train-test split) using the different classification algorithms. Moreover, the experimentation are performed using the two feature selection methods (Chi-Square and IM), and are applied on different categories of attributes (SAT, MAT) and their combination. The results are summarized in the following subsections.

Results Using Cross-Validation. The experimental results using 10-folds cross-validation are presented in the Table 2.

Table 2. Accuracy results using 10-fold cross validation

K	Best attributes	best classifiers	**AC**
Chi-square based feature selection			
10	MAT + SAT-OH	LR, Voting	0.980
20	**SAT-OH + SAT-FH, MAT + SAT-FH**	**Voting**	**0.989**
30	**SAT-OH + SAT-FH, MAT + SAT-FH**	**Voting**	**0.989**
All	MAT	LDA, NB, SVM	0.98
Mutual Information based feature selection			
10	**MAT + SAT-OH**	**Voting**	**0.991**
20	**MAT + SAT-FH**	**Voting**	**0.991**
30	MAT + SAT-FH	Voting	0.985
All	MAT	LDA, NB, SVM	0.98

From the results presented in Table 2, we can see that the feature selection method, which is based on mutual information, offers the highest accuracy (99,1%) with an improvement of 0.2% compared with chi-square test. This result is achieved using a combination of ten (10) MAT and SAT-OH, and applying the voting ensemble classifier.

Results Using Train-Test Split. The experimental results using train-test split validation method are presented in Table 3. In this case, we use the best configuration from the cross-validation experiment namely K=10 (K = 6 for MAT), and Mutual information as feature selection method.

By applying the Train-Test split validation, our system achieves an accuracy of 100% using MAT and the classifiers KNN, CART, and Voting. We recall that MAT are composed of six (06) attributes only. The same accuracy is reached using other combinations such as MAT + SAT-OH, and MAT + SAT-FH.

Table 3. Accuracy results using train-test split and mutual information

Attributes	LR	LDA	KNN	CART	NB	SVM	Voting
SAT-OH	0.964	0.946	0.97	0.97	0.775	0.9645	0.97
SAT-FH	0.847	0.766	0.86	0.87	0.586	0.82	0.87
SAT-OH + SAT-FH	0.964	0.946	0.945	0.945	0.802	0.964	0.964
MAT	0.973	0.982	**1.0**	**1.0**	0.973	0.982	**1.0**
MAT + SAT-OH	0.991	0.937	0.982	**1.0**	0.90	0.973	0.991
MAT + SAT-FH	0.991	0.964	**1.0**	0.973	0.892	0.973	0.991
MAT + SAT-OH + SAT-FH	0.991	0.964	0.84	0.83	0.892	0.973	0.991

4.4 Detection Time

The detection time (DT) is calculated by considering the extraction time and the classification time. Table 4 presents both extraction and classification time for the different types of attributes and classifiers.

Table 4. Experimental results for execution time

Attributes	Extraction time	Classifier	Classif. time
SAT-OH	0.33 s	LR	0.0001
SAT-FH	0.15 s	KNN	0.0003
MAT	0.25 s	NB	0.0004
MAT + SAT-FH	0.38 s	Voting	0.007

In our work, we choose the MAT attributes as the main features, as well as the voting classifier, since it achieved the maximum accuracy in all experiments. Therefore, the detection time (DT) of our system is represented as follows:

DT = MAT extraction time + Voting classification time
$DT = 0.25 + 0.007 = 0.257$ s/file
The system is able to categorize of file (i.e., malicious or benign) in **0.257s**.

4.5 Comparison with Previous Works

In this subsection we compare our obtained results with those of the previous works with respect to accuracy and number of attributes used to achieve that accuracy. Results are presented in Fig. 10.

As shown in Fig. 10, our proposed system for malware detection has the highest accuracy with an improvement of 2.75% compared to the results of Belaoued and Mazouzi [3], and 1.6% compared to the results of Kumar et al. [8].

As mentioned previously, our systems needs only 6 attributes to reach the maximum accuracy, which represents more than the half of the attributes required by the system proposed by Belaoued and Mazouzi [3], and 90 % less than those required by the system proposed by Kumar et al. [8] (Fig. 11).

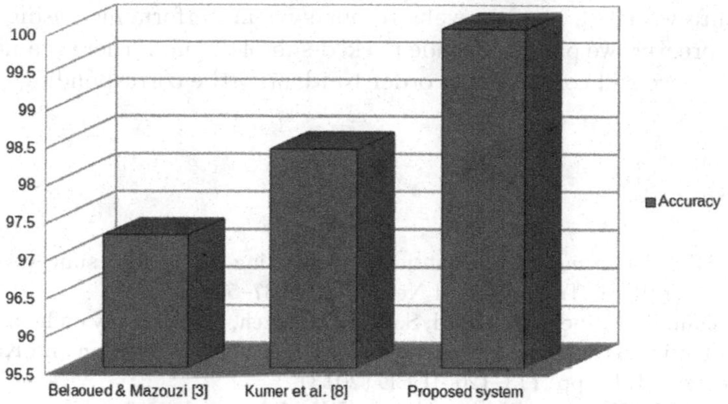

Fig. 10. Comparison with previous works (accuracy)

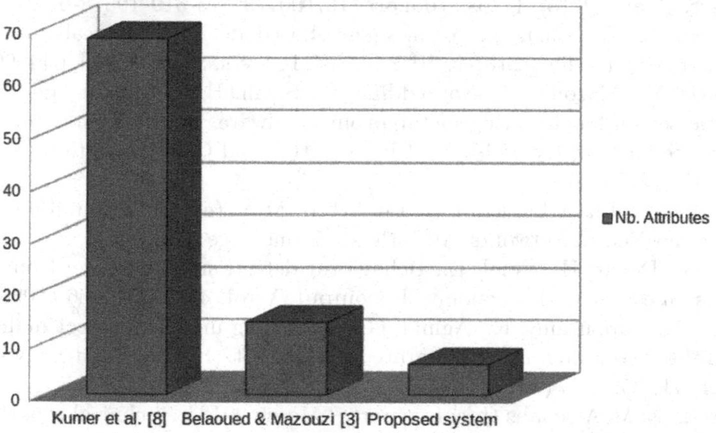

Fig. 11. Comparison with previous works (Nb. of attributes)

5 Conclusions and Future Work

In this work, we have proposed a fast and a highly accurate PE malware detection system. The proposed system was able to detect unknown malware with an accuracy that varies between 99.1% and 100% using ten (10) and six (06) attributes respectively, which represents less than 1% of the initial set of 1329 extracted attributes. The use of voting ensemble classifier that combined the best performing ones made it possible to reach the maximum accuracy under different experimentation settings. Our system has the best performance with respect to accuracy and number of used features, compared to two state-of-the-art methods.

As future work, we plan to evaluate our system performance using a larger dataset. Moreover, we plan to include packed samples, and if there is a need for a new analysis, we will conduct it in order to identify the corresponding adequate features.

References

1. Battiti, R.: Using mutual information for selecting features in supervised neural net learning. IEEE Trans. Neural Netw. **5**(4), 537–550 (1994)
2. Bazrafshan, Z., Hashemi, H., Fard, S.M.H., Hamzeh, A.: A survey on heuristic malware detection techniques. In: 2013 5th Conference on Information and Knowledge Technology (IKT), pp. 113–120. IEEE (2013)
3. Belaoued, M., Mazouzi, S.: A real-time PE-Malware detection system based on CHI-square test and PE-file features. In: Amine, A., Bellatreche, L., Elberrichi, Z., Neuhold, E.J., Wrembel, R. (eds.) CIIA 2015. IAICT, vol. 456, pp. 416–425. Springer, Cham (2015). https://doi.org/10.1007/978-3-319-19578-0_34
4. Belaoued, M., Mazouzi, S.: A chi-square-based decision for real-time malware detection using PE-file features. JIPS (J. Inf. Process. Syst.) **12**(4), 644–660 (2016)
5. Belaoued, M., Mazouzi, S., Noureddine, S., Salah, B.: Using chi-square test and heuristic search for detecting metamorphic malware. In: 2015 First International Conference on New Technologies of Information and Communication (NTIC), pp. 1–4. IEEE (2015)
6. Dietterich, T.: Ensemble learning. In: Arbib, M.A. (ed.) The Handbook of Brain Theory and Neural Networks. MIT Press, Cambridge (2002)
7. Jacob, G., Debar, H., Filiol, E.: Behavioral detection of malware: from a survey towards an established taxonomy. J. Comput. Virol. **4**(3), 251–266 (2008)
8. Kumar, A., Kuppusamy, K., Aghila, G.: A learning model to detect maliciousness of portable executable using integrated feature set. J. King Saud Univ.-Comput. Inf. Sci. **31**, 252–265 (2017)
9. McAfee-labs: McAfee labs threats report, "March 2018". Technical report, McAfee labs, January 2018
10. Moore, D.S.: Chi-square tests. Technical report, Purdue University Lafayette Indiana Department of Statistics (1976)
11. Pietrek, M.: Peering inside the PE: a tour of the Win32 (r) portable executable file format. Microsoft Syst. J.-US Ed. **9**, 15–38 (1994)
12. Schultz, M.G., Eskin, E., Zadok, F., Stolfo, S.J.: Data mining methods for detection of new malicious executables. In: Proceedings of 2001 IEEE Symposium on Security and Privacy, S&P 2001, pp. 38–49. IEEE (2001)
13. Shabtai, A., Moskovitch, R., Elovici, Y., Glezer, C.: Detection of malicious code by applying machine learning classifiers on static features: a state-of-the-art survey. Inf. Secur. Tech. Rep. **14**(1), 16–29 (2009)

DNS Traffic Forecasting Using Deep Neural Networks

Diego Madariaga(✉), Martín Panza, and Javier Bustos-Jiménez

NIC Chile Research Labs, Universidad de Chile, Santiago, Chile
{diego,martin,jbustos}@niclabs.cl

Abstract. With the continuous growth of Internet usage, the importance of DNS has also increased, and the large amount of data collected by DNS servers from users' queries becomes a very valuable data source, since it reveals user patterns and how their Internet usage changes through time. The periodicity in human behavior is also reflected in how users use the Internet and therefore in the DNS queries they generate. Thus, in this paper we propose the use of Machine Learning models in order to capture these Internet usage patterns for predicting DNS traffic, which has a huge relevance since a big difference between the expected DNS traffic and the real one, could be a sign of an anomaly in the data stream caused by an attack or a failure. To the best of the authors' knowledge this is the first attempt of forecasting DNS traffic using Neural Networks models, in order to propose an unsupervised and lightweight method to perform fast detection of anomalies in DNS data streams observed in DNS servers.

Keywords: DNS traffic · Forecasting · Machine Learning

1 Introduction

Internet has experienced a huge growth during the last decade undeniably, as well as DNS system along with it, which has become a fundamental part of Internet working. Both systems, the whole Internet and DNS, have also become more complex due to the constant technological development, emerging of new needs, and the study and successful correction of failures caused by either errors or by the design of varied attacks against the vulnerabilities of the systems' infrastructure. Unfortunately, it is impossible to assure a complete protection of the systems as new unknown attacks and failures are always possible. However, due to its importance it is indispensable to keep the correct working of DNS, since if this is not achieved, the users would be seriously affected by the system's failures. This is especially important for DNS operators, who are responsible for responding to any existing failure.

The early detection of possible failures or attacks is a powerful aid for keeping the correct working of any system. That is why the automatic detection of anomalous events on computer networks issue has taken big relevance in the last years.

E. Renault et al. (Eds.): MLN 2018, LNCS 11407, pp. 181–192, 2019.
https://doi.org/10.1007/978-3-030-19945-6_12

However, in addition to the innate difficulty of this problem, when we refer to real world DNS we are referring to large amount of data, given the huge and growing number of users who are constantly querying the system's servers. This demands automatism and efficiency to the systems that perform anomaly detection in real-time.

Forecasting the system's behavior would contribute to the solution to detect anomalies in network traffic, as finding big differences between what is expected, given the system's past information, and the encountered values would give a quick sign of an anomaly occurring in the data stream. This is specially plausible in DNS, where data behavior is strongly influenced by human patterns which present a strong periodicity.

This work performs DNS traffic forecasting on real DNS data from the Chilean country code top-level domain '.cl' using Machine Learning models. It also proposes the development of an unsupervised and lightweight method, and its usage for fast anomaly detection in DNS data streams of DNS servers.

To the best of the authors' knowledge, this is the first attempt of forecasting DNS traffic using Neural Networks models, with an important future usage for early detection of anomalies on DNS traffic.

2 Related Work

Network traffic prediction is a task that has been continuously studied from a lot of different points of view, focusing on specific portions of the whole network traffic, according to the network protocol, or the target users to study.

Among the important studies about traffic forecast, some that can be mentioned are the prediction of IP backbone traffic using AutoRegressive Integrated Moving Average (ARIMA) models [14], the prediction of data traffic belonging to most popular TCP ports using AutoRegressive Moving Average (ARMA) models [3], and the prediction of TCP/IP traffic using both ARIMA and Neural Networks models [5].

In addition, other research studies have focused on predicting mobile network traffic, implementing models to forecast future call traffic using ARIMA models [1], and based on chaos analysis [9]. Other studies have researched mobile Internet usage, forecasting mobile Internet data traffic using Seasonal ARIMA (SARIMA) models for 2G GSM networks [18], 2G/3G networks [22], and 3G/4G LTE networks [10].

In the area of automatic detection of network traffic anomalies and attacks there exist relevant works, which are closely related to DNS traffic since both general network traffic and DNS traffic are similar, due to how DNS process works. That is why anomalies and attacks are also visible in traffic at DNS level. Most of these studies perform event detection using supervised learning techniques, training the models with specific instances of pre-classified events (attacks or failures) in order to recognize them in future network flows [2].

Also, there are implemented systems that establish certain rules to find different kinds of anomalous events like network intrusion detection [15,17]. Moreover, the problem of detecting network anomalies by using unsupervised learning methods has been studied with the purpose of recognizing and detecting anomalies from unknown threats, by performing outlier detection in network traffic [4].

With respect to DNS, there are studies that analyze and propose supervised methods to detect some specific DNS attacks, such as DoS [7], DDoS [6,11], Domain Fluxing [23], Botnet Domains and Malware [20], and Kaminsky Cache Poisoning [12]. Nevertheless, there are not deep studies with regard to the use of unsupervised learning techniques in DNS data or to the use of big amount of data from real DNS queries. Furthermore, DNS data, that corresponds mostly to UDP packets and uses particular rules, has not yet been exploited to perform analysis, compared to other types of network traffic flows.

Close to the goals of this paper, people from InternetNZ, the registry for the '.nz' ccTLD (Country Code Top-Level Domain) from New Zealand, showed in their blog the use of the Prophet forecasting model [19] to analyze trends on DNS queries for '.nz' domains. Also, they look into their DNS traffic dataset in order to find past anomalies in the stream by visual inspection [16]. Our work proposes the use of Neural Networks models to forecast DNS data traffic and to propose the implementation of an automatic lightweight method to perform early anomaly detection.

3 Data Set Overview

The data used in this work consists of a month of normal operation traffic of one of NIC-Chile's authoritative DNS servers. It starts on 2 October, 2017, until 1 November of the same year. NIC-Chile is the official registry for the '.cl' ccTLD, which is the geographic top-level domain from Chile. Every DNS packet from queries to the server and responses to users are present in the dataset. The server studied belongs to an anycast configuration along with other servers.

A time series of DNS traffic was built by aggregating all the successful server responses into 1-hour intervals. Therefore, each point of the time series corresponds to the amount of DNS packets from server responses with record types 1, 2, 15 or 28 (A, NS, AAAA, MX) obtained in one hour of data. The whole time series represents a total of 249,434,772 DNS packets.

The Fig. 1 below shows a portion of the time series used to train and test the model, corresponding to two weeks of the DNS traffic time series. It is possible to observe some regular patterns on it, represented as periods of one day and one week in the data. This makes perfect sense with the fact that the regularity of human patterns [8,13] is reflected on Internet usage, producing regular patterns as well [21]. Moreover, there are some visible outliers in the time series shown below, and these anomalies in the traffic could be easily detected by comparing the amount of data traffic measured in a real-time flow and the amount expected, obtained by the use of forecast models.

The data was split into training and testing set, leaving the last complete week for testing and the rest for training. That is roughly a 77%/23% distribution.

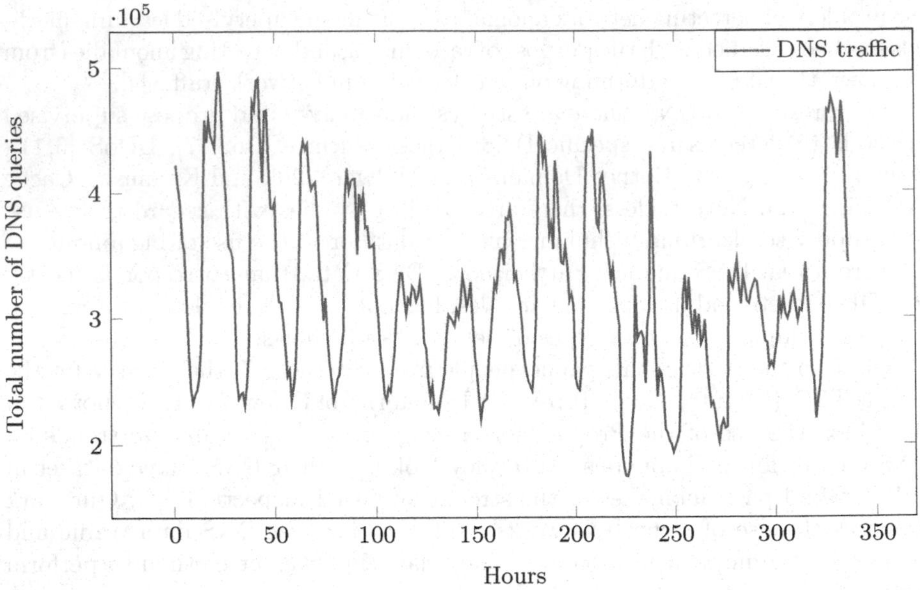

Fig. 1. DNS traffic time series

Since data is real from a normal working of the system, it takes on great importance in the analysis of this work and gives relevance to the results obtained as users patterns are captured in the traffic, showing clear periodicity from day to day.

4 Forecast Models

The selected neural network models follow some commonly used architectures in literature that address time series analysis problems.

To evaluate these forecast models, a more basic forecast model is proposed for establishing a comparison on some forecast errors indicators obtained at predicting DNS traffic. The selected comparison model is Weighted Moving Average.

4.1 Neural Networks

Artificial Neural Network models, based on the natural neural networks of the human brain, consist of the interconnection of several nodes that individually define weights and operations to transmit data over their own connections and thus, over the network. The storage and update of the data perceived on the nodes allow the network training and learning from the input, in order to give logical predictions to future information. The different configuration of the connections between the nodes leads to different types of networks, some of which are commonly used at time series forecasting.

LSTM. Long Short-Term Memory is a type of recurrent neural network. That is, a network whose connections contain loops in order to keep information, passing it through the steps of the network. An LSTM-layer's special feature is that it can retain long-term information and learn when to get or not get it into account. The implementation of an LSTM unit consists of three gates: input gate, output gate and forget gate that are related according to the following equations:

$$c_t = i_t \circ tanh(W_c x_t + V_c y_{t-1} + b_c) + f_t \circ c_{t-1} \tag{1}$$

$$y_t = o_t \circ tanh(c_t), \tag{2}$$

where i_t, f_t, o_t are the respective activation functions of each gate:

$$g_t = sigmoid(W_g x_t + V_g y_{t-1} + b_g), \tag{3}$$

where g is the corresponding gate, W and V are weight matrices, b is a bias vector, x_t and y_t are input and output vectors of the step t, and $\circ$ corresponds to the entry-wise product between two matrices.

CNN-LSTM. Convolutional Neural Network (CNN) differs from a normal artificial neural network at its use of kernels to apply a convolutional operation over the input data in order to transform it and obtain specific information to focus on. The equation of a convolutional layer for a 2-dimensional input is described below:

$$X_{ij} * K_{ij} = \sum_{a=0}^{m-1} \sum_{b=0}^{n-1} w_{ab} X_{(i+a)(j+b)} \cdot K_{ab} + b \tag{4}$$

where X is the input and K is the $m \times n$ kernel. w and b are weight and bias respectively.

CNN-LSTM would be the combination of the two networks mentioned above. It will consist of a convolutional layer before an LSTM layer.

4.2 WMA

Weighted Moving Average, commonly used to smooth out short-term fluctuations and highlight longer-term trends in time series, can be defined as a forecasting model where the next forecast values are calculated as a weighted average of past values, where the weights decrease in arithmetical progression. A weighted moving average with linear decreasing weights and a window k is defined as follows:

$$WMA_n = \frac{y_n \times k + y_{n-1} \times (k-1) + ... + y_{n-k} \times 1}{k + (k-1) + ... + 1} \tag{5}$$

5 Accuracy Measures

In order to compare the obtained results by the different forecast models proposed, it is necessary to establish the metrics to consider during the evaluation process. The selected metrics are divided in two groups:

5.1 Prediction Errors

Firstly, is necessary to evaluate the forecast results with regard to the difference between real and predicted values, for which the following measures are considered:

1. **Root-Mean-Square-Error** (RMSE): Square root of the average of the square of vertical distance between each real value and its forecast.

$$RMSE = \sum_{i=1}^{n} \sqrt{\frac{(r_i - p_i)^2}{n}}$$

2. **Mean Absolute Error** (MAE): Mean of the vertical distance between each real value and its forecast.

$$MAE = \sum_{i=1}^{n} \frac{|r_i - p_i|}{n}$$

5.2 Time Series Distances

In addition to the prediction errors, it is important to take into account other factors when performing an evaluation of the forecast models, mainly because the prediction errors mentioned above do not consider important characteristics of the nature of a time series when comparing them, such as the fact that the points of time series have a logical order, given by the time. Therefore, another way to measure accuracy of forecast results is needed, in order to capture the similarity between the shape of real data and predicted curves or the data distribution. The following distance measures, commonly used for time series clustering, are considered:

1. **Edit Distance for Real Sequences** (EDR): Compares the two time series in terms of how many edit operations (delete, insert or replace) are necessary to transform one curve to the other. The distance between two points is reduced to 0 or 1, where if the distance between two points r_i and p_j is less than a given ϵ, then the points are considered equal.
2. **Dynamic Time Warping** (DTW): It is a time series alignment algorithm that aims at aligning two sequences by warping the time axis iteratively until an optimal match between the two sequences is found, which minimize the sum of absolute differences for each matched pair of indices. An acceptable match must follow four conditions:

- Every index from the first sequence must be matched with one or more indices from the other sequence, and vice versa.
- The first index from the first sequence must be matched with the first index from the other sequence.
- The last index from the first sequence must be matched with the last index from the other sequence.
- The mapping of the indices from the first to the second sequence must be monotonically increasing, and vice versa.

6 Experimental Results

As mentioned in Sect. 3, the DNS traffic time series with information about 1-hour aggregated data is forecasted sequentially hour by hour until completing one week of predicted data, i.e. 168 forecast operations.

Figures 2 and 3 show the forecast results of performing DNS traffic forecast using LSTM and CNN-LSTM methods described in Sect. 4. Also, Fig. 4 shows the results obtained by using the Weighted Moving Average as basic forecast model for comparison.

The accuracy measures obtained for each forecast model, are presented in Table 1.

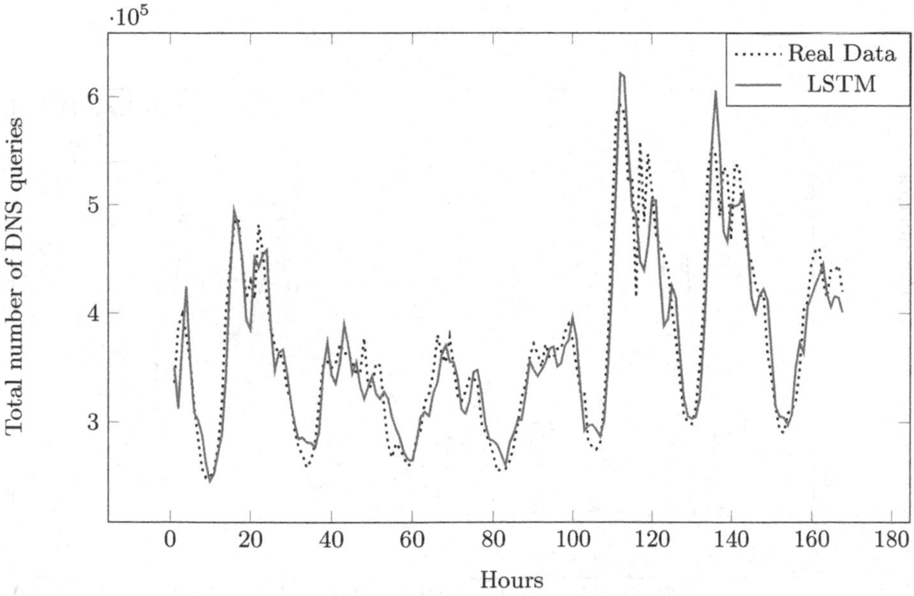

Fig. 2. DNS traffic forecast using LSTM network

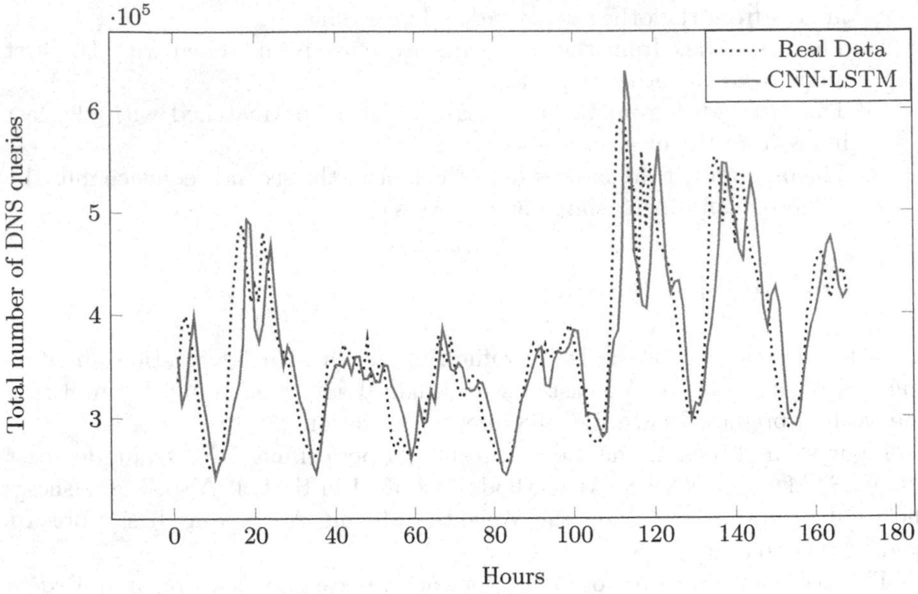

Fig. 3. DNS traffic forecast using CNN-LSTM network

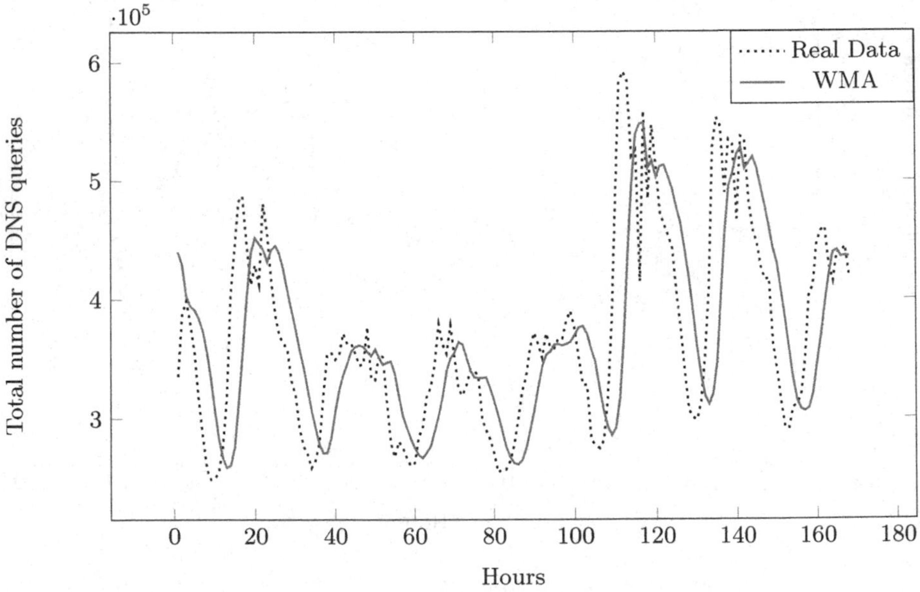

Fig. 4. DNS traffic forecast using Weighted Moving Average

Table 1. DNS traffic forecast errors

Method	RMSE	MAE	EDR	DTW
WMA	68330	50952	68	3910580
LSTM	**28980**	**21902**	60	3710717
CNN-LSTM	48752	34432	**53**	**3531212**

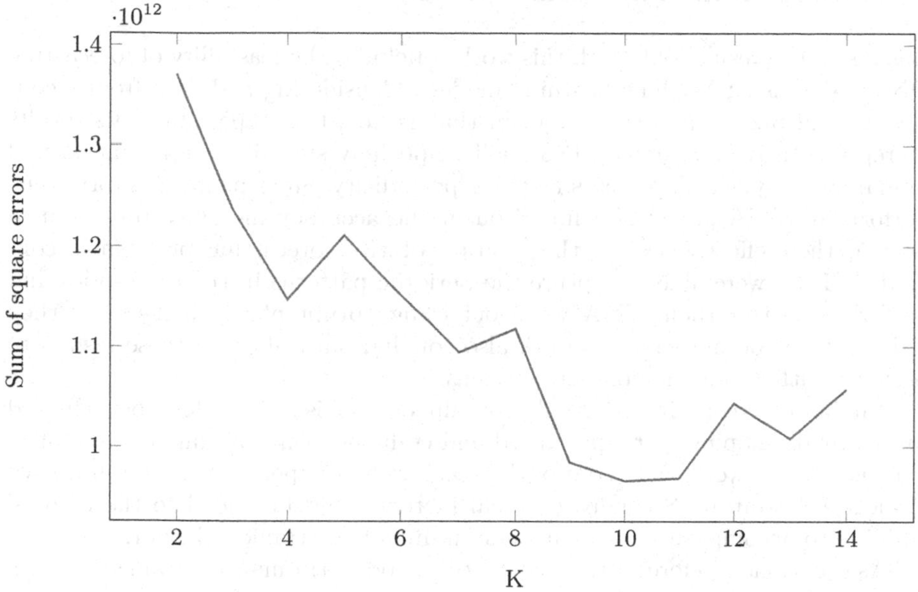

Fig. 5. Elbow method for DNS traffic clustering

7 Discussion

As Figures at Sect. 6 illustrate, all models show positive visual results, as the predicted curves captures the basic periodic pattern present in data. Both LSTM and CNN-LSTM models are capable of predict almost all sudden changes in real data curve thanks to their more risky behavior at forecasting, obtaining results quite fit the real values. In the other hand, WMA results are basically presented as a smoothing of the real curve, without foreseeing the most interesting points of real data. Also, it is also remarkable the fact that WMA model takes a couple of points before detecting the change of phase in the day periodicity, presenting clearly a big delay with respect to the real curve. Both neural network models show a similar behavior, as observed in Figs. 2 and 3, where LSTM demonstrates a slight improve over CNN-LSTM when detecting changes in the periodicity phase.

With regard to the accuracy measures presented in Table 1, neural network models outperformed WMA model in every distance and error measure, proving that results obtained by LSTM and CNN-LSTM have a better accuracy in average and also, that their curves are more similar in shape to the real one. At comparing both neural network models, LSTM obtains lower prediction error results (RMSE and MAE), while CNN-LSTM obtains lower time series distances (DTW and EDR).

8 Conclusions and Future Work

Basing on the results obtained, this work concludes the feasibility of forecasting DNS traffic using Machine Learning models. Considering real data from a constantly and massively queried domain that is '.cl', this paper states its results as representative for similar DNS traffic. Specially since it comes from normal human activity, which presents a strong periodicity. Since neural networks outperformed WMA model of comparison in the accuracy measures presented in Sect. 5, their effectiveness to the solution of this forecasting problem is concluded. They were able to capture the periodic patterns in the time series and performed better than WMA when detecting abrupt phase changes. As they did not focus on outliers, this work also concludes a utility for these results to perform outlier detection on data streams.

An improvement in the results obtained in this work could be achieved firstly, by developing more specialized and complex deep learning architectures. This is a tough work that is out of the reach of this paper as it used well-known models in literature. Secondly, external factors could be added to the models' training to accomplish a better understanding of the traffic behavior.

As mentioned before in this paper, the good performance obtained at forecasting DNS traffic is a very valuable result especially for DNS providers since the forecast models can be easily used to develop a lightweight monitoring tool to continuously analyze traffic in DNS servers, where a big difference between the real amount of DNS data experienced and the forecast levels of traffic could be an early sign of the presence of an anomaly in the data stream produced by an attack or a failure.

Also, as future work it is important to carry out deeper analysis about DNS traffic data, performing smart data aggregation to try to infer more detailed reasons of unexpected traffic anomalies. Possible ways of performing this aggregation is by grouping DNS queries according to their query type (A, AAAA, MX, NS), by applying subnet mask to the sender IP addresses or by grouping queries for similar domains, where domains are considered similar if they are requested in similar way, i.e. their DNS traffic time series are close in distance. By doing this, it could be possible to perform forecasts of different portions of the complete DNS traffic, giving to DNS providers the possibility to detect in which specific portion of the whole stream there is an anomalous event.

In relation to the aforementioned, the first steps have been taken in order to group queries for similar domains, by creating time series with the portion of the DNS traffic according to each requested domain and performing clustering by using time series K-means algorithm with Dynamic Time Warping distance.

Figure 5 shows the results obtained by performing *Elbow method* to find an appropriate number of clusters. According to these results, a good approach for a future work would be to use nine clusters, analyze their content, and to use the forecast models described in this paper to perform more accurate and understandable anomalous event detection.

References

1. Akinaga, Y., Kaneda, S., Shinagawa, N., Miura, A.: A proposal for a mobile communication traffic forecasting method using time-series analysis for multi-variate data. In: IEEE Global Telecommunications Conference, GLOBECOM 2005, vol. 2, pp. 6-pp. IEEE (2005)
2. Alsirhani, A., Sampalli, S., Bodorik, P.: DDoS attack detection system: utilizing classification algorithms with apache spark. In: 2018 9th IFIP International Conference on New Technologies, Mobility and Security (NTMS), pp. 1–7. IEEE (2018)
3. Basu, S., Mukherjee, A., Klivansky, S.: Time series models for internet traffic. In: Fifteenth Annual Joint Conference of the IEEE Computer Societies. Networking the Next Generation. Proceedings IEEE INFOCOM 1996, vol. 2, pp. 611–620. IEEE (1996)
4. Casas, P., Mazel, J., Owezarski, P.: Unsupervised network intrusion detection systems: detecting the unknown without knowledge. Comput. Commun. **35**(7), 772–783 (2012)
5. Cortez, P., Rio, M., Rocha, M., Sousa, P.: Multi-scale internet traffic forecasting using neural networks and time series methods. Expert Syst. **29**(2), 143–155 (2012)
6. Douligeris, C., Mitrokotsa, A.: Ddos attacks and defense mechanisms: classification and state-of-the-art. Comput. Netw. **44**(5), 643–666 (2004)
7. Feinstein, L., Schnackenberg, D., Balupari, R., Kindred, D.: Statistical approaches to DDoS attack detection and response. In: Null, p. 303. IEEE (2003)
8. Gonzalez, M.C., Hidalgo, C.A., Barabasi, A.L.: Understanding individual human mobility patterns. Nature **453**(7196), 779 (2008)
9. Hu, X., Wu, J.: Traffic forecasting based on chaos analysis in GSM communication network. In: International Conference on Computational Intelligence and Security Workshops, CISW 2007, pp. 829–833. IEEE (2007)
10. Miao, D., Qin, X., Wang, W.: The periodic data traffic modeling based on multiplicative seasonal arima model. In: 2014 Sixth International Conference on Wireless Communications and Signal Processing (WCSP), pp. 1–5. IEEE (2014)
11. Mirkovic, J., Reiher, P.: A taxonomy of DDoS attack and DDoS defense mechanisms. ACM SIGCOMM Comput. Commun. Rev. **34**(2), 39–53 (2004)
12. Musashi, Y., Kumagai, M., Kubota, S., Sugitani, K.: Detection of Kaminsky DNS cache poisoning attack. In: 2011 4th International Conference on Intelligent Networks and Intelligent Systems (ICINIS), pp. 121–124. IEEE (2011)
13. Oliveira, E.M.R., Viana, A.C., Sarraute, C., Brea, J., Alvarez-Hamelin, I.: On the regularity of human mobility. Pervasive Mob. Comput. **33**, 73–90 (2016)
14. Papagiannaki, K., Taft, N., Zhang, Z.L., Diot, C.: Long-term forecasting of internet backbone traffic: observations and initial models. In: Twenty-Second Annual Joint Conference of the IEEE Computer and Communications, INFOCOM 2003, IEEE Societies, vol. 2, pp. 1178–1188. IEEE (2003)
15. Paxson, V.: Bro: a system for detecting network intruders in real-time. Comput. Netw. **31**(23–24), 2435–2463 (1999)

16. Qiao, J.: .nz DNS traffic: trend and anomalies (2017). https://blog.nzrs.net.nz/nz-dns-traffic-trend-and-anomalies/
17. Roesch, M., et al.: Snort: lightweight intrusion detection for networks. In: Lisa, vol. 99, pp. 229–238 (1999)
18. Shu, Y., Yu, M., Liu, J., Yang, O.W.: Wireless traffic modeling and prediction using seasonal arima models. In: IEEE International Conference on Communications, ICC 2003, vol. 3, pp. 1675–1679. IEEE (2003)
19. Taylor, S.J., Letham, B.: Forecasting at scale. Am. Stat. **72**(1), 37–45 (2018)
20. Thomas, M., Mohaisen, A.: Kindred domains: detecting and clustering botnet domains using DNS traffic. In: Proceedings of the 23rd International Conference on World Wide Web, pp. 707–712. ACM (2014)
21. Wang, H., Xu, F., Li, Y., Zhang, P., Jin, D.: Understanding mobile traffic patterns of large scale cellular towers in urban environment. In: Proceedings of the 2015 Internet Measurement Conference, pp. 225–238. ACM (2015)
22. Xu, F., et al.: Big data driven mobile traffic understanding and forecasting: a time series approach. IEEE Trans. Serv. Comput. **9**(5), 796–805 (2016)
23. Yadav, S., Reddy, A.K.K., Reddy, A.N., Ranjan, S.: Detecting algorithmically generated domain-flux attacks with DNS traffic analysis. IEEE/ACM Trans. Netw. **20**(5), 1663–1677 (2012)

Energy-Based Connected Dominating Set for Data Aggregation for Intelligent Wireless Sensor Networks

Basem Abid[1], Sarra Messai[1,2(✉)], and Hamida Seba[1]

[1] Université de Lyon, CNRS Université Lyon 1, LIRIS, UMR5205, 69622 Lyon, France
{basem.abid,sarra.messai,hamida.seba}@liris.cnrs.fr

[2] Département d'informatique, Faculté des sciences, Université Ferhat Abbas Sétif 1, 19000 Sétif, Algérie

Abstract. The main mission of deploying sensors is data collection and the main sensor resource to save is energy. For this reason, data aggregation is an important method to maximize sensors' lifetime. Aggregating sensed data from multiple sensors eliminates the redundant transmissions and provides fused information to the sink. It has been proved in the literature that a structure based data aggregation gives better results in terms of packet delivery and energy saving which prolong the network lifetime. In this paper, we propose a novel approach called Distributed Connected Dominating Set for Data Aggregation (DCDSDA) to construct our network topology. The sensors of the network compute in a distributed way and based on the residual energy of each sensor, a connected dominating set to form a virtual backbone. This backbone forms a tree topology and as it is computed and maintained in a distributed way based on predefined energy constraints, it represents an intelligent fault tolerance mechanism to maintain our network and to deal with packet loss. The simulation results show that our proposed method outperforms existing methods.

Keywords: Sensor networks · Data aggregation · Virtual backbone network · Connected dominating set (CDS)

1 Introduction

In the last few years, networks of wireless sensors are used in diverse domains and have gained much attention within the research community. A sensor network consist of a large set of highly-resource constrained sensors (called also nodes). Such a network is able to support various real world applications [16,25] like habitat monitoring, building monitoring, military survival lances and target tracking, ... etc. Generally, for most of these applications, sensors are required to

This work was supported by PHC TASSILI 17MDU984.

E. Renault et al. (Eds.): MLN 2018, LNCS 11407, pp. 193–211, 2019.
https://doi.org/10.1007/978-3-030-19945-6_13

sense their environment, detect some events and report sensed data to a special node called a sink.

The collection of these data must be efficient as it is an important issue in sensor networks due to the resource-constraint of sensors. In this paper, we propose several algorithms to achieve this requirement.

In the literature, several proposals for better power efficiency have been suggested. Some of them organize the channel access by the sensor nodes to reduce the packet retransmission due to the packet collision [3,26], others propose some compression techniques [6,23] to reduce the amount of transmitted data. Finally, some works propose to reduce the number of transmissions by aggregating packets which is the scope of our work.

Data aggregation is defined as the process that aims to reduce the energy consumption by lowering the number of transmissions [9,14], this task is achieved by some intermediate nodes in the network. These nodes may perform some operations (e.g. sum, maximum, minimum, mean, etc.) or basically concatenate the received data packets before forwarding it to an upper level in the network architecture.

In this paper, we propose an intelligent structure-based data aggregation approach in which the sensors build and maintain the ctructure without centrelized supervision. Therefore, our algorithm computes a Connecting Dominating Set (CDS) based on the residual energy of each sensor.

Our solution is built in two steps. The first step constructs the dominating set (DS) in a distributed way. A graph $G = (V, E)$ is a set of vertices V and edges E. A dominating set of graph G is a subset $S \subseteq V$ of its nodes such that for all nodes $v \in V$, either $v \in S$ or a neighbor u of v in S. It is well known that computing a dominating set of minimal size is $NP-hard$ [18]. We therefore provide an approximation algorithm. The second step of our algorithm selects additional nodes to construct a connected dominating set (CDS). The CDS of a graph $G = (V, E)$, is a subset S of vertices such that the subgraph induced by S is connected. This means that any node in S can reach any other node in S by a path that is entirely in S. The second property that must be verified as well is the following: The subset S forms a DS. This problem is known to be $NP-hard$ [18]. If the CDS is with a minimal cardinality it is called a minimum connected dominating set (MCDS). The final CDS constructed by our approach forms a tree routed at the Sink.

The set of nodes in the CDS (aggregator nodes) are those with the best energy level and all the sensors contribute together to elect this set of aggregator nodes. Therefore, if an aggregator node fails (no more energy is available) it will never be elected during the next setup phases. The main advantage of developing a distributed solution rather than a centralized one at the sink node is that the sink has to gather new topology information from the network to build the CDS and the gathering tree at each round time. When the size of the CDS is small, we simplify the network control operations and confines routing operations to a few nodes set leading to advantages such as energy efficiency and low latency.

In this work we minimize the number of nodes in the CDS based on the degree of each node as we will see later, thus we combine two constraints to build our network architecture, residual energy and node degree. To the best of our knowledge, most of the previous algorithms are centralized or needs special nodes to set up the network. Unlike other methods, our algorithm constructs the network topology without any knowledge of the node distribution inside the area and without any leader node to initiate the construction of the CDS or the tree. The sink has just the role to trigger the construction of the network topology and discover the first level nodes with whom it may receive the aggregated data. The gathering tree is computed as a second step based on the CDS result where the aggregator nodes compose the main nodes of the tree and we attach a level to each of them, thus we guarantee that each node is placed as near as possible to the sink, therefore, the packet-delay is reduced.

The rest of the paper is organized as follows: Sect. 2 is about some related work. The proposed solution is explained in Sect. 3. The performance of our algorithms under various network configurations are discussed in Sect. 4. Section 5 concludes the paper.

2 Related Work

In this section we present some data-aggregation protocols and some algorithms used to construct a CDS and gathering tree.

2.1 Data Aggregation

Several protocols were proposed for data aggregation and most of them implement a static structure, such as cluster-based approaches and tree-based approaches. For the first approach, some nodes in the network are elected as cluster-heads, they aggregate data and transmit the result to some intermediate nodes or directly to the sink. Low Energy Adaptive Clustering Hierarchy (LEACH) [9] was the first clustering algorithm in sensor networks. Nodes in the network randomly determine whether they are cluster-heads or not. The communication between the cluster members and the cluster-head is scheduled using TDMA protocol (Time Division Multiple Access). LEACH protocol provides a balancing of energy usage by random rotation of cluster-heads.

Two-level LEACH (TL-LEACH) [15] is an improvement of LEACH protocol. It processes primary and secondary-level cluster-heads selection to reduce the energy consumption. The HEED protocol [27], further improves the algorithm by considering the residual energy of sensors and intra-cluster communication costs while making the selection of the cluster-heads in multi-hop sensor networks.

In [11], the authors propose an energy efficient multi-level clustering (EEMC) which organizes the nodes into a hierarchy of clusters with the objective of minimizing the total energy spent in the network. Part of the cluster-heads election is centralized at the sink which collects location information and the residual energy collected from the nodes in the network. In [12], Khamfroush et al.

propose a tree-structure for data aggregation. Nodes inside a cluster are organized into a tree and the tree structure is used as the transmission path from the cluster-heads to the sink.

Tan et al. propose a power efficient data gathering and aggregation protocol (PEDAP) [19]. PEDAP is a minimum spanning tree based protocol computed at the sink. The goal of PEDAP is to maximize the lifetime of the network in terms of number of rounds.

A part of our proposed topology is considered as a cluster-based with a tree structure to connect the different cluster-heads to the sink. However, most of the proposed approaches suffer from the centralized way to elect the cluster heads and to compute the different paths to the sink with the energy constraint and the dynamic topology of the sensor networks.

2.2 Connected Dominating Set

In the past few years, many CDS-related algorithms have been proposed. These algorithms can be classified as follows: (1) the centralized approach which it starts with a full knowledge of the network topology. This assumption is not practical for sensors networks. (2) the decentralized approaches: it includes a set of distributed algorithms where the decision is decentralized and/or depends on information from a leader node.

Using a CDS approach in sensor networks *reduces the routing overhead* [24]: The non-backbone nodes conserve their links only with the backbone-node and ignore all the others, therefore, the maintenance cost of the routing table is reduced and by using only the backbone node to forward broadcast packets, the excessive broadcast redundancy can be avoided. In addition, the node's *energy consumption can be significantly reduced* by putting the non-backbone nodes in sleep mode and the network connectivity is maintained by backbone nodes [2].

In [4,13], the authors propose a centralized algorithm to construct a CDS, however, these algorithms require a global information of the network, thus, these approaches are applicable only in static networks. In [13], authors, investigated the problem of constructing quality CDS in terms of size, diameter, and average backbone path length, and propose two centralized algorithms with constant performance ratios for the size and diameter of the constructed CDS.

In the literature, we found many decentralized algorithms either localized or distributed. For a localized algorithm, the decision process is distributed, and requires only a constant number of communication rounds. The localized algorithms are divided in two main categories: probabilistic or deterministic. In this kind of algorithms the status of a given node in the network (backbone or non-backbone) depends on the neighborhood information. A typical probabilistic algorithm is the gossip-based algorithm [10]. In [10], the selection of the set of backbone nodes depends on a probability p. When p is larger than a threshold, the backbone nodes form a CDS with high probability. This threshold is determined based on experimental data. The produced backbone is usually large because the selection of p is conservative due to the unpredictable network conditions.

In [1], authors propose an algorithm used as a topology control, but it can be extended for virtual backbone construction. This algorithm is a grid based, where the network is with non-uniform node distribution and every node has β backbone neighbors, then all backbone nodes form a CDS with high probability. The value of β is determined based on experimental data. In [22], the authors propose a Local algorithm (L-DCDS) to select forwarding nodes and forwarding edges at the same time. The algorithm uses a pruning rule to obtain a minimum directional connected dominating set. In [20], the authors propose a distributed algorithm to construct a CDS. It constructs a rooted spanning tree, then a maximal independent set (MIS) and finally a dominating tree based on the MIS. The maximal independent set is an independent set such that adding any node not in the set breaks the independence property of the set. Thus, any node not in the MIS must be adjacent to some nodes in the set.

In the above discussed algorithms, the authors propose different techniques to construct the CDS without considering the energy level of each node who becomes a backbone-node and taking into account the different energy consuming tasks of these kinds of nodes, thus, during the election of the special nodes our proposed algorithm gives a high priority to the residual energy of sensors without expanding the CDS size while retaining decentralization which is the main characteristic of intelligent systems.

3 Our Proposal

This section mainly focuses on the intelligent formation of a connected dominating set by the sensors. The main objective of this paper is to combine the features of the CDS with the data aggregation technique. Some desirable features are: (1) The formation process should be distributed. (2) The resulting CDS should be close to minimum and forms a tree. (3) The CDS nodes are the aggregator nodes.

We assume a connected sensor networks and all the nodes have the same transmission range. Hence, we model a sensor network as an non oriented graph $G = (V, E)$, where V is the set of n sensors denoted by v_i $(i \leq i \leq n)$, i is the node ID of v_i; E represents the link set, $\forall\ u, v \in V, u \neq v$, there exists a link (u, v) in E if and only if u and v are in each other transmission range. We also assume that the links are bidirectional, which means that two linked nodes are able to transmit and receive data from each other. We divide the time frame into different time intervals as described in Fig. 1 and for each time interval a specific task is executed.

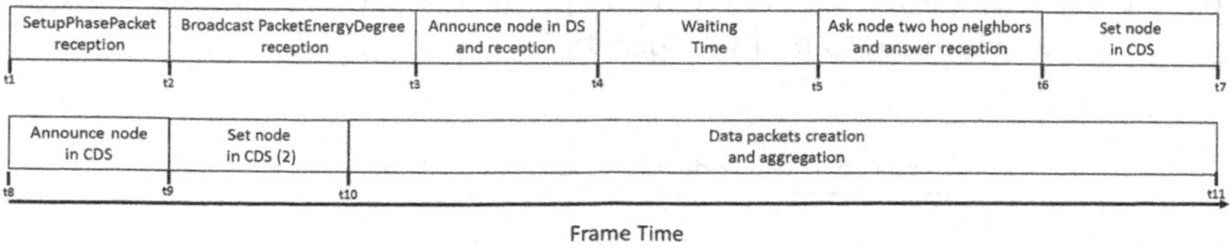

Fig. 1. Frame time composition

3.1 Distributed DS with Energy Constraint (DDSEC)

In this section we describe our algorithm to compute the DS in a connected graph $G(V, E)$ with energy constraint. We assume that each sensor in G has an initial energy greater than a threshold E_{th}. The sink node is without energy constraint. The residual energy of each sensor is an important factor to decide whether that sensor could belong to the DS or not:

$$u_i \in DS \; iff \; E_{R_{u_i}} \geq E_{th}$$

($E_{R_{u_i}}$ *is the residual energy of sensor* u_i $1 \leq i \leq n$)

To build the network, we suppose that the sink broadcasts a *StartSetupPhase* control packet and all the sensors are initially WHITE. The steps of the algorithm are as follows:

1. During the *SetupPhasePacket reception* period (Fig. 1), if a node $v \in V$ receives the *StartSetupPhase* packet directly from the sink and its residual energy is greater that E_{th} it becomes a BLACK node (a dominator), then it creates its own *StartSetupPhase* packet and forwards it to the other nodes. The *StartSetupPhase* packet fields are described in Fig. 2. We note that in this case the node v directly linked to the sink receives a $NodeLevel = 0$, therefore, its level becomes 1.

Node addr	Node level

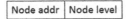

Fig. 2. StartSetupPhase packet

2. If node $v \in V$ receives the *StartSetupPhase* packet from a node $v' \in V$, the node v changes its level if the following criterion is satisfied:

$$CurrentL > R_{cv}L + 1$$

   - $CurrentL$ is the current level of node v
   - $R_{cv}L$ is the received node level of node v'

   When the above criteria are satisfied the node v creates its own *StartSetupPhase* packet and forward it to the other nodes.
3. During the *Broadcast PacketEnergyDegree reception* period (Fig. 1), each node $v \in V$ computes its degree based on the different *StartSetupPhase* packet received during the previous step, then it creates an *EnergyDegree* packet as shown in Fig. 3 and broadcasts it.

Node addr	Node degree	Node residual energy

Fig. 3. EnergyDegree packet

4. If the node v has a sufficient E_R, it compares its degree $d(v)$ with the different one-hop neighbors. The node v behaves as follows:
   - If v has $E_R \geq E_{th}$ and $d(v) > \delta(N[v])$, it becomes a BLACK node ($\delta(N[v])$ is the maximum degree of all the closed neighborhood of N of v with sufficient energy).
   - if v has $E_R \geq E_{th}$ and $d(v) = \delta(N[v])$, v becomes a BLACK node only if $E_R(v_i) > E_R(N[v]), 1 \leq i \leq n$. ($E_R(N[v])$ is the maximum residual energy of the nodes with the same degree as v).
   - if $E_R < E_{th}$ the node v becomes a GRAY node.
5. During the *Announce Node In DS and reception* period (Fig. 1), all the BLACK nodes announce their new situation to their one-hop neighbors by sending an *AnnounceNodeInDs* packet.
6. If v is not a BLACK node and receives the *AnnounceNodeInDs* packet it becomes GRAY node.
7. If the node v does not receive an *AnnounceNodeInDs* packet, during the *waiting time period* (see Fig. 1) it waits for a random time, then it becomes a BLACK node and sends an *AnnounceNodeInDs* packet. This waiting event is interrupted if any announcement is received in meanwhile and v becomes GRAY. We interrupt this process to minimize the final number of nodes in the DS.

To illustrate the DDSEC algorithm, Fig. 4 gives an example of how the DS nodes are elected:

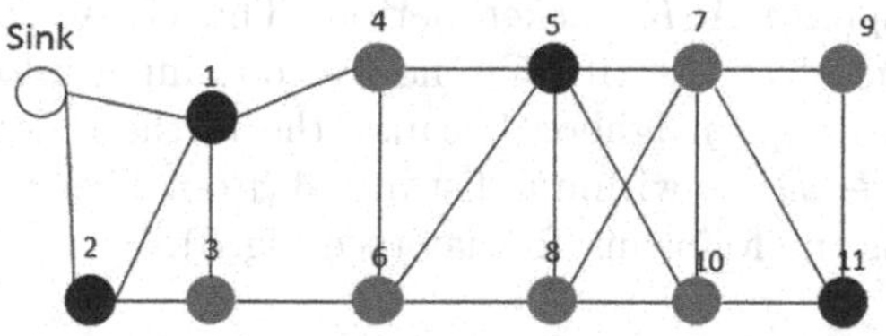

Fig. 4. 1, 2, 5 and 11 form the DS

1. Node 1 and 2 become BLACK because they are in direct link with the sink and their level is equal 1. The remaining nodes receive the *StartSetupPhase* packets and compute their level.
2. The different nodes exchange the *EnergyDegree* packet.
3. We suppose that node 5 has a residual energy $E_R = 50$ *units* and node 7 has a residual energy $E_R = 40$ *units* and the $E_{th} = 20$ *units*, therefore, both nodes have the possibility to become a BLACK node. In this case, 5 becomes a BLACK node as it has the best residual energy with respect to 7.
4. 3, 4, 6, 7, 8 and 10 will receive the *AnnounceNodeInDs* packet and they become GRAY nodes.

5. 9 and 11 will not receive the announcement so during the *waiting time* period (see Fig. 1) they wait a random time, then they generate a *AnnounceNodeInDs* packet. In our example we suppose that 11 generates the packet first (its random waiting time is smaller than the waiting time of 9), so it becomes a BLACK node and 9 get interrupted and became GRAY node.

Nodes 1, 2, 5 and 11 form the dominating set.

3.2 Distributed CDS with Energy Constraint (DCDSEC)

In this section we present our algorithm to construct the CDS in a connected graph G with energy constraint. The set of nodes elected by DDSEC algorithm are automatically included in the final CDS, therefore, the main task of DCDSEC is to add several nodes to the initial DS such that all nodes in DS are connected with at least one path.

To select these nodes our method is mainly based on two criteria:

- The node degree.
- The node residual energy E_R.

The steps of our algorithm are as follows:

1. During the *Ask node two hop neighbors and answer reception* period (Fig. 1), each BLACK node (i.e. node in DS) broadcast a packet control called *AskNodeForNeighbors*. The GRAY node will store all the received packets and wait for the *Response ASK packet* period (see Fig. 1) to answer each of them.
2. During the *Response ASK packet* period, The GRAY node must send a *ResponseAsk* packet to the BLACK nodes containing information about its one hop and two hops neighbor because the farthest neighbor in DS of a BLACK node $v \in DS$ is within a distance 3 (from Theorem 1). The *ResponseAsk* packet has the following format (see Fig. 5).

Node addr	Node degree	Node residual energy	Addr destination	List of BLACK neighbors and the hop distance

Fig. 5. Response Ask packet

3. If the GRAY node has a direct BLACK neighbor its answer is as in Fig. 6(a), otherwise the "*List of BLACK neighbors and the hop distance*" field of the *ResponseAsk* packet is more complex as represented in Fig. 6(b).
4. The BLACK nodes store all the answers and wait for the *Set Node In CDS* period to decide which nodes should be elected in the final CDS. This decision procedure will be explained later in this paper (Sect. 3.2).

5. During the *Set Node In CDS* period, the BLACK nodes send *AddNodeInCDS* packet to the set of nodes elected by the algorithm of the previous point. For each selected node α, the received *AddNodeInCDS* packet contains the list of BLACK node that α must connect. If this list contains BLACK nodes 2-hop far from α (from Theorem 1), therefore, during the *Announce node in CDS* period, α waits for the announcement packet coming from a one-hop neighbor and connecting α to the BLACK node δ. Otherwise, if α does not receive the announcement, it waits for the *Set Node In CDS (2)* period and selects a node β from its one-hop neighbor list that may connect α to β and sends a *AddNodeInCDS* packet. The selection algorithm is the same as in point (4) and explained in Sect. 3.2.
6. If a GRAY node receives a *AddNodeInCDS* packet it changes its status to BLACK and announce it to its neighbors during the *Announce node in CDS* period.

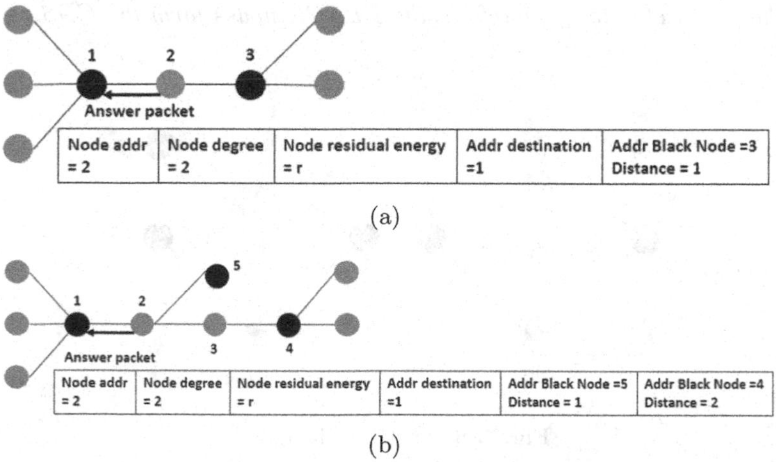

Fig. 6. Example: Response Ask packet

Select Node in CDS. This algorithm aims to select the nodes to be integrated in the CDS as well as minimizing the CDS size. We explain this algorithm with an example shown in Fig. 7, so if w, x and y are GRAY nodes and u, v and z are BLACK nodes distributed as in Fig. 7(a) and (b). The same code is executed in all the BLACK nodes, thus the selected node in CDS will be the same by any BLACK node (u, v, z).

In Fig. 7(a) node u, will get a *ResponseAsk* packets from w, x and y (the same for v). Between these three nodes, the node u will select one of them based on the following criteria:

1. The node with sufficient energy is selected.

2. If many nodes satisfy the above condition we select the one with the max number of BLACK nodes as neighbors (The node w in Fig. 7(b)).
3. If the above conditions are satisfied by many nodes, the one with the best degree is selected.
4. Finally, when all the competitor nodes have the same properties described above, the node with the lowest Id will be selected.

Theorem 1. *If the original network is connected, the BLACK nodes form a CDS of the network.*

Proof 1. *According to Theorem 1, for each BLACK node α, there is at least one BLACK node β where the $dist(\alpha, \beta) \leq 3$, therefore, to connect α to β we need at most two sensors, which means that α selects the first node δ and β selects the second node ω. If ω is not a one hop neighbor of δ, which is the worst case, δ will not receive the announcement packet, thus, it selects the appropriate node during the set node in CDS (2) period to construct a path to β (the same for ω), hence, $\forall\ \alpha, \beta$ BLACK nodes and $dist(\alpha, \beta) \leq 3$ there is at most two BLACK nodes that connect α to β, therefore, the BLACK nodes form the CDS.*

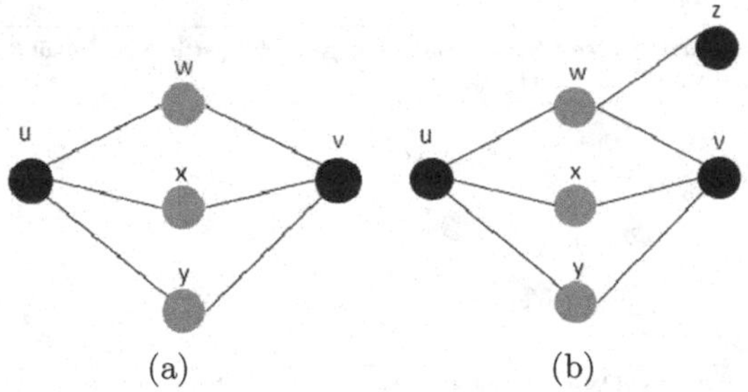

Fig. 7. Select BLACK node

The above criteria to select the node in the CDS represent our strategy to minimize the number of nodes in this set, as choosing the node with the max degree and with the max BLACK neighbors will avoid the selection of additional useless nodes.

For a more complex example where the nearest BLACK neighbor to v ($v \in DS$) is within 3-hop distance as represented in Fig. 8(a), the final result will be as shown in Fig. 8(b).

Node 2 and 3 are competitors and both have the same degree and the same number of BLACK neighbors, but node 2 has the lowest Id, thus, 2 is elected. Additionally, 2 is the unique node who can connect node 1 to node 6. Node 3 and 8 are competitors to connect node 1 to node 5. Node 3 is elected as it has the best degree with respect to node 8. Node 4 is elected by 5.

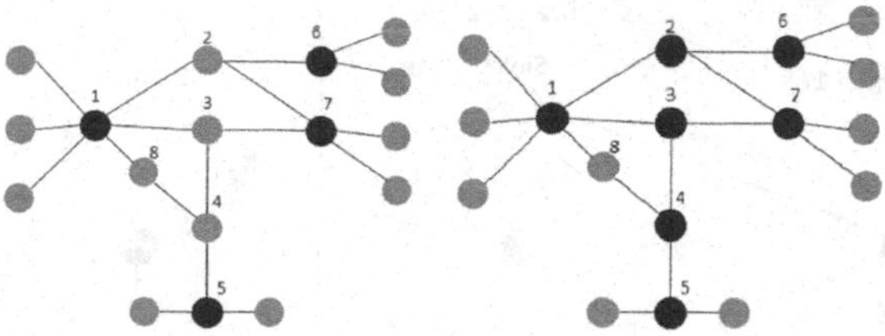

Fig. 8. Example: select BLACK nodes

3.3 Tree Construction

This task consists of computing the parent of each node. During the DS and CDS computation each node holds information about its neighbors (status, level, degree, E_R). For each node $v \in V$ in $G(V, E)$, we locally compute the parent of v noted $P(v)$.

$P(v)$ must satisfy the following conditions:

- $E(P(v)) > E_{th}$ ($E(P(v))$ is the residual energy of $P(v)$)
- $P(v)$ is a BLACK node
- $L(P(v)) < L(v)$ (L is the node level and $L(P(v))$ is as minimum as possible)

We attribute the node v to a parent $P(v)$ with a minimum level in order to minimize the packet delay. We note that all the nodes directly linked to the sink ($dist(v, sink) \leq R_c$, R_c is the communication range) have $L(v) = 1$ and they are BLACK nodes, therefore, we construct the branches of the tree as described in Fig. 9. Any node in the tree (except nodes at level 1) can change their branch during the *data transmission* period as we will see in the next section.

3.4 Data Aggregation

The aggregation task starts during the *Data packet creation and aggregation* period (see Fig. 1). BLACK nodes represent the aggregator sensors. They aggregate the maximum of data and they forward it hop by hop to the sink node. The aggregation period is divided into two repetitive periods as described in Fig. 10. Our approach is for event-driven application, thus, if the node v detects an event, it creates a data packet, and forward it to its parent. During the *random time for aggregation*, the GRAY node sends the data packet to its upper level parent using a CSMA/CA as Medium Access Control protocol (MAC).

Each node $v \in V$ has three trials one at each *random time for aggregation* period for GRAY nodes and one at each *forward data to the upper level* period for BLACK nodes to send their data packets to an upper level. The second and third trials are used only if the ACK (Acknowledgement) packet is not received by the nodes who send the data packet. After the third trial, the node $v \in V$ tries to find a new parent in its neighbor list. First, it tries to find a Parent with

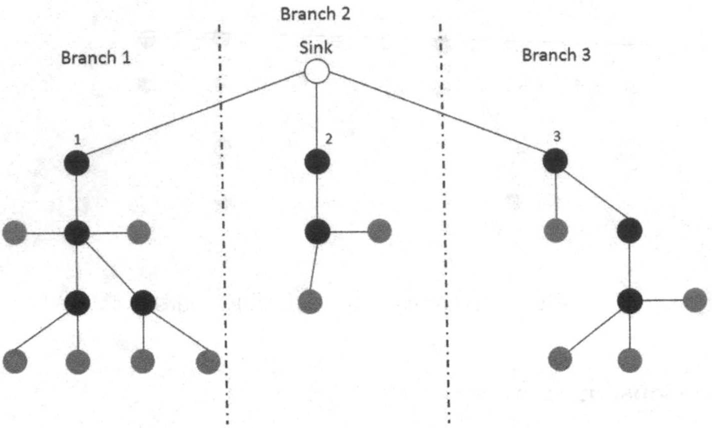

Fig. 9. Tree structure

$L(P(v))$ $L(v)$ to keep the packet delay low, otherwise, it tries to find a new parent with $(L(P(v)) \geq L(v)$ keeping the parent level as minimum as possible. We apply this strategy of parent redistribution to overcome the problem of radio link instability and to get a fault tolerant network with dynamic tree topology. With this approach we increase the packet delivery ratio to the sink.

Random time for aggregation 1	Forward data to upper level 1	Random time for aggregation 2	Forward data to upper level 2	...	Random time for aggregation n	Forward data to upper level n

Fig. 10. Aggregation frame time

In the example shown in Fig. 11(a), nodes 7 and 8 sends their data packets during *random time for aggregation 1* period to their parent (node 4). During the same period, nodes 3, 5 and 6 send also their packet to their parents as shown in Fig. 11(b). During the next period of *forward data to upper level 1*, each BLACK node aggregates its own data with the received packets during the previous period and sends the result aggregated packet to its BLACK parent as shown in Fig. 11(b) and (d). In Fig. 11(c) nothing happens as there is no data packet to send from a GRAY node to a BLACK node.

4 Evaluation

In this section, we evaluate the performance of our solution via Cooja cross-level network simulator [17]. Cooja is an open discrete event simulation package for the Contiki operating system [5].

The sensors are randomly deployed in a square area with side length (L). The results of our proposed solution are compared with the MCDS construction

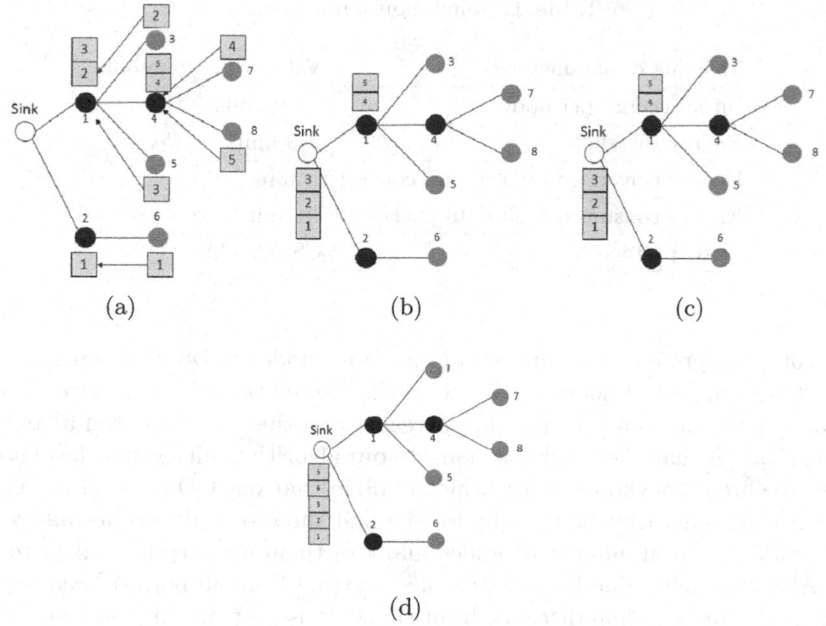

Fig. 11. Sending packets to the sink

algorithm [21] denoted by MCDS, the LBCDS-GA algorithm proposed in [7] denoted by GA and the LBVB algorithm proposed in [8] denoted by LBVB. To compare these algorithms, we use the following metrics: the CDS size and the remaining energy over the whole network. For the last simulation we run our algorithm under different number of generated events and we compute the delivery ratio for our aggregation solution.

Before the simulation begins, we make sure that all the sensors in the network are connected to avoid some isolated sensors which affect the simulation results.

Table 1 summarizes all the network configuration used in simulation. The MAC protocol used in our simulation for both CDS creation and the data transmission period is the CSMA/CA protocol. For both packet control and data packet the energy consumed is the same for both transmission and reception phase.

Three parameters are considered to build the different simulation scenarios: the sensor transmission range, the total number of deployed sensor in the square area, and the side length of the square area.

4.1 CDS Size and Remaining Energy with Different Number of Nodes

The transmission range of the different nodes is equal to 50 m and they are randomly deployed in a square area of 300 m × 300 m. We increment the number of nodes from 50 to 100 by 10.

Table 1. Simulation parameters

Meaning of parameter	Value	Symbol
Initial energy per node	1000 units	E
Energy threshold	200 units	E_{th}
Energy consumed by a packet control	1 unit	
Energy consumed by a data packet	10 units	
MAC protocol	CSMA/CA	

Figure 12 represents the number of backbone nodes (CDS size) with respect to the total number of nodes in the network. We see that the CDS size slightly increases when the number of nodes increases and that our proposed algorithm performs better than the other solutions. In our algorithm, all the first level nodes of the tree are a backbone nodes, which explain that our CDS size grows faster than the other algorithms especially for the first three simulations because when we increase the total number of nodes many of them are directly linked to the sink. Although this value becomes stable (starting from 80 nodes) because less nodes are within one-hop distance from the sink as its transmission range does not change in this scenario. Figure 13 shows the remaining energy over the whole

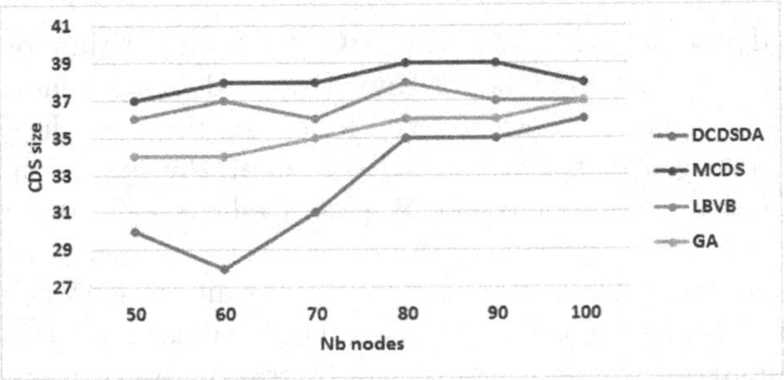

Fig. 12. CDS size vs Number of nodes

network. When the number of nodes increases the remaining energy increases as well for all the algorithms. For the different number of nodes we observe that our approach performs better than MCDS, LBVB, and GA because in DCDSDA initially we select the node with the best remaining energy to be a BLACK node in the dominating set, hence during the second step of our algorithm most of the task is executed by nodes with the best energy to elect the nodes in CDS which conserve energy in the whole network.

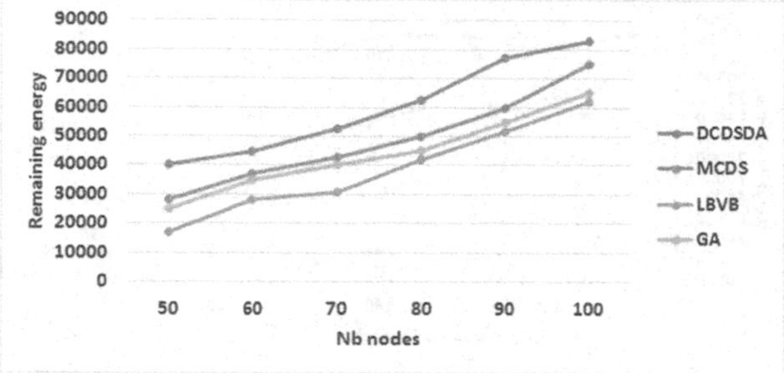

Fig. 13. Remaining energy vs Number of nodes

4.2 CDS Size and Remaining Energy with Different Side Length of the Square Area

In this scenario, the side length of the square area is ranging from 100 to 150 and incremented by 10. We attribute the same transmission range of all the sensors ($R = 20$) and we randomly deploy 100 nodes in a square area. In Fig. 14, we see that, for all the algorithms, the total number of nodes in the CDS increases when the area side length increases because when the deployment region becomes larger, more nodes are needed to maintain the connectivity of the CDS. The CDS size of our DCDSDA approach is close to the CDS size obtained with MCDS, LBVB and GA approaches. We do not obtain better results because as explained in the previous scenario, all the sensors at level one are BLACK nodes.

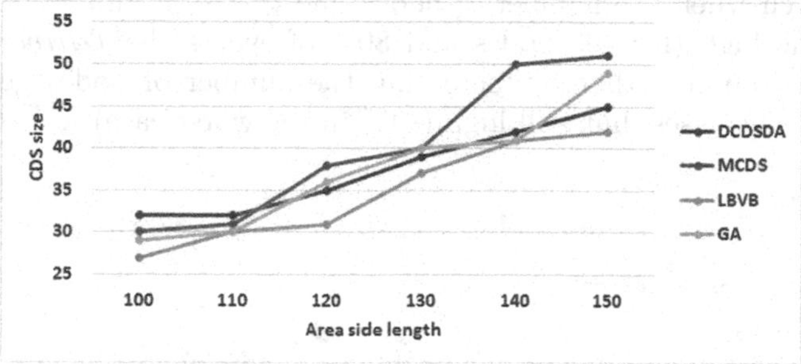

Fig. 14. CDS size vs Area side length

Figure 15 shows that the remaining energy in the whole network decreases when the area side length increases because more packets are sent to select the backbone nodes. We also observe that our method performs better than the three other methods and the remaining energy decreases, but still stable compared to MCDS, LBVB and GA. For all the simulations the DCDSDA has better remaining energy with respect to the other methods for the same reasons of the previous scenario.

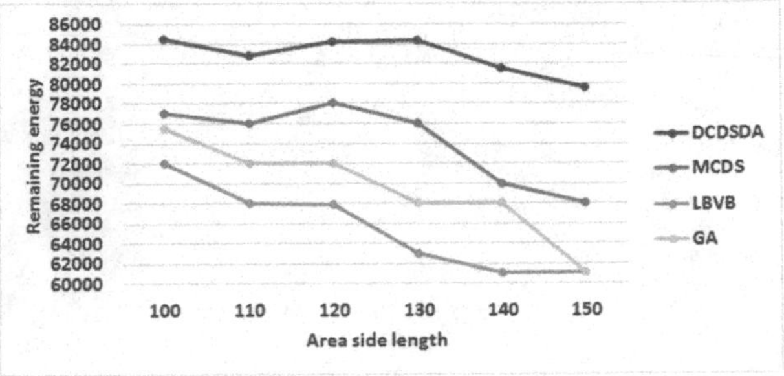

Fig. 15. Remaining energy vs Area side length

4.3 Delivery Ratio

We evaluate the delivery ratio of our protocol with different number of nodes ranging from 50 to 100. The delivery ratio shows the percentage of data packets transmitted and that were successfully received by the sink. The Delivery Ratio is only considered for data packets and computed as follows:

$$DeliveryRatio = \frac{nb_{received}}{nb_{sent}} \times 100$$

The nb_{sent} is the number of events generated by the sensors and it is proportional to the number of nodes deployed in the network, hence if we deploy 50 nodes and we generate 20% of events it means that $nb_{sent} = 10$.

Figure 16 illustrates the obtained results. We can see that our protocol performs well even when the number of nodes and the total number of generated events become high (for 100 nodes and 80% of events the $DeliveryRatio$ = 85%). We also observe that by increasing the number of nodes generally the delivery ratio decreases, but still high (84% in the worst case).

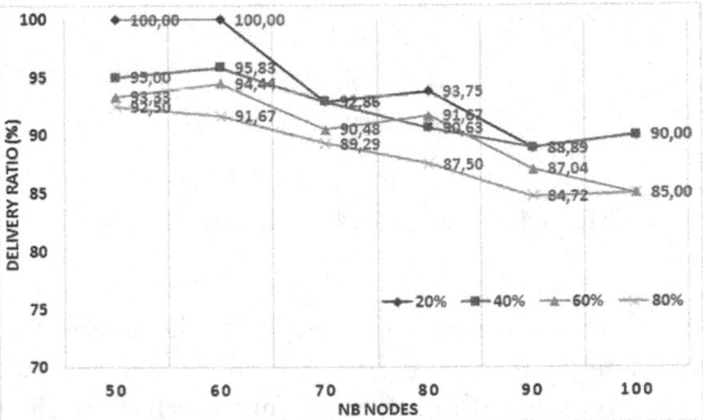

Fig. 16. Delivery ratio vs % generated envents

5 Conclusion

The use of a backbone in a sensor network reduces the communication overhead and decreases the overall energy consumption. However, existing backbone-based solutions are centralized and depend on some particular nodes. In this paper, we provide an intelligent CDS-based backbone construction where the sensors construct the network structure in a distributed manner relying on their energy budget as a key element. Our prime conclusion is that our proposed algorithm is a promising solution for data aggregation in sensor networks in terms of energy consumption and delivery ration. We compare our method with others in the literature and we observe that it performs better especially in terms of energy consumption which proves that considering the remaining energy at each sensor while selecting the CDS nodes reduces the energy consumption. We also show that we obtain a good delivery ratio with our method.

References

1. Blough, D.M., Leoncini, M., Resta, G., Santi, P.: The k-neigh protocol for symmetric topology control in ad hoc networks. In: Proceedings of the 4th ACM International Symposium on Mobile Ad Hoc Networking & Computing, MobiHoc 2003, pp. 141–152. ACM, New York (2003)
2. Chen, B., Jamieson, K., Balakrishnan, H., Morris, R.: Span: an energy-efficient coordination algorithm for topology maintenance in ad hoc wireless networks. Wirel. Netw. **8**(5), 481–494 (2002)
3. van Dam, T., Langendoen, K.: An adaptive energy-efficient MAC protocol for wireless sensor networks. In: Proceedings of the 1st International Conference on Embedded Networked Sensor Systems, SenSys 2003, pp. 171–180. ACM, New York (2003). https://doi.org/10.1145/958491.958512
4. Das, B., Sivakumar, R., Bharghavan, V.: Routing in ad hoc networks using a spine, pp. 1–20 (1997)
5. Dunkels, A., Gronvall, B., Voigt, T.: Contiki - a lightweight and flexible operating system for tiny networked sensors. In: 2004 29th Annual IEEE International Conference on Local Computer Networks, pp. 455–462, November 2004
6. Ganesan, D., Greenstein, B., Estrin, D., Heidemann, J., Govindan, R.: Multiresolution storage and search in sensor networks. Trans. Storage **1**(3), 277–315 (2005). https://doi.org/10.1145/1084779.1084780
7. He, J., Ji, S., Yan, M., Pan, Y., Li, Y.: Load-balanced CDS construction in wireless sensor networks via genetic algorithm. Int. J. Sen. Netw. **11**(3), 166–178 (2012)
8. He, J.S., Ji, S., Pan, Y., Cai, Z.: Approximation algorithms for load-balanced virtual backbone construction in wireless sensor networks. Theor. Comput. Sci. **507**, 2–16 (2013)
9. Heinzelman, W.R., Chandrakasan, A., Balakrishnan, H.: Energy-efficient communication protocol for wireless microsensor networks. In: Proceedings of the 33rd Hawaii International Conference on System Sciences, HICSS 2000, vol. 8, p. 8020. IEEE Computer Society, Washington (2000). http://dl.acm.org/citation.cfm?id=820264.820485
10. Hou, X., Tipper, D.: Gossip-based sleep protocol (GSP) for energy efficient routing in wireless ad hoc networks. In: 2004 IEEE Wireless Communications and Networking Conference, WCNC 2004, vol. 3, pp. 1305–1310, March 2004

11. Jin, Y., Wang, L., Kim, Y., Yang, X.: EEMC: an energy-efficient multi-level clustering algorithm for large-scale wireless sensor networks. Comput. Netw. **52**(3), 542–562 (2008). https://doi.org/10.1016/j.comnet.2007.10.005
12. Khamfroush, H., Saadat, R., Heshmati, S.: A new tree-based routing algorithm for energy reduction in wireless sensor networks. In: 2009 International Conference on Signal Processing Systems, pp. 116–120, May 2009. https://doi.org/10.1109/ICSPS.2009.38
13. Kim, D., Wu, Y., Li, Y., Zou, F., Du, D.Z.: Constructing minimum connected dominating sets with bounded diameters in wireless networks. IEEE Trans. Parallel Distrib. Syst. **20**(2), 147–157 (2009)
14. Li, D., Cao, J., Liu, M., Zheng, Y.: Construction of optimal data aggregation trees for wireless sensor networks. In: 2006 Proceedings of 15th International Conference on Computer Communications and Networks, ICCCN 2006, pp. 475–480, October 2006. https://doi.org/10.1109/ICCCN.2006.286323
15. Loscri, V., Morabito, G., Marano, S.: A two-levels hierarchy for low-energy adaptive clustering hierarchy (TL-LEACH). In: 2005 IEEE 62nd Conference on Vehicular Technology Conference, VTC-2005-Fall, vol. 3, pp. 1809–1813, September 2005. https://doi.org/10.1109/VETECF.2005.1558418
16. Mainwaring, A., Culler, D., Polastre, J., Szewczyk, R., Anderson, J.: Wireless sensor networks for habitat monitoring. In: Proceedings of the 1st ACM International Workshop on Wireless Sensor Networks and Applications, WSNA 2002, pp. 88–97. ACM, New York (2002). https://doi.org/10.1145/570738.570751
17. Osterlind, F., Dunkels, A., Eriksson, J., Finne, N., Voigt, T.: Cross-level sensor network simulation with COOJA. In: Proceedings 2006 31st IEEE Conference on Local Computer Networks, pp. 641–648, November 2006
18. Schmid, S., Wattenhofer, R.: Algorithmic models for sensor networks. In: 2006 20th International Conference on Parallel and Distributed Processing Symposium, IPDPS 2006, p. 11, April 2006
19. Tan, H.O., Körpeoğlu, I.: Power efficient data gathering and aggregation in wireless sensor networks. SIGMOD Rec. **32**(4), 66–71 (2003). https://doi.org/10.1145/959060.959072
20. Wan, P.J., Alzoubi, K., Frieder, O.: Distributed construction of connected dominating set in wireless ad hoc networks. In: Proceedings of Twenty-First Annual Joint Conference of the IEEE Computer and Communications Societies, INFOCOM 2002, vol. 3, pp. 1597–1604. IEEE (2002)
21. Wan, P.J., Huang, S.C.H., Wang, L., Wan, Z., Jia, X.: Minimum-latency aggregation scheduling in multihop wireless networks. In: Proceedings of the Tenth ACM International Symposium on Mobile Ad Hoc Networking and Computing, MobiHoc 2009, pp. 185–194. ACM, New York (2009)
22. Wang, N., Yu, J., Li, G.: A localized algorithm for constructing directional connected dominating sets in ad hoc networks. Comput. Eng. Appl. 102–106 (2012)
23. Welch, T.A.: A technique for high-performance data compression. Computer **17**(6), 8–19 (1984). https://doi.org/10.1109/MC.1984.1659158
24. Wu, J., Li, H.: On calculating connected dominating set for efficient routing in ad hoc wireless networks. In: Proceedings of the 3rd International Workshop on Discrete Algorithms and Methods for Mobile Computing and Communications, DIALM 1999, pp. 7–14. ACM, New York (1999)
25. Xu, N., et al.: A wireless sensor network for structural monitoring. In: Proceedings of the 2nd International Conference on Embedded Networked Sensor Systems, SenSys 2004, pp. 13–24. ACM, New York (2004). https://doi.org/10.1145/1031495.1031498

26. Ye, W., Heidemann, J.S., Estrin, D.: An energy-efficient MAC protocol for wireless sensor networks. In: INFOCOM (2002)
27. Younis, O., Fahmy, S.: Heed: A hybrid, energy-efficient, distributed clustering approach for ad hoc sensor networks. IEEE Trans. Mob. Comput. **3**(4), 366–379 (2004). https://doi.org/10.1109/TMC.2004.41

Touchless Recognition of Hand Gesture Digits and English Characters Using Convolutional Neural Networks

Ujjwal Peshin[1,2(✉)], Tanay Jha[1,3], and Shivam Kantival[1,4]

[1] Netaji Subas Institute of Technology, New Delhi 10078, India
jha.tanay@gmail.com,shivam.kantival@gmail.com
[2] Columbia University, New York, NY 10027, USA
up2138@columbia.edu
[3] Sumo Logic, 3rd Fl, Tower-1, Highway Towers, Block A, Industrial Area, Sector 62, Noida 201309, Uttar Pradesh, India
[4] Sprinklr, 902 Iris Tech Park, Sector 48, Gurgaon, India

Abstract. Computer technology has changed the way humans define success. Gesture recognition is a new and intuitive interaction method, as it does not require any mechanical interface for interaction. The model proposed takes input from the user using an infrared emitter and a web camera and this is used to generate a 28×28 image contains the drawn character. This is passed through a convolutional neural network which detects the character drawn. The benefits include reliability and cost-effective nature of the proposed system. Such an approach can help people with disabilities.

Keywords: Machine learning · Deep learning · Convolutional neural networks · Computer vision · Human Computer Interaction · Image analysis

1 Introduction

Advancements in computer technology have become the forefront metric of human success. Self-sufficiency of machines is the primary focus of computer engineers and scientists to reduce human effort and need for expertise in each field. Automation requires interaction between humans and computers, giving rise to a field known as Human-Computer Interaction, in which it is required to have several interfaces between machines and humans.

Gesture recognition is a new and intuitive interaction method in Human-Computer Interaction, and it does not require any mechanical device for interaction with machines. Hand gesture recognition has become an important and vital part of Human-Computer Interaction due to its value in real-life applications like in Virtual Reality, teaching, and learning, recognizing Sign Language and in gaming. There has been a lot of traditional work that has been carried out in the field of vision-based hand gesture recognition systems, although they have

É. Renault et al. (Eds.): MLN 2018, LNCS 11407, pp. 212–221, 2019.
https://doi.org/10.1007/978-3-030-19945-6_14

not provided satisfactory results. Some of the limitations include susceptibility to lighting conditions and backgrounds or using gesture-based gloves, which need calibration and are often more expensive than complex vision based sensors [1].

All this requires extensive recognition of hand movements and motion-based pattern. While a pattern is easily discernible to humans, for any machine, it is only input data. Hence, a machine has to learn to discern between different noise and data to recognize patterns and generalize those patterns to other similar data input. Hence, human designed the architecture of artificial neural networks which can be used to recognize patterns in a fashion similar to the biological neural network. It is used to predict values on the basis of some input data, which is passed through a system of interconnected 'neurons'.

Artificial Neural Networks consists of input cases, hidden layer(s) and the output layer. There are a varying number of neurons in each layer. The output of a neuron is a function applied to the weighted sum of inputs with a bias.

An artificial neural network can be trained with a training set, which is a set of an output of known cases of a problem before it is tested how well it can generalize to unknown instances of the problem. Their adaptive nature leads to learning by examples.

The model proposed takes input from the user using an infrared emitter and a webcam which can detect infrared and this is used to generate a 28×28 image contains the drawn character. This is passed through a convolutional neural network which detects the character drawn. The benefits over present systems include reliability and cost-effective nature of the proposed system.

An approach like this can help the disabled people who knew to write, but later on were unable to write on paper by using hand because of some difficulties.

The outline of this paper is as follows: Sect. 2 describes existing character recognition techniques. Section 3 presents the proposed model. Experimental studies have been discussed in Sect. 4. Finally, concluding remarks are explained in Sect. 5.

2 Literature Survey

Several efforts have been made designed to provide a means to make hand gesture recognition useful for recognizing digits and English characters. Some of the research in this field has been mentioned below:

2.1 English Sentence Recognition Using Artificial Neural Network Through Mouse-Based Gestures

Parwej [2] used mouse-based gestures in real-time for continuous English sentence recognition. Artificial neural networks were used which had been trained using the traditional backpropagation algorithm. The English Sentence was given as an input to the system, which was recognized by the system and the same sentence was displayed on the screen after recognition. It required an input of a mouse for the usability of the system.

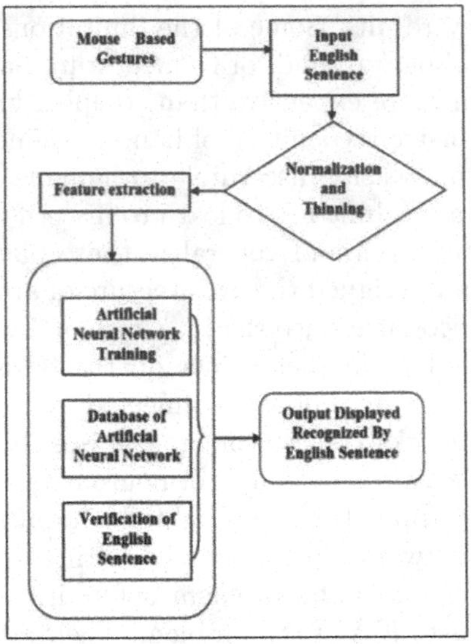

Fig. 1. Mouse based English sentence recognition

As can be observed in Fig. 1, it consisted of two major modules, Preprocessing and Feature Extraction. Preprocessing includes segmentation, thinning (skeletonization), contour marking, normalization, filtration etc. Feature Extraction is done using Pixel Delivery method.

2.2 Neural Network Based Approach for Recognition of Text Images

Kumar and Bhatia [3] devised a method of recognizing the characters present on the text images. The image was read in a binary format. Preprocessing was carried out using thresholding, skeletonization operations, and normalization. Feature extraction was carried out using Fourier Descriptor method. A two layer feed forward neural network of 26 input neurons with a learning rate of 0.01 was trained for 2000 epochs after which testing was carried out.

2.3 Finger Writing in the Air Using a Kinect Sensor

Zeng et al. [4] formulated and tested a system which proposed to use the fingertip as a 'virtual pen'. The use of the Kinect allowed them to incorporate the depth and visual information to track the characters which were being traced in real time. The hand was differentiated from the background using color and depth sequences. A depth-skin-background mixture model(DSB MM) was proposed to solve various issues related to hand recognition, like uneven illumination and hand-face overlap issues (Fig. 2).

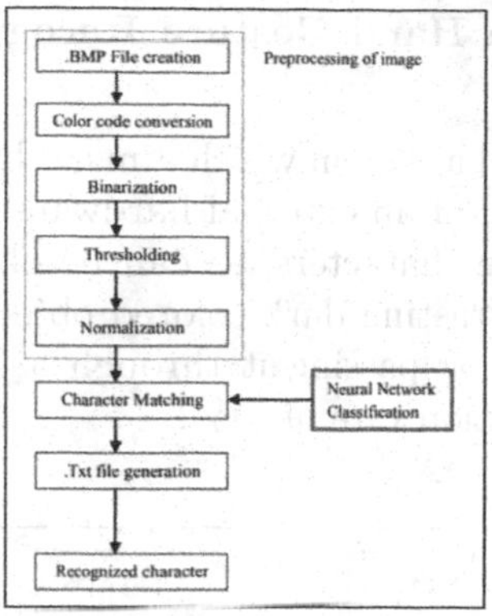

Fig. 2. Handwritten character recognition

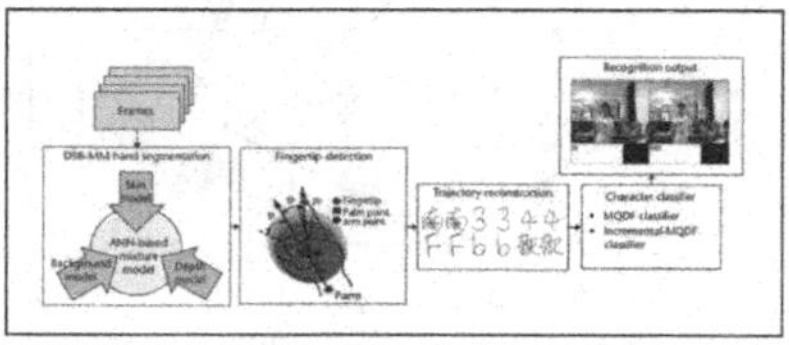

Fig. 3. Fingertip writing using Kinect as a depth sensor

As seen in Fig. 3, a dual-mode switching algorithm was used to detect the fingertip from various hand poses. The trajectory of the fingertip was extracted and linked and then reconstructed as an inkless character. A state-of-the-art handwriting character recognition method was used to produce the final output. The system requires expensive hardware but deals with various poses of the hand. The system produced an accuracy of 94.62% for the recognition of uppercase English letter and 86.15% for the recognition of lowercase English letter.

2.4 Handwritten English Character Recognition Using Neural Network

In the above-mentioned paper, Pal and Singh [5] have made efforts towards developing automatic handwritten character recognition system for the English language, which will have a high accuracy and minimized training and classification time. The sample is obtained by scanning 32×32 images and skeletonized and normalized into thin strokes. Boundary Detection Feature Extraction was applied for feature extraction. A 2 layer neural network was used for the remaining classification task.

2.5 Touchless Writer a Hand Gesture Recognizer for English Characters

Joshi et al. [6] have proposed a system which aims to be much more economically viable and without the need of an external hardware device. It was designed to recognize English uppercase characters, as can be observed in Fig. 4. It uses a webcam to recognize a contrasting dark colored object which is further used to draw an English character by passing it through a 3 layer neural network. It provides an average recognition rate of 91%.

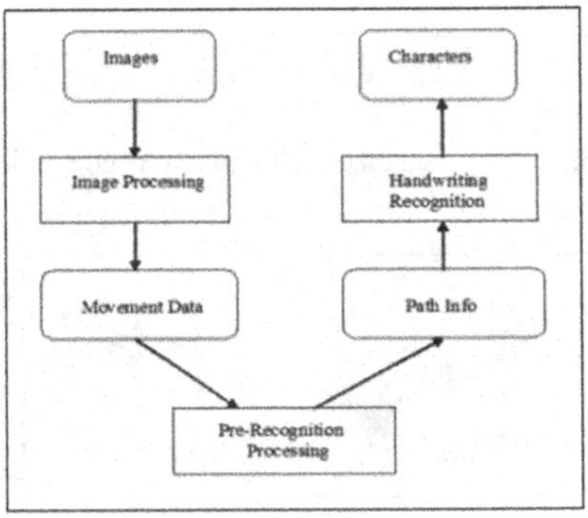

Fig. 4. Hand gesture recognition for English uppercase characters

3 System Architecture

The system proposed comprises 4 major modules as shown in Fig. 5. The description of each module is described below:

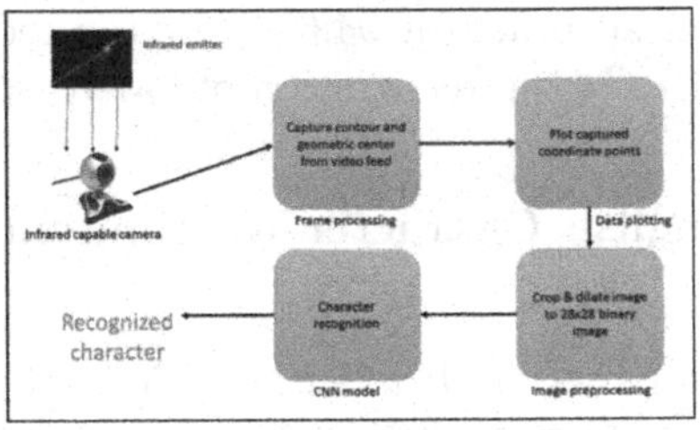

Fig. 5. Proposed system architecture

3.1 Frame Processing Module

This module receives the video frames from the camera, detects the region in which infrared is present and draws a contour around it. It then finds the minimum enclosing rectangle around this contour and finds the geometric center of the contour.

An empty deque is maintained and the camera frame is read in and flipped, as input provided by the camera is inverted in nature. The image is then converted to greyscale as the color information does not provide any additional information to us to help in detecting features and edges and to provide fast real-time processing of character.

Binary thresholding is applied to focus on the edges of the contours and then the image was dilated to focus on the image edges and remove the extra detected infrared portion.

Next, the contours were detected and their centers of contours found are added to a deque. Once this module detects that no contours have been found for 5 simultaneous frames, it automatically passes this deque to next module for plotting.

3.2 Data Plotting

This module uses the 'pts' deque passed to it by the Frame Processing module. This module uses matplotlib library for plotting the tracked points to an 800×600 (wxh) white canvas. This plot is saved as a temporary file named "foo1.jpeg", which is further used by modules.

The output plot of this module is as shown in the figure, Fig. 6, we can see the dimensions of the image generated and the tracked points plotted to a blank background.

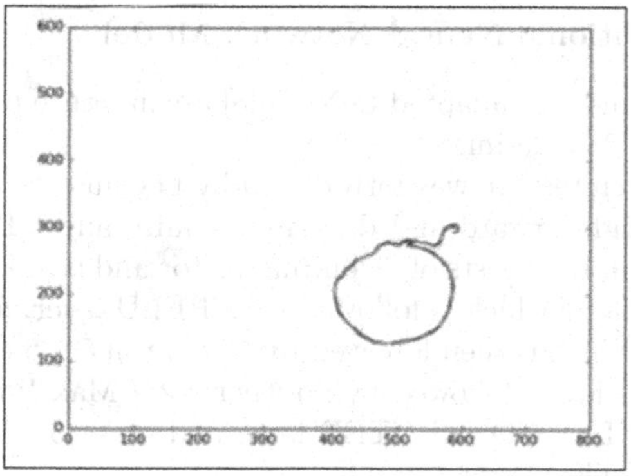

Fig. 6. Output of Data Plotting module

3.3 Image Preprocessing

The image needs to be preprocessed before it can be passed through the neural network to decrease noise an normalize the shape of objects. This module itself is called by the Data Plotting module. No argument is passed to this module as an input, instead, it uses the temporary file updated by the previous module as an input. The plot in image "foo1.jpeg" is read in and further processed to create a binary image that is accepted by the convolutional neural network classifier.

The module first converts the image to a binary space using thresholding function and dilates the image to focus on the black boundaries of the generated plot. Then it finds the area of interest by finding a minimal enclosing rectangle around the plotted pixels. This sub spaced image of the original canvas is then resized to a 28×28 pixel image, as can be seen in Fig. 7. This is done because the convolutional neural network is trained using dataset [7] and so as to ensure that the training and test set is picked from similar distributions, the image is resized to dimensions similar to that of the training set. The figure below shows the output generated by this module.

Fig. 7. Output of image processing module

3.4 The Convolutional Neural Network Model

This module uses a modern adapted LeNet5 [8] architecture to predict the character present in the 28×28 image.

The LeNet [8] architecture was introduced by LeCun et al., as seen in Fig. 8. It is small and straightforward and does not require any GPU and hence was used for classification. It consists of 784 input vector and is followed by 20 neuron 5×5 convolutional layer which is followed by a RELU layer, followed by a 2×2 Max Pooling layer, which is then followed by 50 neuron 5×5 convolutional layer followed by a RELU layer, followed by another 2×2 Max Pooling layer, which is connected to a FULLY CONNECTED layer and then to a RELU and then to a FULLY CONNECTED layer.

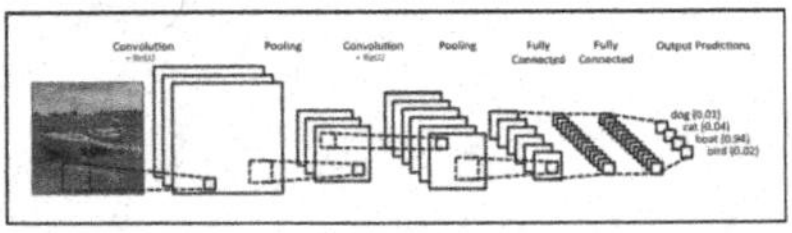

Fig. 8. LeNet architecture [8]

The neural network was trained using the MNIST dataset [7]. The goal of this dataset is to classify the handwritten digits 0–9. The dataset contains a total of 70,000 images, which were used completely for training the neural network. Each digit is represented as a 28×28 grayscale image with the pixel intensities falling in the range [0, 255]. All digits are placed on a black background with a light foreground.

The binary image is fed to a convolutional neural network which has already been trained on the MNIST dataset [7], and it outputs the predicted character along with the prediction confidence value.

4 Experimental Setup and Results

The experiment required a webcam which was infrared detection enabled. Hence, a normal webcam had to go through a few changes to detect infrared from an emitter, which was as follows:

1. The webcam casing was unscrewed.
2. The infrared filter (tiny color film) was removed with help of cutter or scissors.
3. The infrared filter was replaced with a clean unused negative film.
4. Finally, to make this an infrared-only camera, the visible light needed to be filter-out. To do that, a filter was prepared for the front of the lens out of several layers of fully-exposed 35 mm color negative film.
5. The setup was screwed back and the focus was adjusted back to as it was.

A simple infrared emitter can be a TV remote. Nowadays, several mobile devices come equipped with an infrared emitter.

The user has to just draw a digit or an English lowercase character in front of the camera and the character would be recognized by the proposed system.

The experimental result shows an acceptable recognition rate for digits and English characters (Tables 1 and 2).

Table 1. Accuracy achieved by LeNet architecture trained on MNIST [7]: 98.6%

Character	Size of test set	Correctly predicted
0	100	99
1	100	98
2	100	99
3	100	100
4	100	98
5	100	100
6	100	96
7	100	100
8	100	99
9	100	97

Table 2. Accuracy achieved by inception model trained on MNIST [7]: 98.0%

Character	Size of test set	Correctly predicted
0	20	20
1	20	20
2	20	18
3	20	19
4	20	20
5	20	20
6	20	20
7	20	20
8	20	20
9	20	19

4.1 Results

The final system was successfully created which can take input from the infrared emitting device and predict the corresponding digit. The prediction was done using convolutional neural networks. Convolutional neural networks can have different architectures. Some of the most common architectures used with convolutional neural networks are LeNet [8], AlexNet [9], GoogLeNet [10] etc. We tried the LeNet and GoogLeNet architecture in our project using transfer learning.

5 Applications and Future Scope

5.1 Applications

1. This software can be used by physically challenged people for writing on displays and providing input to computer systems.
2. It will prove useful in sterile environments like hospitals, where touching hardware is of severe concern.
3. It can have extensive use in gesture-based gaming and virtual reality-based systems.
4. On larger screens, like TVs or desktops, where there is no inherent touch-control system and the device itself is sometimes out of reach, a gesture-based system makes a lot of sense as it provides an intuitive control system.
5. If your hands are dirty or if you really hate smudges, touch-free controls are a huge benefit over conventional means of communication with the device. Touch-free screens would also make using touchscreen devices in the winter much easier when you're wearing gloves.
6. One big future use can be with a 3-dimensional display and projections, as that display can be combined with this project to create immersive experiences with futuristic devices.

5.2 Future Scope

Once the character recognition is included in the project, it can be extended to complete sentence detection (using a combination of Long Short Term Memory and Convolutional Neural Networks). Also, using better computing power we can train our predictor more extensively, to get even better prediction accuracy, for eg. using a Graphics Processing Unit. A better quality camera can be used to make the process of taking input more efficient.

References

1. Li, X.: Hand gesture recognition technology in human-computer interaction, School of Computer Science, University of Birmingham (2016)
2. Parwej, F.: English sentence recognition using artificial neural network through mouse-based gestures. Int. J. Comput. Appl. (IJCA) (2013). arXiv:1301.4659
3. Zeng, W.: A new writing experience: finger writing in the air using a Kinect sensor. IEEE Computer Society (2013)
4. Kumar, G., Bhatia, P.K.: Neural network based approach for recognition of text images. Int. J. Comput. Appl. **62**(14), 8–13 (2013)
5. Pal, A., Singh, D.: Handwritten English character recognition using neural network. Int. J. Comput. Sci. Commun. **1**(2), 141–144 (2010)
6. Joshi, A., et al.: Touchless writer a hand gesture recognizer for English characters. Int. J. Adv. Comput. Eng. Network. (IJACEN) **3**, 1–5 (2015)
7. Lecun, Y., Cortes, C.: The MNIST database of handwritten digits
8. Lecun, Y., Bottou, L., Bengio, Y., Haffner, P.: Gradient-based learning applied to document recognition. Proc. IEEE **86**(11), 2278–2324 (1998). https://doi.org/10.1109/5.726791
9. Krizhevsky, A., Sutskever, I., Hinton, G.E.: ImageNet classification with deep convolutional neural networks 84–90 (2017). https://doi.org/10.1145/3065386
10. Szegedy, C., et al.: Going deeper with convolutions 1–9 (2015). https://doi.org/10.1109/CVPR.2015.7298594

LSTM Recurrent Neural Network (RNN) for Anomaly Detection in Cellular Mobile Networks

S. M. Abdullah Al Mamun[1(✉)] and Mehmet Beyaz[2(✉)]

[1] Research and Development Specialist, TTG International Ltd., 34799 Istanbul, Turkey
abdullah.almamun@ttgint.com

[2] CTO and Research Director, TTG International Ltd., 34799 Istanbul, Turkey
mehmet.beyaz@ttgint.com

Abstract. Anomaly detection can show significant behavior changes in the cellular mobile network. It can explain much important missing information and which can be monitored using advanced AI (Artificial Intelligent) applications/tools. In this paper, we have proposed LSTM (Long Short-Term Memory) based RNN (Recurrent Neural Network) which can model a time series profile for LTE network based on cell KPI values. We have shown in this paper that the dynamic behavior of a single cell can be simplified using a combination of a set for neighbor cells. We can predict the profile and anomalous behavior using this method. According to the best of our knowledge this approach is applied here for the first time for cell level performance profile generation and anomaly detection. In a related work, they have proposed ensemble method to compare different KPIs and cell performance using machine learning algorithm. We have applied DNN (Deep Neural Network) to generate a profile on KPI features from historical data. It gave us deeper insight into how the cell is performing over time and can connect with the root causes or hidden fault of a major failure in the cellular network.

Keywords: Anomaly detection · Deep Neural Network · Cell performance degradation · Recurrent Neural Network (RNN) · Cell diagnostics

1 Introduction

Telecommunication Network Operators face the challenge to minimize technical problems using different software tools and network analytics report. For this reason, the HW (Hardware) of each individual NE (Network Element) needs to be controlled before transport to the deployment site. Visual inspection at the target site is performed to detect any transport damage. The deployment on the field requires typically external passive components like cables and connectors without any self- diagnostics. Therefore, the complete installation needs to be tested once more for basic functionality using measurement equipment and solving any root causes behind possible alarms generated by the NE self-diagnostics. In this phase, the NE should still be as isolated as possible

É. Renault et al. (Eds.): MLN 2018, LNCS 11407, pp. 222–237, 2019.
https://doi.org/10.1007/978-3-030-19945-6_15

to minimize any remaining problems affecting normal subscribers in the area. The remaining technical challenge a Telecommunications, Especially Cellular/Wireless Network Operator/Carrier faces is to detect and solve new technical problems as soon as possible. For example, an excavator might cut or damage underground cables, a water leakage in the cell-site building or outdoor container could reach the NE causing damage (oxidation, corrosion, etc.), a random HW component may fail as they won't last forever or a storm might turn some antenna to wrong direction.

A part of such faults can be detected automatically from an internal alarm raised by the NE self-diagnostics or from a configured external alarm from sensors at the site [6]. The NE contains a panel with connectors for external sensors to enable external/environmental alarms. An external alarm could be raised by e.g. a flood detector, a fire detector or a room temperature sensor. The NE sends new Alarm events and Cancel/Ceasing events of old Alarms to a remote OSS in the Operation and Maintenance (O&M) center of the operator/carrier. The most difficult cases to detect are Faults Without Alarm & Without Downtime. We named them Hiding HW Faults/Hiding SW-Bugs [1]. Such problems are not detected by the self-diagnostics of the NE. They exist as the NE cannot have diagnostics for all possible unexpected fault scenarios without becoming prohibitively complex, slow to develop and expensive. Faults in passive components without any intelligence are also typical cases (cables, connectors, etc.).

Performance Measurement (PM) counters that the NE periodically uploads to the OSS in the O&M center are providing important site information. In this paper, we concentrate on the detection of Hiding Faults using both the raw PM counter values and the values of derived formulas Key Performance Indicators (KPI) as input time series. Time series based anomaly detection is studied for various performance measurements [9, 11]. Correlation in anomalies is also studied [2]. Historically 60 min is a typical duration of PM statistics collection before upload of the counter data sets to the OSS and resetting the counters for the next interval. Smaller intervals like 5 min, 15 min, and 30 min have also become a configurable option. They are useful especially for an upper-level NE connected to a lot of base station cell sites under it. A problem in such an NE would affect a large number of subscribers and/or a large geographical area. But a short interval does not make sense if the rate of HW-Fault detection exceeds the daily amount field engineers are able to solve. The local road traffic and distance conditions need to be considered time from office/warehouse to a typical NE site. Otherwise, the short measurement upload interval would only waste storage disk space at OSS.

In this study, we have proposed LSTM RNN based individual cell profile generation technique which is able to demonstrate Root Cause Analysis (RCA) for the anomaly in the cell level. This technique shows possible anomaly candidates and also the real anomaly in the site. We can detect both short interval and long interval anomalous behavior in the network using our proposed method.

1.1 Related Works

In literature, we have found that anomaly detection have been studied mostly for network flow, anomaly symptom detection in 5G networks [23] and classification for Network Intrusion Detection [7, 12, 21, 22].

Ciocarlie et al. mentioned that, the key challenge for defining the more general problem of cell degradation is creating a robust method for modeling cell behavior [14]. Using KPI values, this paper proposes a novel method for modeling cell behavior to address these problems. KPIs are highly dynamic measurements of cell performance. In this study they have tried to model normal cell behavior using robust models. They detect anomalies in the cell behavior based on the KPI measurements. They have shown univariate, time series analysis, multivariate time series analysis and finally ensemble method for cell anomaly detection. They also proposed an adaptive ensemble method for modeling cell behavior with KPI values [4]. This cell anomaly detection framework aims to determine the relevant features required for detection of anomalies in cell behavior based on the KPI measurements [8]. This ADT (Anomaly Detection Tool) also helps to explore KPI data, visualizes anomalies and reduces the time required to find the actual anomaly [3]. In this study they have used Support Vector Machine (SVM) algorithm to build KPI models. Then using a predictive approach, KPI behavior is captured by autoregressive, integrated moving average (ARIMA) models.

Maimo et al. proposed a novel 5G oriented cyber defense architecture to identify cyberthreats in 5G mobile networks efficient and quickly. The architecture uses deep learning techniques to analyze network traffic by extracting features from network flows [17]. Another study proposed a MEC-oriented solution in 5G networks to detect network anomalies in real-time in automatic way. They proposed a deep learning technique to analyze anomalous network flows [18].

Wu et al. showed how to accurately detect Key Performance Indicator (KPI) anomalies is a critical issue in cellular network management. They presented CELLPAD a unified performance anomaly detection framework for KPI time series data and demonstrated two types of anomalies of practical interest, namely sudden drops and correlation changes [20].

Cheung et al. proposed a new algorithm, which is designed to detect system performance degradations and paving the way to more mature fault prediction strategies. Detecting degradations is a precursor to fault predictions, as degradations are often early signatures of potentially catastrophic faults [21].

1.2 Preparation of Data Sets

We have prepared the training dataset according to our proposed method for applying on LTE networks. Firstly, we have selected a region or site where we have chosen a set of neighbor cells to collect 7 weeks of data with hourly time interval [15–17]. The datasets are collected from 26-02-2018 to 15-04-2018. In this study, we have collected rate based KPI values and profile based KPI values to prepare our feature sets for RNN models to compare real anomaly in the cell behavior.

It is clear that our proposal to go through cell diagnostics from top to bottom using eNodeB first check. The method gave us more effective results to find out cell level anomalies. There will be two different steps in diagnostics, firstly eNodeB level diagnostics for the set of cell combinations. If it says there is no anomaly then it won't go for the next step. The second step is only for individual cell diagnostics.

We have developed this concept based on KPI values degradation observations. We noticed that randomly selected set of the cell or a single cell cannot be expected to produce any repeating profile pattern. So, even if we apply this concept to an eNodeB level that won't perform well because of the randomness. Dynamic behavior of cell traffic pattern was discussed by Ciocarlie et al. in their work [13]. In our study, we noticed that low traffic cells out of hot spots have a lot of randomness in their profile of events. There is no certain reason for telecom subscribers to repeat cell phone activities at the same time on the same days.

1.3 LSTM RNN for Anomaly Detection

We have applied LSTM RNN to generate a model for anomaly detection in the KPI values. It is found that LSTM RNN can get better results in Network Intrusion detection [7, 22]. We have realized that we can apply this model to generate a certain predictive profile for network cells. LSTM performs well where RNN fails to generate profiles for time lags in the data. LSTM networks have been successful for the demonstration in learning sequences containing long-term patterns of unknown length due to their ability to maintain long-term memory.

Deep Learning-based model increased the possibility to predict more accurate results in terms of long-term profile generations [19]. Here to the best of our knowledge, First time we have applied this method successfully in cellular network data to model its cell behavior. We have also demonstrated that the eNodeB level shows the total impact of the anomalous behavior of the selected neighbor cells.

This kind of ML (Machine Learning) model can generate better results for RCA (Root Cause Analysis). This kind of study can support to find out false alarms. Kushnir et al. showed that the statistical method can be used to predict days before the outages of Radio networks with alarm data [24]. We are hopeful that our current study can be associated with the alarm related studies to analyze fault detection and prediction in the future. Our main goal is to find correlated results and make a relation of the real failure causes.

2 System Architecture

We have focused on finding out the solution for anomaly detection from cell level. We have collected and prepared our training set to train our LSTM RNN model. It contains only part of a dataset from a certain region for the concept validation. We have shown that this model can give us expected RCA using ML-based profile generation.

Cellular Network can create a very large scale big data which should be handled through our design process [10]. During our preparation, we observed that to train a data set and test our architecture with LSTM RNN models, we need to find a site or region in the network where anomaly may happen. To understand it, preparation of data is very important. The following diagram gives an idea of how the process applied to generate the profiles for individual cells. Once we are ready to apply this model, it will give predictions of the general behavior of the cell for a certain time period. We can prepare the training sets using different trends or seasonal calendar can be applied.

In this architecture, LSTM RNN model uses 90 epochs, batch size 45 and 4 layers. In training, RMSE is calculated as 14.31 and 15.72 for test RMSE. This profile generation method helped us to recreate the actual profile and predicted profile of any cell for a given time series. It is finally related to eNodeB total profile. We have mentioned that how they are affecting from a dynamic behavior to more stable understandable results in the eNodeB level. Unexpected behavior of this network cell diagnostics provides us with more specific results to reach the Root Cause Analysis (RCA) in cell performance degradation.

Certain selected features related to this diagnostics can produce more accurate results. Machine learning based system should be trained through those features which can give us more predictable profile. Here we have used total call attempted, KPI values related to call dropped number and their failure rates in the network. It is possible to add more related KPI values to generate features for LSTM RNN model.

A fault in a cellular network is a consequence of hardware or software level bugs which significantly degrades network performance. Such kind of problems almost always exist in cellular networks and we can define them as permanent and temporary cell performance degradation. Temporary cell degradation can recover without external support where a longer period cell degradation causes the real problem in the network [25]. Generally, this kind of degradations is not easy to detect because they might not trigger alarms even when users are affected seriously (Fig. 1).

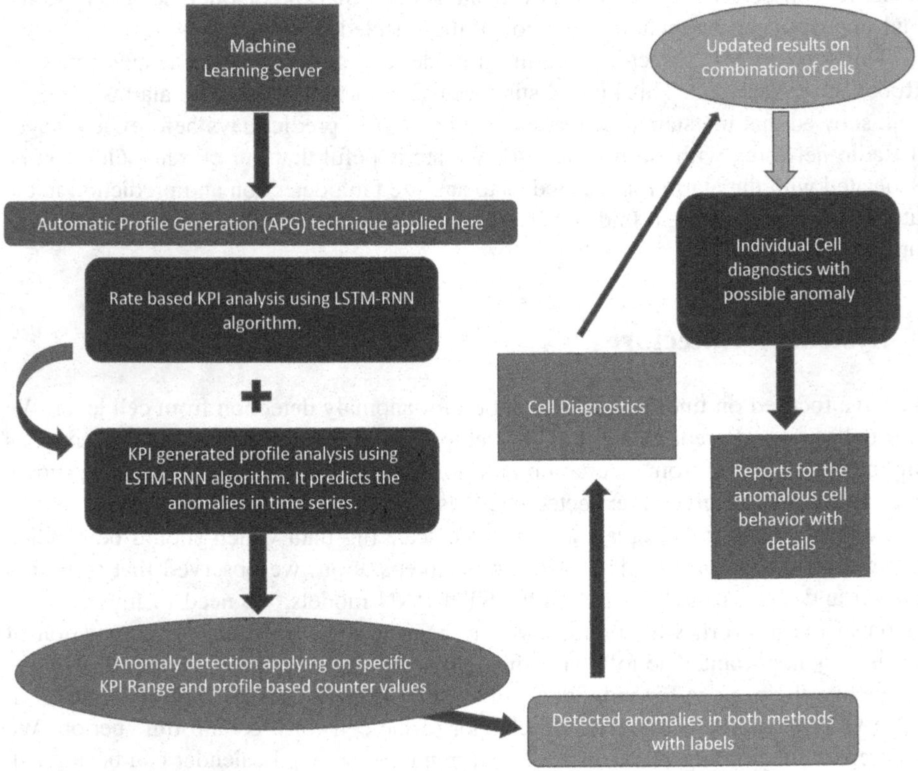

Fig. 1. ML implementation flow for cell level profile generation

We can understand from the figure that in this process we have applied LSTM RNN for both rate and profile based KPI values. It is possible to compare different features from the cell KPI to compare these performance results over a certain period. In this result section, we have shown one of the features collected from LTE cellular networks.

2.1 Results

a. Combination of the results from neighbor cells (eNodeB). Results for LTE networks from the selected cell group

We have prepared this section to describe our idea more elaborately. We have shown the combination of the set of group cells and it's effect on the performance. It is possible to monitor them focusing on the regular formation changes from the eNodeB level and to generate a profile which can explain exact root causes for the faulty cell.

This section is prepared from our sample data sets and specific on two different features which can explain the performance KPI values in a significant way. RNN model made it possible to generate a regular profile using these features and which explains the cell level behavior for the cellular mobile network. Marked with the red line represents the baseline for the KPI feature.

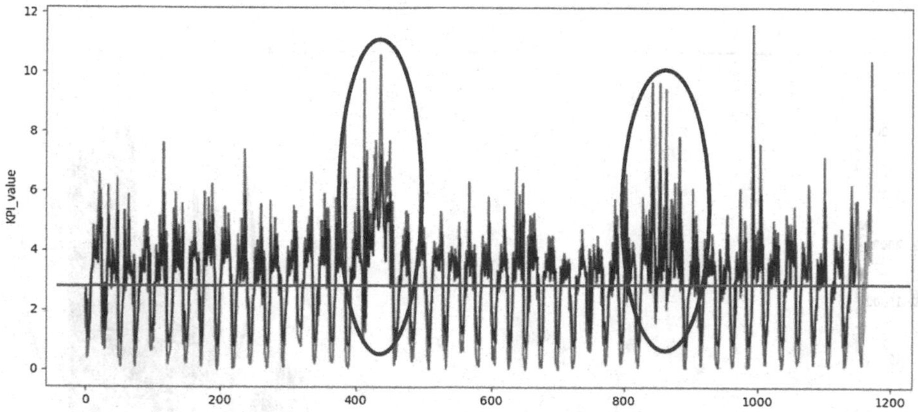

Fig. 2. Combination of rate based KPI values from 9 different cells [hourly] (Color figure online)

Here Fig. 2 describes the total rate KPI value from the set of cells. Here we have used 9 cells to calculate it. This major changes in the profile give us an idea that there might be an anomaly in that set of cells. Marked places are possible anomaly candidates. It will appear when there will be a major change in the KPI values. We have used CDR (Call Drop Rate) as KPI value here. As the drop rate is shown as an aggregated value with changes in the normal profile are marked with rings. This kind of behavior shows the potential anomaly candidate in that selected part. We can then go for more details of individual cell diagnostics to understand why this might happen.

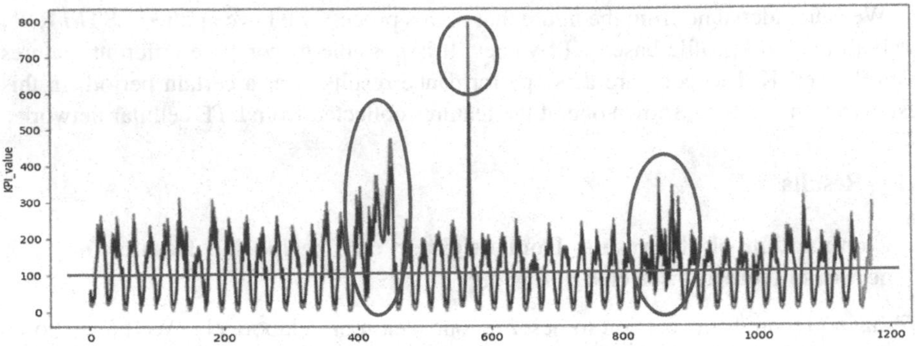

Fig. 3. Combination of numerical KPI values from 9 different cells [hourly]

In Fig. 3 there is another anomaly candidate point appeared which was not available in the rate based KPI value feature. In this part, we have used the real number of calls dropped feature. It was a combination of 9 different cells. So, we can go into individual cell level profile check to find out the real cause of anomalies from given set of cells. Marked circles are possible anomaly candidates here. Both of the features were trained and predicted with LSTM RNN, where deep blue is predicted value and cyan is real KPI value.

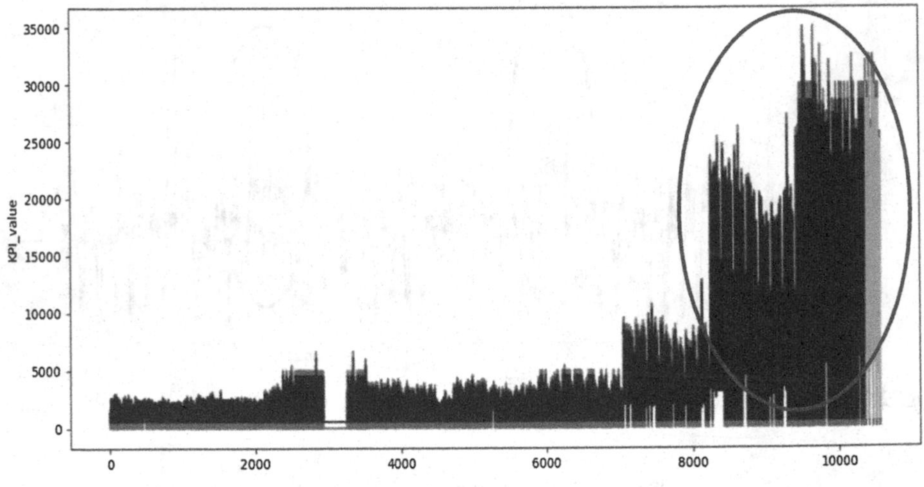

Fig. 4. Combination of total call attempts from 9 different cells [hourly]

In Fig. 4 we can see the total call attempts for those specific 9 different cells. This information is needed when there will be an unusual case of high call attempts and significant call drop rate. The marked area is most probably the highly affected area. Expected anomaly candidates can be found in that area. It is numerically a series of the selected set of cells which are represented here. In the result, we observed exactly the expected event in cell level diagnostics.

b. Individual cell anomaly analysis using profile generation

In this section, we have demonstrated our observation of individual cell details for cell degradation reasons. It has cleared that short duration and long duration cell degradation events can be detected using LSTM RNN profile generation. This kind of event observation can eventually lead to RCA.

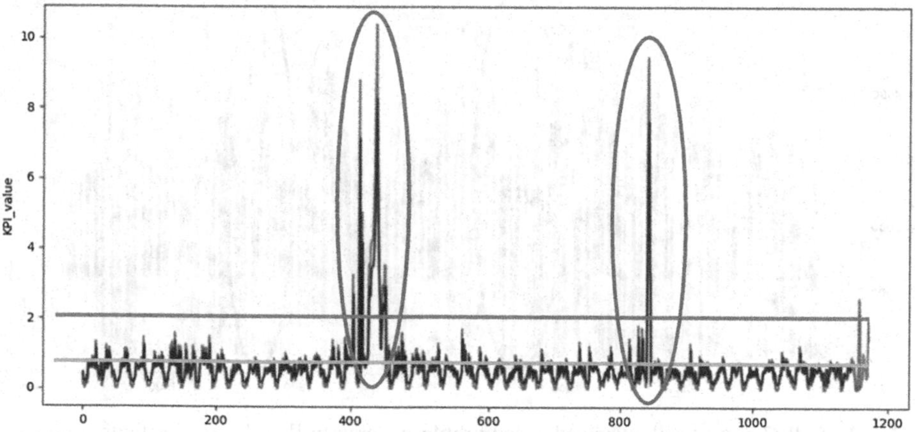

Fig. 5. Cell#7 rate KPI rate based profile generation [hourly] (Color figure online)

In Fig. 5 rate based KPI value is shown. CDR (Call Drop Rate) is selected as a feature. Marked area is a possible anomaly candidate which is also out of the normal profile trend. Here x-axis represents the time tags of those events. The green line is indicating a safe threshold and the red line is boarder for a hard line to the acceptable threshold. A long duration of this event marked as a real anomalous candidate according to our proposal.

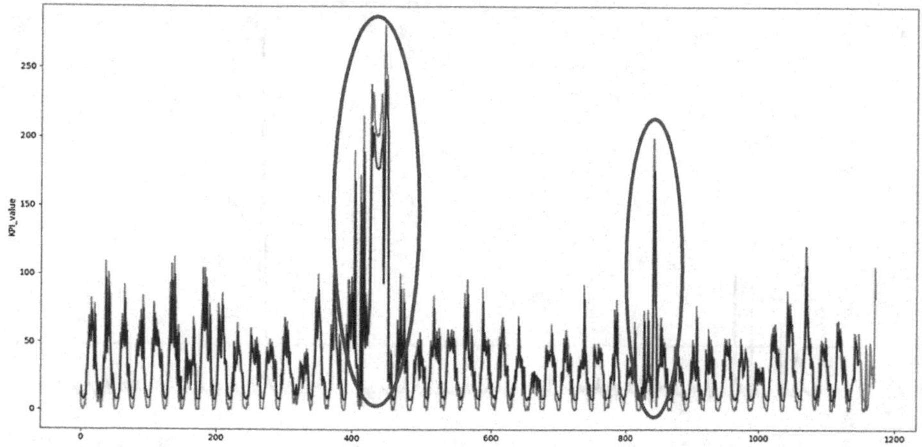

Fig. 6. Cell#7 numerical KPI value based profile generation [hourly] (Color figure online)

In Fig. 6 value based KPI feature is presented. The number of dropped calls is used as a feature and it is hourly total call drops for that selected cell. We have identified an area which is also marked in rate based KPI value. We can understand the long duration and short duration changes in profile from the LSTM RNN prediction model. Figures 5 and 6 of cell number 7 clearly agrees with eNodeB level anomaly predictions.

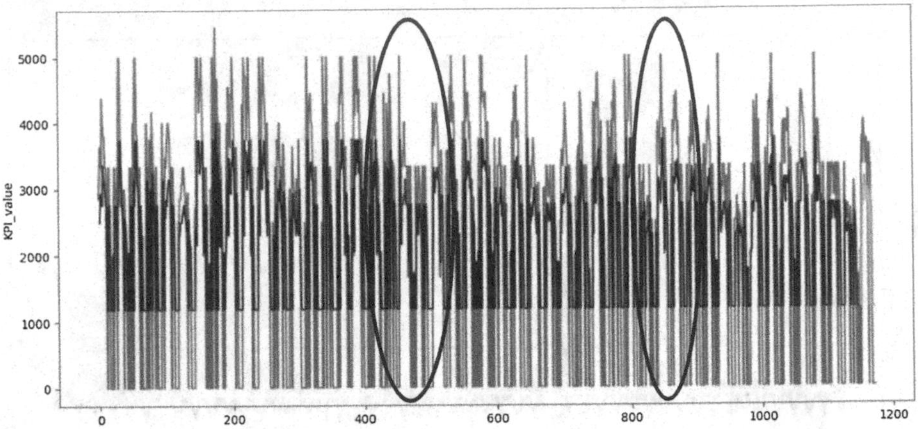

Fig. 7. Cell#7 hourly call attempted based profile generation [hourly] (Color figure online)

In Fig. 7 hourly call attempts have been projected. We can check the real call attempt number in cyan color. Deep blue is predicted values in LSTM RNN model. It is significantly low in affected hours. We can see that real attempts were less or normal but the problem continued for a long duration. So, this is one of the symptoms we have expected in the probable anomalous candidates. It can be marked as a major anomaly because the duration for the cell performance degradation was too long than a sudden problem.

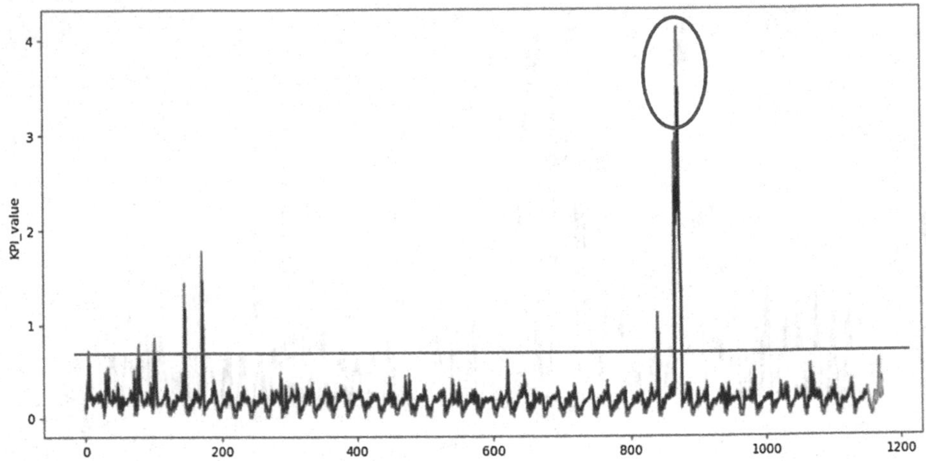

Fig. 8. Cell#8 rate KPI value based profile generation [hourly] (Color figure online)

In Fig. 8 cell number eight is shown with CDR based features. We can mark the cell degradation was affected by short duration. There is no effect of other major cell degradation part. It is marked and a red line threshold is shown to compare the general profile of the cell behavior. This sudden change is also important for many faults to have occurred.

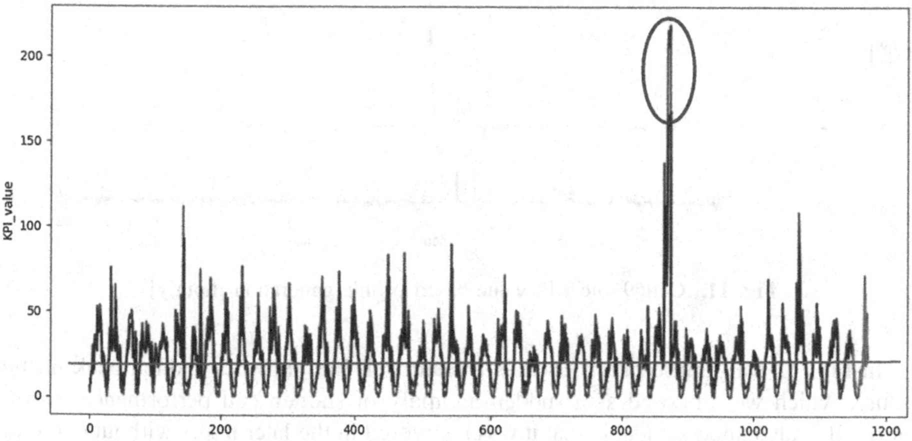

Fig. 9. Cell#8 numerical KPI value based profile generation [hourly] (Color figure online)

Figures 8 and 9 of cell number 8 clearly agrees with eNodeB level anomaly predictions. In Fig. 9 the number of calls dropped for cell 8 is generated. It is also clear from the red line threshold that this kind of event didn't take place before for this cell.

Figure 10 is total call attempted in hours as a feature for cell number 8. It is clearly the highest peak value for this cell. We can understand that it has crossed the general profile for hourly call attempts. It has an impact on temporary cell performance degradation.

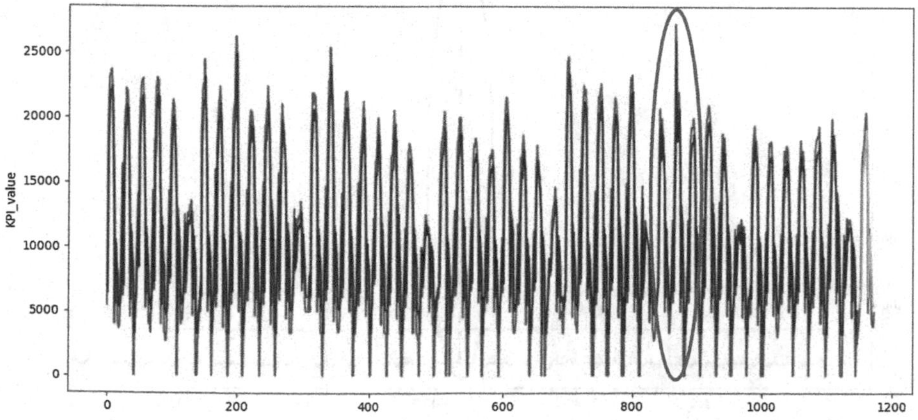

Fig. 10. Cell#8 hourly call attempted based profile generation [hourly]

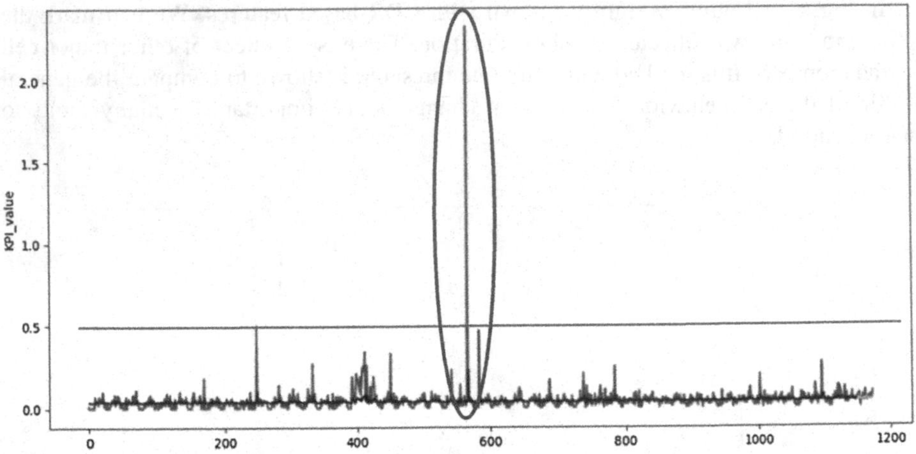

Fig. 11. Cell#9 rate KPI value based profile generation [hourly]

In Fig. 11 call failure rate is used as a feature. It has shown a sudden peak in the failure, which we proposed as a sudden anomaly or sudden cell performance degradation. It is also understandable that it was recovered in the later hours without external hand. This kind of sudden failure can emerge a very big impact on other cellular network performance monitoring tools. When we can check the cell behavior it is obviously unexpected but this sudden peak was also affected in the eNodeB level combination results. It is also in the list of anomaly candidates when we run the cell level diagnostics we found that it won't cause a major issue for the subscriber's dissatisfaction. Long duration cell performance degradation needs to be detected by the tools. Our proposed LSTM RNN model based profiles can satisfy that.

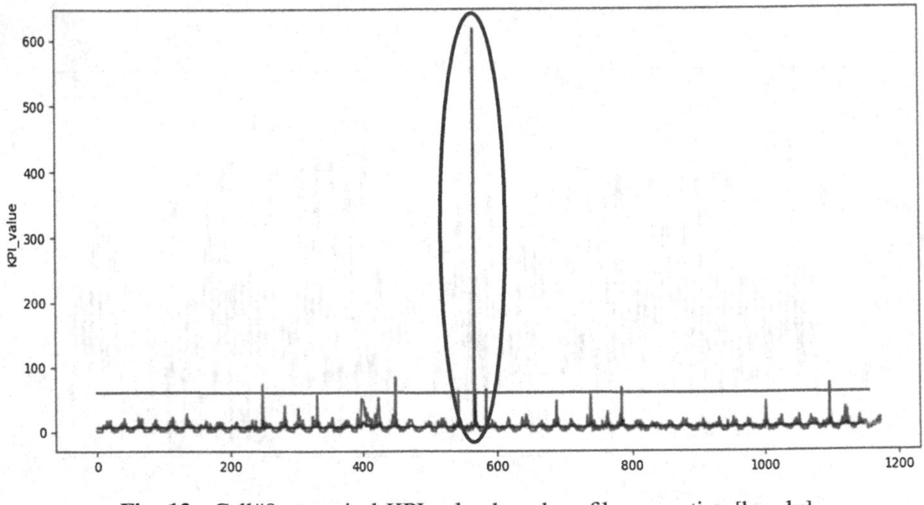

Fig. 12. Cell#9 numerical KPI value based profile generation [hourly]

In Fig. 12 the number of failures is used as a feature. It is also a high peak in the profile but the number crossed anything took place earlier. So this kind of anomaly should be marked in the RCA.

In Fig. 12 the number of failures is used as a feature. It is also a high peak in the profile but the number crossed anything took place earlier. So this kind of anomaly should be marked in the RCA.

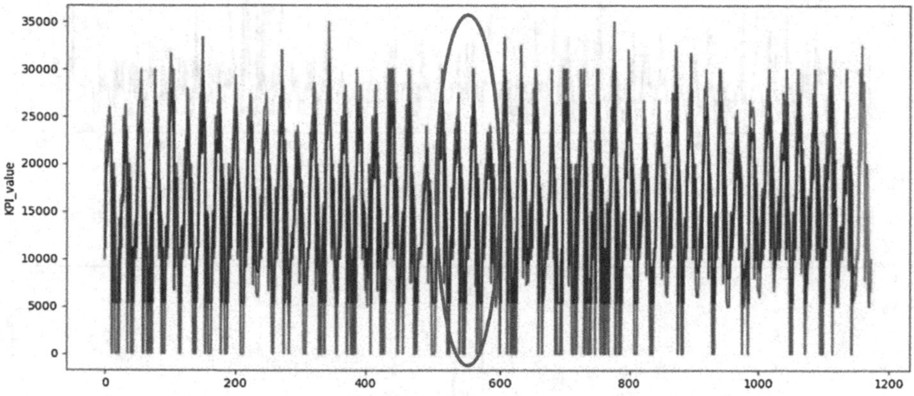

Fig. 13. Cell#9 hourly call attempted based profile generation [hourly]

In Fig. 13 cell number 9 is shown with the hourly call attempts here. Marked hours with high failure rate is marked here. We can understand that the call numbers are not so different than other hours or daily profile. So, these sudden failure numbers are the real effect of any bug or hidden bugs in hardware or software system. This kind of profile based decision will lead to the success for Root Cause Analysis (RCA).

In the next part, we will review our proposal for anomaly candidates. We marked two important events in same hours where the number of dropped calls and call failure rate mismatch in cell number 6. It's possible when the total number of call attempts are lower when the drop rate is high. Another case when dropped call numbers are high but the failure rate is not that high where call attempts are normal. We can mark this cell event as neighbor affected one. Here numerical KPI value is the number of calls drooped hourly feature.

Figures 14 and 15 of cell number 6 clearly agrees with eNodeB level anomaly predictions. This cell is indicating the two different cases in the profile. This event regularly happens in most of the cells for the random nature of the subscribers [4]. So our proposed method to the random nature eNodeB level combination of a set of cells is going to minimize it.

In Fig. 16 hourly call attempts are presented for cell number 6. It is discussed in the earlier section that this kind of cases may be normal or affected cell behavior in a region as a neighbor cell.

In this next section, we are going to demonstrate the 3 different cases that might happen in the mobile cellular networks for cell performance degradation. In this table, we will see possible combinations of features like call drop number, call drop rate and

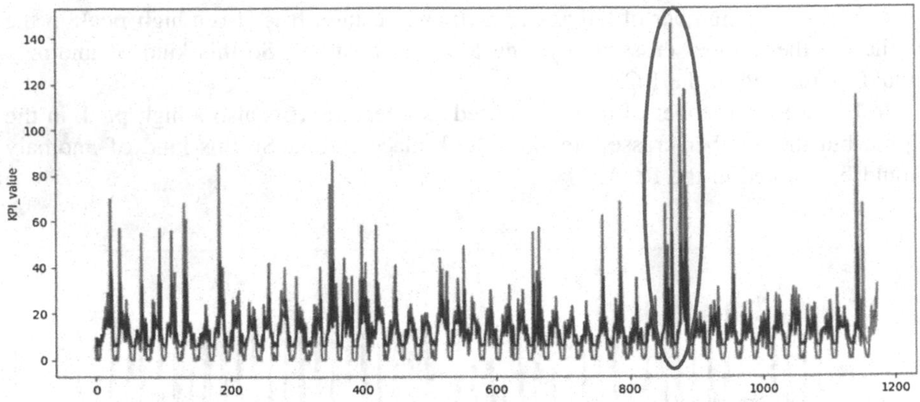

Fig. 14. Cell#6 numerical KPI value based profile generation [hourly]

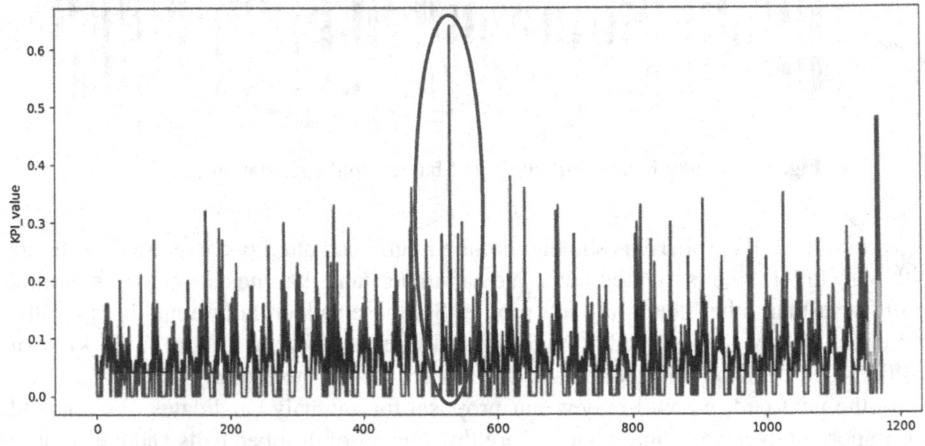

Fig. 15. Cell#6 rate KPI value based profile generation [hourly]

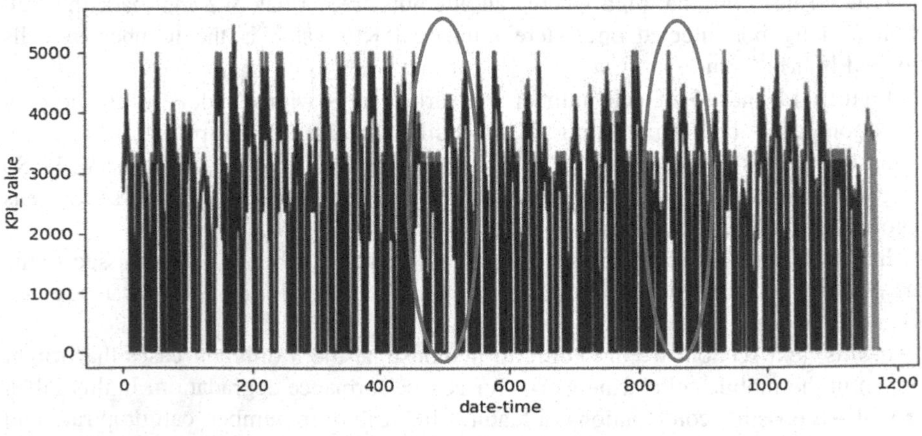

Fig. 16. Cell#6 hourly call attempted based profile generation [hourly]

call attempts are measured and marked with signs to relate them in making a decision. LSTM RNN based generated profile can successfully detect following cases and make decisions for it. This table has all signs and its description with comments what kind of event that might be. We are able to find out more insights of the cell degradation and which is also important to find out RCA (Table 1).

Table 1. LSTM RNN profile based decision for different cases

3 Cases for Anomalous Event Detection (AED)	Sign	Sign details	LSTM RNN Comments
Call drop Number	↑	High drop Number	Sudden cell performance degradation / Short duration cell performance degradation
Call drop Rate	↑	Very High drop Rate	
Call attempt	↑	Number of High Call attempt hourly	
Call drop Number	↑	High drop Number	Long duration cell performance degradation
Call drop Rate	↑	High drop Rate	
Call attempt	↑	Number of High Call attempt hourly	
Call drop Number	↓	Low drop Number	Below/ around threshold cell performance degradation
Call drop Rate	↑	High drop Rate	
Call attempt	↓	Number of expected Call attempt hourly	

Data preparation is very important in the whole process because the correct data sample can be verified for anomaly detection using LSTM RNN models. We can simply understand that for mobile network operators this kind of solution will bring more easy operation in the sites. Sometimes it hard to find out the main reason and some affected cells are also considered for not knowing the exact reason. So we believe that our proposed solution will help the research community and telecom industry for the major solution of RCA.

2.2 Conclusion

Our study will be able to find the unpredictable behavior of the cells in the operational cellular networks. It will be a more efficient way to understand the individual cell level anomaly and its relation with upper network layer like eNodeB. We can understand that a combination of results from neighbor cells can simplify the total scenario by decreasing the randomness in behavior. It can provide us with a faster solution that if there is really any kind of anomaly in the network or not.

The next step of the cell diagnostic can exactly show that which cells are affected and how it was created through all those cells. This solution eventually will guide us to find the RCA in the cellular networks.

Acknowledgments. We would like to thank TTG International Ltd. R&D Istanbul, Turkey and our General Manager Mrs. Emine Beyaz for the support in this research.

References

1. Bouillard, A., Junier, A., Ronot, B.: Hidden anomaly detection in telecommunication networks. In: 2012 8th International Conference on Network and Service Management (CNSM) and 2012 Workshop on Systems Virtualiztion Management (SVM), pp. 82–90 (2012)
2. Qiu, H., Liu, Y., Subrahmanya, N.A., Li, W.: Granger causality for time-series anomaly detection. In: 2012 IEEE 12th International Conference on Data Mining, pp. 1074–1079 (2012)
3. Ciocarlie, G.F., Lindqvist, U., Nitz, K., Nováczki, S., Sanneck, H.: On the feasibility of deploying cell anomaly detection in operational cellular networks. In: 2014 IEEE Network Operations and Management Symposium (NOMS), pp. 1–6 (2014)
4. Ciocarlie, G.F., Lindqvist, U., Nitz, K., Nováczki, S., Sanneck, H.: DCAD: Dynamic cell anomaly detection for operational cellular networks. In: 2014 IEEE Network Operations and Management Symposium (NOMS), pp. 1–2 (2014)
5. Chernov, S., Cochez, M., Ristaniemi, T.: Anomaly detection algorithms for the sleeping cell detection in LTE networks. In: 2015 IEEE 81st Vehicular Technology Conference (VTC Spring), pp. 1–5 (2015)
6. Slimen, Y. B., Allio, S., Jacques, J.: Anomaly prevision in radio access networks using functional data analysis. In: GLOBECOM 2017 - 2017 IEEE Global Communications Conference, pp. 1–6 (2017)
7. Shon, T., Moon, J.: A hybrid machine learning approach to network anomaly detection. Inf. Sci. **177**(18), 3799–3821 (2007)

8. Ciocarlie, G.F., et al.: Demo: SONVer: SON verification for operational cellular networks. In: 2014 11th International Symposium on Wireless Communications Systems (ISWCS), pp. 611–612 (2014)
9. Karatepe, I.A., Zeydan, E.: Anomaly detection in cellular network data using big data analytics. In: Proceedings of 20th European Wireless Conference on European Wireless 2014, pp. 1–5 (2014)
10. Shipmon, D.T., Gurevitch, J.M., Piselli, P.M., Edwards, S.T.: Time Series Anomaly Detection: Detection of Anomalous Drops with Limited Features and Sparse Examples in Noisy Periodic Data. ArXiv Preprint ArXiv:1708.03665 (2017)
11. Brutlag, J.D.: Aberrant behavior detection in time series for network monitoring. In: LISA 2000 Proceedings of the 14th USENIX Conference on System Administration, pp. 139–146 (2000)
12. Himura, Y., Fukuda, K., Cho, K., Esaki, H.: An automatic and dynamic parameter tuning of a statistics-based anomaly detection algorithm. In: 2009 IEEE International Conference on Communications, pp. 1–6 (2009)
13. Ciocarlie, G.F., Lindqvist, U., Nováczki, S., Sanneck, H.: Detecting anomalies in cellular networks using an ensemble method. In: Proceedings of the 9th International Conference on Network and Service Management (CNSM 2013), pp. 171–174 (2013)
14. LTE; Telecommunication management; Key Performance Indicators (KPI) for the Evolved Packet Core (EPC) (3GPP TS 32.455 version 10.0.0 Release 10)
15. Anomaly Detection "automate watching dashboard" by Mr. Nathaniel Cook, Influxdata, SLC DevOpsDays 2016. https://www.youtube.com/watch?v=3swnsoydKTI
16. KPI targets and KPI ranges, IBM Knowledge Center, IBM Business Process Manager Standard 8.5.7, https://www.ibm.com/support/knowledgecenter/SSFTDH_8.5.7/com.ibm.wbpm.wid.tkit.doc/model/kpitargets.html
17. Maimo, L.F., Gomez, A.L.P., Clemente, F.J.G., Perez, M.G., Perez, G.M.: A self-adaptive deep learning-based system for anomaly detection in 5G networks. IEEE Access **6**, 7700–7712 (2018)
18. Maimó, L.F., Celdrán, A.H., Pérez, M.G., Clemente, F.J.G., Pérez, G.M.: Dynamic management of a deep learning-based anomaly detection system for 5G networks. J. Ambient Intell. Humanized Comput. 1–15 (2018)
19. Zhu, M., Ye, K., Xu, C.-Z.: Network anomaly detection and identification based on deep learning methods. In: CLOUD, pp. 219–234 (2018)
20. Wu, J., Lee, P. C., Li, Q., Pan, L., Zhang, J.: CellPAD: Detecting Performance Anomalies in Cellular Networks via Regression Analysis. In: Proceedings of IFIP Networking (Networking 2018), Zurich, Switzerland, May 2018 (2018)
21. Casas, P., D'Alconzo, A., Fiadino, P., Callegari, C.: Detecting and diagnosing anomalies in cellular networks using Random Neural Networks. In: 2016 International Wireless Communications and Mobile Computing Conference (IWCMC), pp. 351–356 (2016)
22. Kim, J., Kim, J., Thu, H.L.T., Kim, H.: Long short term memory recurrent neural network classifier for intrusion detection. In: 2016 International Conference on Platform Technology and Service (PlatCon), pp. 1–5 (2016)
23. Li, J., Zhao, Z., Li, R.: A machine learning based intrusion detection system for software defined 5G network. Let Networks (2017)
24. Kushnir, D., Gohil, G., Sayeed, Z., Uzunalioglu, H.: Predicting outages in radio networks with alarm data. In: IEEE/ACM International Symposium on Quality of Service (2018)
25. Asghar, M., Nieminen, P., Hämäläinen, S., Ristaniemi, T., Imran, M.A., Hämäläinen, T.: Cell degradation detection based on an inter-cell approach. Int. J. Dig. Content Technol. Appl. **11** (2017)
26. Cheung, B., Kumar, G., Rao, S.A.: Statistical algorithms in fault detection and prediction: Toward a healthier network. Bell Labs Tech. J. **9**(4), 171–185 (2005)

Towards a Better Compromise Between Shallow and Deep CNN for Binary Classification Problems of Unstructured Data

Khadoudja Ghanem(✉)

MISC Laboratory, Constantine 2 University, Constantine, Algeria
gkhadoudja@yahoo.fr

Abstract. Deep Neural Network is a large scale neural network. Deep Learning, refers to training very large Neural Networks in order to discover good representations, at multiple levels, with higher-level learned features. The rise of deep learning is especially due to the technological evolution and huge amounts of data. Since that, it becomes a powerful tool that everyone can use specifically on supervised learning, because it's by far the dominant form of deep learning today. Many works based on Deep learning have already been proposed. However, these works have not given any explanation on the choice of the number of the network layers. This makes it difficult to decide on the appropriate deep of the network and its performances for a specific classification problem. In this paper the objective is threefold. The first objective was to study the effect of facial expressions on facial features deformations and its consequences on gender recognition. The second objective is to evaluate the use of Deep learning in the form of transfer learning for binary classification on small datasets (containing images with different Facial expressions). Our third goal is then to find a compromise between too much capacity and not enough capacity of the used deep Neural Network in order to don't over fit nor under fit. Three different architectures were tested: a shallow convolutional neural network (CNN) with 6 layers, a deep CNN VGG16 (16 layers) and very deep CNN RESNET50 (50 Layers). Many conclusions have been drawn.

Keywords: Convolutional neural network · Deep learning · VGG16 · RESNET50 · Gender classification

1 Introduction

Deep learning has reached a level of public attention and industry investment never before seen in the history of Artificial Intelligence; it is used in many fields especially in computer vision, speech recognition, machine translation, web search, advertising and many other fields.

Deep Neural Network is a large scale neural network. Deep Learning, refers to training very large Neural Networks in order to discover good representations, at multiple levels, with higher-level learned features.

In this paper the objective is threefold. Our first objective is to study the effect of facial expressions on facial features deformations and its consequences on gender

É. Renault et al. (Eds.): MLN 2018, LNCS 11407, pp. 238–246, 2019.
https://doi.org/10.1007/978-3-030-19945-6_16

recognition. In order to realize this study, we have built a facial image database that brings together all facial images of the most known datasets used by the research community in the field of facial expression recognition (Datasets are indicated in Sect. 4). The new built dataset is composed by many universal facial expressions especially: Joy, Anger, Disgust, Sad, Fear, Surprise and Neutral. It contains as many common variations, and counts 6400 images with different: Facial expressions, ethnicity, Color, resolution, lighting, age, pose, background and with and without occlusions.

To recognize gender and avoid the classical feature engineering techniques for potential feature extraction, we used deep learning. The problem is that deep learning is generally used with very large datasets. And as we have a small built Dataset for training (few thousands) at hand, we have set our second goal which is: Evaluate the use of Deep learning in the form of transfer learning for binary classification on small datasets. We take an image-classification model trained on the large-scale dataset ImageNet [1] and reuse it on our specific problem (gender recognition) on our specific dataset. This model consists of classifying high resolution color images into 1,000 different categories after training on 1.4 million images of the specified dataset (ImageNet).

Our third and last goal, is to find the minimal depth of a convolutional neural network needed to perform a binary classification task with high accuracy. We are looking for a possibility to draw new conclusions. In order to find an appropriate model size for data, we start with relatively few layers and parameters, and then we add new layers until we see diminishing returns with regard to validation loss (the metric used to evaluate performances). Once we obtained a simple model, the question becomes, is our model sufficiently powerful? Does it have enough layers and parameters to properly model the problem at hand? The ideal model is one that stands right at the border between under fitting and over fitting; between under capacity and over capacity [2]. To figure out where this border lies and to figure out how big a model we'll need, we develop a model that overfits. So we add layers and train for more epochs. We should use models that have enough parameters that they don't over fit nor under fit. There is a compromise to be found between too much capacity and not enough capacity. Unfortunately, there is no magical formula to determine the right number of layers or the right size for each layer [2].

So in this paper we are interested by only recent works based on pre-trained deep and very deep models for binary classification, especially for gender recognition.

In [3] a CNN is trained to perform gender recognition by fine-tuning a pre-trained network, the network is composed by 8 layers, and then an SVM is trained using the deep features computed by the CNN. The proposed model is evaluated on Feret Dataset and on Adience dataset.

In [4] authors used AlexNet-like and domain specific VGG-Face CNN models. SVM are used to classify CNN features. The top accuracy score obtained in gender classification on the Adience dataset is 92.0%.

In [5] A feed forward CNN pipeline which incorporates an attention mechanism for automatic age and gender recognition for face analysis has been proposed. they used the VGG16 Model to extract features, then, like in [3], an SVM is trained using the deep features computed by the CNN to classify age and gender. The proposed system is evaluated on Adience, Images of Groups, and MORPH II benchmarks

In [6] authors used The VGG16 and RESNET 50 models to recognize gender and age, they design the state-of-the-art gender recognition and age estimation models according to three popular benchmarks: LFW, MORPH-II and FGNET. They get high performances on LFW dataset with both tasks. But they demonstrate that, Combining the two approaches Gender and Age Recognition together does not improve age performances and even leads to a slight decrease of gender accuracy.

In the next section, gender recognition is evaluated using recent deep models. The publicly available Deep models VGG16 and Resnet50 are used out of our specific data.

2 Deep Convolutional Neural Network [2]

Deep convolutional neural networks do a mapping inputs (images) to targets (the label), by observing many examples of input and targets via a deep sequence of simple data transformations, these data transformations are learned by exposure to examples. The specification of what a layer does to its input data is stored in the layers weights. The first layer acts as a collection of various edge detectors. At that stage, the activations retain almost all of the information present in the initial image. With the higher layers, the activations become increasingly abstract and less visually interpretable. They begin to encode higher-level concepts. Higher presentations carry increasingly less information about the visual contents of the image, and increasingly more information related to the class of the image.

2.1 Shallow Neural Network

A basic convolutional neutral network is a stack of convolution layers and pooling layers, they end with a densely connected layer. Convolution layers learn local patterns and the role of pooling is to aggressively down sample feature maps. The last (full connected layer) represents the classifier.

2.2 VGG16

VGG16 architecture, developed by Karen Simonyan and Andrew Zisserman in 2014 [7]; it's a simple and widely used Convolutional Network architecture for ImageNet. Although it's an older model, far from the current state of the art and somewhat heavier than many other recent models. It is composed of stacked blocks of Convolutional layers and Pooling steps, a densely connected classifier is staked at the top of these blocks.

2.3 ResNet50

ResNet50 is a powerful model for image classification. It is a very deep Residual Convolutional Network introduced by He et al. [8], it is a plain network mainly inspired by the philosophy of VGG nets with an inserted shortcut connections. It allows to train much deeper networks than were previously practically feasible. It can also learn features at many different levels of abstraction, from edges (at the lower layers) to very complex features (at the deeper layers). Deep Residual Networks are built by stacking together two type of blocks: The identity block and the convolutional block. The identity block is the standard block used in ResNets, this block is composed by a Convolution, a BatchNorm step and a ReLU activation steps in each layer. The Convolutional block is the other type of block. The difference with the identity block is that there is a convolution layer in the shortcut path. In ResNets, a "shortcut" or a "skip connection" allows the gradient to be directly back-propagated to earlier layers.

3 Deep Learning for Gender Recognition

This section analyses the use of shallow, deep and very deep neural networks for gender recognition. To use a pre-trained deep learning model (VGG16 or Resnet50), we start by a feature extraction process then, we fine-tune the pre-trained network. Feature extraction consists of using the representations learned by a previous network to extract interesting features from new samples. These features are then run through a new classifier, which is trained from scratch.

3.1 Key Choices and Hypothesis

We make three key choices to build our three proposed working models: the first one is about the last-layer activation, as we are dealing with binary classification problems we used "sigmoid function". The second one is about the "Loss function (objective function)", it captures how well the network is doing, it takes the predictions of the network and the true target and computes a distance score, this score is used as a feed back to adjust the weights in order to lower the loss. We used "binary_crossentropy" as an objective function. This adjustment is done by an optimizer which is the last key choice, this optimizer implements the Backpropagation algorithm (the central algorithm in deep learning). We used "RMSProp" and its default learning rate to govern the use of gradient descent.

In order to evaluate the performances of the used model, we use the Fold-Validation protocol. To reduce time training and exploit minimum of layers in the extraction step of more general facial features of the three convolutional models, we proceed to face detection using Viola and Jones Algorithm [9], after that, data is preprocessed and formatted into appropriately form to be fed into the network.

The framework Keras [10] which is a library (API), is used to facilitate the use of the three models.

3.2 Models Architectures and Configurations

Our first used model is simple CNN and it is composed of 6 layers, 4 convolutional layers, alternated with maxpooling steps and ends by 2 Full Connected (FC) layers.

When considering the two Deep models VGG16 and Resnet50, two experiments were conducted, in the first experiment, all convolutional layers were frozen, the output of the considered network with shape (4, 4, 512) is flatten so the new shape is (8192), the two full connected layers were ignored and the last layer was replaced by the Sigmoid layer, this one represent the classifier (Logistic regression).

In the second experimentation, we have unfreeze the last convolution Block, we get its precedent layer outputs, then add a FC layer with Relu activation, this new layer produces a vector with shape (256). Finally, we add a last FC layer with Sigmoid activation, this last layer represent the classifier, its output is a scalar (0, 1): 0 for Female, and 1 for Male).

The input and the output of all changed layers are the same for the two models and for the two experiments, the differences are given in the following Table 1:

Table 1. Used CNNs configurations.

Model	Layers	Learned parameters
Simple CNN	7 layers	452 806
VGG16	16 layers	16 812 353
RESNET50	50 layers	24 112 513

In our first experiment, earlier obtained results start to over fit due to the high number of learned parameters and much training data for that, and to fight overfitting, we apply some regularization techniques such as: Dropout, Add and remove layers, data augmentation (to increase training data: Double of trained images). Once we get a satisfactory model we put validation and training data together then we proceed to a test step.

3.3 Used Algorithm (for VGG16 and Resnet50 Models)

Begin

a. *Instantiation of the convolutional base of both models;*
b. *Feature extraction of the new datasets;*
c. *If Feature extraction without data Augmentation then:*
    i. *Define the FC classifier;*
    ii. *Train the FC classifier;*
d. *Else (with Data augmentation)*
    i. *Add a new FC classifier at the top of the convolutional base;*
    ii. *Train the model end to end with convolutional base freezed*
e. *Unfreeze some layers at the top of the convolutional base used in feature extraction;*
f. *Train the joined the new added layers and unfreezed layers;*
g. *Fine tuning the model with RMSProp optimizer, and a small training rate;*

End

4 Experimental Analysis

4.1 Datasets

1. FERET Facial Dataset [11]: It counts 920 images (460 Female, 460 Male).
2. Expression-Gender dataset [11]: A new facial data set composed of all the images of nine universal facial expression data sets: Yale, Jaffe, Caplier, Bioid, Dafex, EEbd, Cohn & Kanade datasets. The new constructed data set counts 6400 images (3200 female, 3200 Male) with different: Facial expressions, ethnicity, Color, resolution, lighting, age, pose, background and with or without occlusions. FERET data set, is also added to all these datasets. The new constructed dataset is well-balanced, the ratio of men and women is 50%, 50%.

Each considered dataset is split to three partitions: training, validation and test. After the best model configuration is found, the two partitions; training and validation are fused then, the models are tested on the unseen images.

All results reported are calculated according to the Fold-validation protocol.

4.2 Obtained Results and Used Parameters

In the following subsection, we present most values of parameters and hyperparameters used to get our best results for the three models (Table 2).

Table 2. Used parameters.

Model	Batch_size	Step_per_epoch	Validation_steps	Input_shape	Final characteristic form
Simple CNN	10	100	50	(150, 150, 3)	(4, 4, 512)
VGG16	10	100	50	(150, 150, 3)	(4, 4, 512)
RESNET50	10	100	50	(197, 197, 3)	(1,1,2048)

To evaluate models in term of performances, we used: Classification Accuracy and Loss.

From Tables 3 and 4, we can observe that the test-set accuracy is lower than the training set accuracy. This means that the model's performance on the validation data begins to degrade, we've achieved an overfitting. This can be explained by the use of a large number of learned parameters and a little number of training examples. This is why, we explored the use of deep learning with small datasets (transfer learning).

In the following, our best performing Gender Recognition Results are reported on the new constructed dataset (Expression-Gender dataset) (Tables 5 and 6).

In this work, we employ VGG-16, and obtain an accuracy of 97.2% on the new constructed data set. We have pre-trained ResNet-50 following the same training strategy and the resulting CNN reaches 97.37%. Hence, ResNet-50 has proven to be much more effective.

Table 3. Obtained results for simple CNN model on Feret Dataset.

	Epochs	Train acc	Train (loss)	Validation acc	Validation (loss)	Train time (s)
Simple CNN	10	99.5%	1.14%	93.2%	32.6%	2317.23
	30	100%	0.7%	94.2%	48.91%	6668.25
	50	100%	0.1%	92.6%	68.27%	13030.66
	100	100%	0.1%	93.4%	57.83%	22457.93

Table 4. Obtained results for simple CNN model on Expression-Gender Dataset.

	Epochs	Train acc	Train (loss)	Validation acc	Validation (loss)	Train time (s)
Simple CNN	10	97.6%	6.8%	97.2%	13.21%	2279.61
	30	99%	3.01%	96%	25.11%	6271.30
	50	99.8%	1.4%	98.4%	6.12%	10647.04
	100	99.9%	0.4%	98.4%	18.01%	15141.43

Table 5. Obtained results for VGG model on Expression-Gender Dataset.

	Epochs	Train acc	Train (loss)	Test acc	Train time (s)	Test time (s)
Simple VGG16	100	99.96%	0.45%	**97.11%**	8431.69	0.78
VGG16 with Unfreezed layers	100	87.30%	30.39%	95.99%	202280.4	184.4

Table 6. Obtained results for Resnet50 model on Expression-Gender Dataset

	Epochs	Train acc	Train (loss)	Test acc	Train time (s)	Test time (s)
Simple RresNet50	100	75.39	49.8%	85.99%	566.89	2.91
ResNet50 with Unfreezed layers	100	93.50%	16.19%	**97.39%**	316700.8	487.05

However, we could not compare our results with the state of the art ones during this research for the simple reason that we do not have the same bases. The fact is that our initial objective was to study the influence of the facial deformations due expressions on gender recognition.

4.3 Discussion

During this work, we conducted various experiments. We have tested different dataset sizes (Feret, Expression-Gender **and** the two datasets with augmentation). We have explored different architectures with various depths for each used model (Shallow CNN, VGG16, Resnet50 with freezed and unfreezed layers) and we have tested different values of most hyper parameters having an impact on classification results.

From all these experiments we have drawn some key conclusions which are reported below.

1. The three models were trained with different values of epochs (5, 10, 30, 50, 100, 120, 140). The best results are obtained with a number of epochs equal to 100. It is not always necessary to increase the number of epochs to obtain the best results. 100 is a good compromise in most cases.
2. The best obtained results with the three models are better with the new created Dataset (6400 images). When using transfer learning with small data sets, the size of the dataset at hand should be several thousand or more to obtain acceptable results.
3. Data augmentation and the use of unfreezed layers with VGG16 model have no advantage in this study, on the contrary, they degrade performances.
4. Data augmentation and the use of unfreezed layers with Resnet50 model have an interesting advantage: performances on non seen data are better than on the training data, this means that this model has better capacities to generalize.
5. Obtained results with Resnet 50 outperform those obtained with VGG16 on both datasets(the very little dataset Feret (960 images) and the little dataset Expression-Gender dataset (6400 images).
6. Globally, our models fail to predict the correct gender class of some female images with Anger facial expression (when Anger is expressed with important facial deformations). This corresponds well to the results of researches in the field of facial expression recognition, as well as facial recognition. Anger is the least recognized expression among the six universal expressions [12], and images with Anger expression are facial images which are the least recognized ones.
7. In general, our models succeed to predict the correct gender class of most facial images with Joy expression as well as with neutral expressions.

5 Conclusion and Future Work

In this work our objective is threefold. The first objective is to study the effect of facial expressions on facial features deformations and its consequences on gender recognition. To this end, we have built a facial image database that brings together all facial images of the most known datasets used by the research community in the field of facial expression recognition. The new built dataset contains as many common variations, and counts 6400 images with different: Facial expressions, ethnicity, Color, resolution, lighting, age, pose, background and with and without occlusions. The second objective is to evaluate the use of Deep learning in the form of transfer learning for binary classification on small dataset (containing images with different Facial expressions).

Our final goal is then to find a compromise between too much capacity and not enough capacity of the used deep Neural Network in order to don't over fit nor under fit. We analyze the use of shallow, deep and very deep neural networks for Gender recognition. Three different models were tested: a Shallow Convolutional neural network with 6 layers, a deep CNN: VGG16 and a very deep CNN: RESNET50. Many conclusions have been drawn.

Our perspective is to evaluate other binary classifications in order to find out more conclusions about the appropriate number of necessary layers for binary classification.

References

1. Russakovsky, O., et al.: ImageNet large scale visual recognition challenge. arXiv:1409.0575 (2014). ImageNet https://github.com/itf/imagenet-download
2. Chollet, F.: Deep Learning with Python. Manning Publications, USA (2018)
3. van de Wolfshaar, J., Karaaba, M.F., Wiering, M.A.: Deep convolutional neural networks and support vector machines for gender recognition. In: IEEE Symposium Series on Computational Intelligence (2015)
4. Ozbulak, G., Aytar, Y., Ekenel, H.K.: How transferable are CNN-based features for age and gender classification? In: IEEE International Conference on Biometrics Special Interest Group (BIOSIG), pp. 1–6 (2016)
5. Rodriguez, P., Cucurull, G., Gonfaus, J.M., Roca, F.X,. Gonzalez, J.: Age and gender recognition in the wild with deep attention. Pattern Recogn. (2017). https://doi.org/10.1016/j.patcog.2017.06.028
6. Antipov, G., Baccouche, M., Berrani, S.A., Dugelay, J.-L.: Effective training of convolutional neural networks for face-based gender and age prediction. Pattern Recogn. (2017). https://doi.org/10.1016/j.patcog.2017.06.031
7. Simonyan, K., Zisserman, A.: Very deep convolutional networks for large-scale image recognition. ArXiv https://arxiv.org/abs/1409.1556 (2014)
8. He, K., Zhang, X., Ren, S., Sun, J.: Deep residual learning for image recognition. In: IEEE Conference on Computer Vision and Pattern Recognition (CVPR), USA (2016)
9. Viola, P., Jones, M.: Rapid object detection using a boosted cascade of simple features. Comput. Vis. Pattern Recognit. 511–518 (2001). https://doi.org/10.1109/CVPR.2001.990517
10. Keras. https://keras.io/
11. http://www.face-rec.org/databases/
12. Ghanem, K., Caplier, A.: Towards a full emotional system. Behav. Inf. Technol. J. **32**(8), 783–799 (2013)

Reinforcement Learning Based Routing Protocols Analysis for Mobile Ad-Hoc Networks

Global Routing Versus Local Routing

Redha Mili(✉) and Salim Chikhi

MISC Laboratory, Constantine 2 – Abdelhamid Mehri University, Constantine, Algeria
redha.mili@gmail.com, slchikhi@yahoo.fr

Abstract. Energy consumption and maximize lifetime routing in Mobile Ad hoc Network (MANETs) is one of the most important issues.

In our paper, we compare a global routing approach with a local routing approach both using reinforcement learning to maximize lifetime routing.

We first propose a global routing algorithm based on reinforcement learning algorithm called Q-learning then we compare his results with a local routing algorithm called AODV-SARSA.

Average delivery ratio, End to end delay and Time to Half Energy Depletion are used like metrics to compare both approach.

Keywords: Reinforcement learning · Ad-hoc Network · MANETs · Energy AODV · Q-Learning

1 Introduction

In Mobile Ad-hoc Networks (MANETs) [1] the End to End delay, the delivery Rate, the Network lifetime and the energy consumption are indicators of a good network management and a good offered quality of service. In order to satisfy the strict requirements of these parameters, MANET nodes must deal with routing in an efficient and adaptive way.

Indeed, the routing protocol must perform efficiently in mobile environments; it must be able to adapt automatically to the high mobility, the dynamic network topology and link changes. Simple rules are not enough to extend lifetime of the network.

Hence, Reinforcement learning (RL) [2] methods can be used to control both packet routing decisions and node mobility.

Energy efficient routing is a real challenge and may be the most important design criteria for MANETs since mobile nodes will be powered by batteries with different and limited capacity.

Generally, MANETs routing protocol using reinforcement learning can be classified in two different approaches: Global Routing and Local Routing.

É. Renault et al. (Eds.): MLN 2018, LNCS 11407, pp. 247–256, 2019.
https://doi.org/10.1007/978-3-030-19945-6_17

This paper presents a performances evaluation comparison between the designed Global Routing protocols EQ-AODV (Energy Q-Learning AODV), with AODV-SARSA which is a Local Routing protocol using reinforcement leaning.

The EQ-AODV protocol that we present in this paper is hybridization between AODV (Ad hoc On Demand Distance Vector) [3] and the reinforcement learning algorithm Q-Learning [4].

The remainder of the paper is organizing as follow. In Sect. 2, we discuss the related work covering adaptive energy aware routing in MANETs. In Sect. 3 we give a general description of the protocol EQ-AODV. We present in Sect. 4 a performance evaluation comparison between EQ-AODV and AODV-SARSA. In this simulation we captured several metrics: Lifetime, battery energy, thus, the End to End Delay and Delivery Rate. Finally, Sect. 5 concludes the paper.

2 Energy Aware Routing in MANETs

Maximum lifetime-routing protocols perform energy aware routes discovery in two different ways, namely [5, 6] Global Routing and Local Routing. In this section, we survey related work on modeling routing behavior in ad hoc networks. Most of these papers are intelligent routing based, they combines well-known routing algorithms with well-known learning techniques.

2.1 Global Routing Protocol

In Global Routing, all mobile nodes participate in the route discovery process by forwarding RREQ (Route Request) packets. Subsequently, discovered paths are evaluated according an energy-aware metric either by source or destination nodes.

In [7], the concept is in the time delay route request sent by each node. In fact, a node holds the RREQ packet for some time; this time is inversely proportional to its residual battery energy. Hence, paths with nodes that are poor in energy will have minimal chance to be chosen.

In [8], author aims to maximize the nodes lifetime while minimizing the energy consumption. Every source node runs the first-visit ONMC RL algorithm in order to choose the best path based on three main parameters: The minimum- energy path, the max–min residual battery path, and the minimum-cost path.

Like work in [8], authors in [9] choose also to combine the routing protocol with RL algorithm. First they modeled the issue as a sequential decision making problem, then, they show how to map routing into a reinforcement learning problem involving a partially observable Markov decision process.

[10] This paper presents a new algorithm called Energy-Aware Span Routing Protocol (EASRP) that uses energy-saving approaches such as Span and the Adaptive Fidelity Energy Conservation Algorithm (AFECA) [11]. Energy consumption is further optimized by using a hardware circuit called the Remote Activated Switch (RAS) to wake up sleeping nodes. These energy-saving approaches are well-established in reactive protocols.

However, there are certain issues to be addressed when using EASRP in a hybrid protocol, especially a proactive protocol.

2.2 Local Routing Protocol

In Local Routing, each intermediate node, according to its energy-profile, makes its own decision in order:

- To participate or not in routes-discovery,
- To delay EQ-forwarding, or,
- To eventually adjust its EQ-forwarding rate.

The routing model proposed in [12] gives nodes two possible modes of behavior: to cooperate (forward packets) or to defect (drop packets).

In [13], each node j forwards packets with a probability μj. When a packet is sent, each node computes the current equilibrium strategy and uses it as the forwarding probability. Also, a punishment mechanism is proposed where nodes decrease their forwarding probabilities, when someone deviates from the equilibrium strategy.

Despite the proven effectiveness of these works, authors in [14, 15] offer other efficient routing techniques by applying Reinforcement Learning to enable each node to learn appropriate forwarding rate reflecting its willingness to participate in routes discovery process.

In [16] authors propose a dynamic fuzzy energy state based AODV (DFES-AODV) routing protocol for Mobile Ad-hoc Networks (MANETs) where during route discovery phase, each node uses a Mamdani fuzzy logic system (FLS) to decide its Route Requests (RREQs) forwarding probability.

Unlike work in [16], Fuzzy Logic System (FLS) was used in [17] for adjusting the willingness parameter in OLSR protocol. Decisions made at each mobile node by the FLS take into account its remaining energy and its expected residual lifetime.

Authors in [18] propose a new intelligent routing protocol for MANET based on the combination of Multi Criteria Decision Making (MCDM) technique with an intelligent method, namely, Intuitionistic Fuzzy Soft Set (IFSS) which reduces uncertainty related to the mobile node and offers energy efficient route.

MCDM technique used in this paper is based on entropy and Preference Ranking Organization Method for Enrichment of Evaluations-II (PROMETHEE-II) method to determine efficient route.

3 Proposed Protocol Design

To maximize lifetime of network, we present in this section a reactive protocol called EQ-AODV (Energy Q-learning AODV protocol) using reinforcement learning algorithm.

Based on the original AODV [4], EQ-AODV is an enhanced routing protocol that use Q-learning algorithm [3] to achieve whole network link status information from local communication and change routes preemptively using the information so learned.

In our approach, the network was modeled as a Markov decision processes (MDP) as described in [19] (Fig. 1).

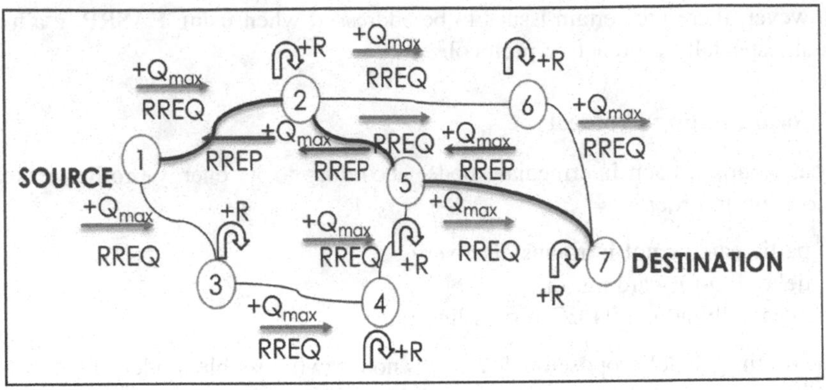

Fig. 1. AODV as a Markov decision process

We see clearly two new values $\mathbf{Q_{max}}$ and **R** added to the original AODV. Respectively, the best Q-value extract from the routing table of neighbor which sends RREQ or RREP, and the reward R calculated at each RREQ and RREP reception based on Energy.

Before gives the proposed RL model, let's do a comparison between AODV-SARSA [14] and EQ-AODV (Table 1):

Table 1. Comparison between AODV-SARSA and EQ-AODV

Comparison criterions	AODV SARSA	EQ-AODV
Global Vs local routing	Local	Global
Set of states	Residual lifetime % (RT)	Nodes
Set of actions	Ratio of RREQs forward	RREQ (route request)
Reward regime	Average of drain rates	Residual lifetime (RT)
Metric	Min-Hop	Q-value

The RL model proposed in this article can be described as follows:

3.1 The Set of States

Each node in the network is considered as a state. The set of all nodes is the state space. Each node:

- Calculates the reward R,
- Calculates Q-value with neighbors,
- Selects the next hop that it should forward packets.

3.2 The Set of Actions

The action can be equivalent to a packet being delivered from one node to its neighbor. The set of neighbor changes due to mobility of nodes. Each node only needs to select its best next hop. The metric used by AODV to choose the best next hop is hop-count. In EQ-AODV, the best next hop is based on the estimation of Q-value from origin to destination based on the Q-Learning algorithm.

3.3 Reward

To calculate our Q-Value for destination, we chose a reward signal based on Drate value (Energy Drain Rate) and the Residual Energy of node. Drate is calculated using the exponential moving average method [20]:

$$DRate_t = \delta \times DRate_{t-1} + (1 - \delta) \times DRate\_Sample_t \quad (1)$$

$DRate_{t-1}$ and $DRate\_Sample_t$ indicate, respectively, the old and the newly calculated energy drain rate values. More priority should be attributed to the current drain rate value using δ weighting factor. To measure the energy drain rate per second, each node monitors its energy consumption during a T seconds sampling interval [20].

We use the result of $DRate_t$ with RE (Residual Energy) to calculate the RT (Residual Lifetime) who is considered like our reward:

$$RT_t = \frac{DRate_t}{RE_t} \quad (2)$$

3.4 RL Algorithm

We chose Q-Learning algorithm [4]. One of the most popular and we define an experience-tuple: (s_t, a_t, R_t, s_{t+1}, a_{t+1}) summarizing a single transition for the RL-agent in its environment. Where:

- s_t is the state before the transition,
- a_t is the taken action,
- R_t is the immediate reward,
- S_{t+1} is the resulting state,
- and a_{t+1} is the chosen action at the next time step t + 1.
- Let α [0, 1] and γ [0, 1] be the learning rate and the discount factor, respectively.

The Action-value function Q(s, a), estimates the expected future reward to the agent when it performs a given action in a given state and following the learned policy π.

```
Algorithm. Q-Learning algorithm
Initializations:
  Initialize Q(s,a);
  Initialize s_t;
Repeat for each time-step
Choose an action a_t using
Take a_t
Observe the reward r_t and the state s_t+1
Update Q(s_t,a_t) : Q(s_t,a_t) ← Q(s_t,a_t)+α(r+γ(s_t+1,a_t+1)-(s_t,a_t))

Until the terminal state is reached
```

In this paper, we assume that the Q-Learning is distributed and each node has a part of Q(s, a) table with Neighbors.

4 Experiments Results and Discussion

In this section, we first describe the simulation environment used in our study and then discuss the results in detail. Our simulations are implemented in Network Simulator (NS-2) [21]. At this level of our study, we discuss the results of both EQ-AODV and AODV-SARSA.

In brief, simulation parameters were set as illustrated in Table 2.

Table 2. Simulation parameters setting

Simulation parameter	Value
Network scale	800 × 800
Simulation time	900 s
Number of nodes	50
Mobility model	Random way point
Pause time	0 s
Traffic type	CBR
Connections number	10, 20, 30
Packets transmission rate	4 packets/s
Initial energy	10 J
Transmission power	0,6 W
Reception power	0,3 W
T sampling interval	6 s
Learning rate	0,9
Discount factor	0,1

To evaluate performance of EQ-AODV, we compare the EQ-AODV algorithm with AODV-SARSA, using the following metrics:

- **Delivery Rate:** the ratio of packets reaching the destination node to the total packets generated at the source node.
- **Average End-to-End Delay:** the interval time between sending by the source node and receiving by the destination node, which includes the processing time and queuing time.
- **The Time Half Nodes Depletion:** the time at which the network see 50% of its nodes exhausting all their batteries [14].

Tables 3, 4 and 5 shows the performances of each protocol EQ-AODV and AODV-SARSA using 50 nodes and Maximum Velocity 10 m/s in low, medium and high traffic.

Table 3. Simulation results for delivery rate

Delivery rate	AODV-SARSA	EQ-AODV
Low traffic-10 connections	82,45798333	85,34965
Medium traffic-20 connections	68,74455517	71,30487333
High traffic-30 connections	54,77744	57,02822

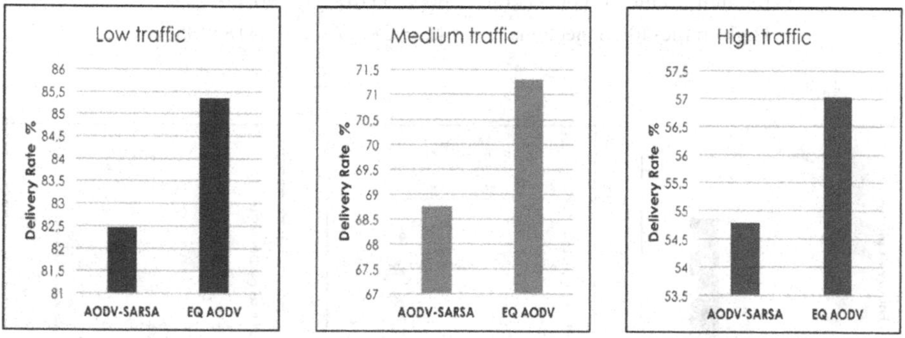

Fig. 2. Average delivery ratio

Results (Table 3 and Fig. 2) show that EQ-AODV has the best Delivery Ration. The ration of packets reaching destination is higher in Low traffic, Medium traffic and High traffic.

Table 4. Simulation results for end to end delay

End to End Delay	AODV-SARSA	EQ-AODV
Low traffic-10 connections	0,046278313	0,085645713
Medium traffic-20 connections	0,079130017	0,095483247
High traffic-30 connections	0,152210963	0,21413287

By changing the hop-count metric of AODV that represents the shortest path, we expected to degrade the End to End Delay. Results (Table 4 and Fig. 3) show that the End to End delay is clearly the weakness point of EQ-AODV. AODV-SARSA is better in Low traffic, Medium traffic and High traffic.

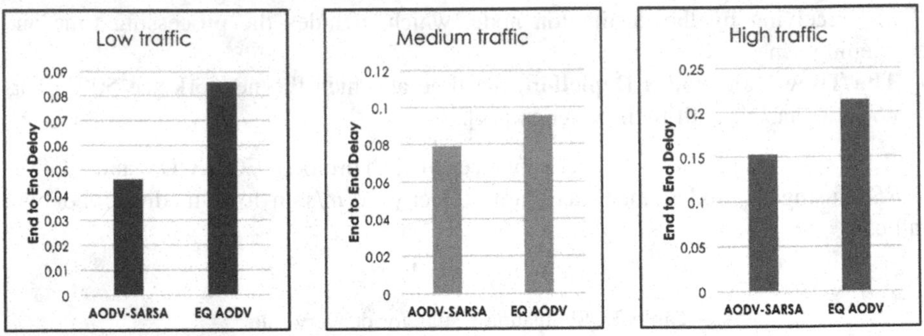

Fig. 3. End to End Delay

Table 5. Simulation results for time half energy deplation

Delivery rate	AODV-SARSA	EQ-AODV
Low traffic-10 connections	118,2996865	122,0999305
Medium traffic-20 connections	86,9253407	90,1144598
High traffic-30 connections	74,99828117	76,31837853

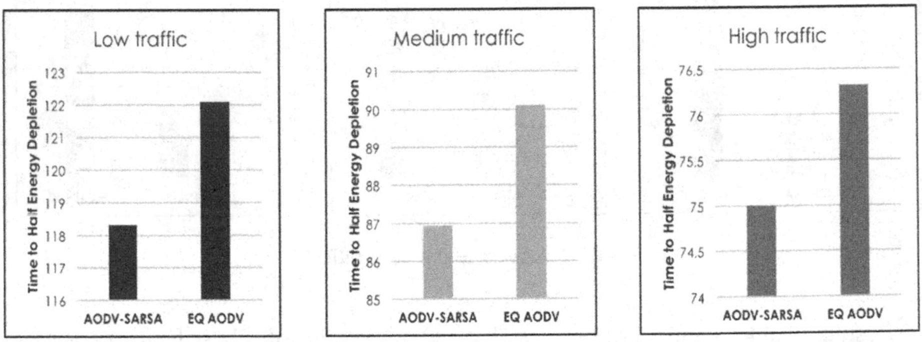

Fig. 4. Time to half energy depletion

About consuming energy, results (Table 5 and Fig. 4) show that EQ-AODV is better than AODV-SARSA in all simulations. The Time to Half Energy Depletion is clearly better in Low traffic, Medium traffic and High traffic.

The more Time Half Energy Depletion, high lifetime will be.

5 Conclusion

In this paper we have raised the issue of Energy Aware Routing while maximizing the Network lifetime in MANET.

Using simulation, we chose to compare two types of routing algorithms: Global and Local Routing. Both algorithms are based on reinforcement learning techniques.

The results show that both algorithms have encouraging performance for MANET networks. The EQ-AODV gives better performances than AODV-SARSA in most metrics, as the packet delivery ratio, and energy consumption. However, AODV-SARSA End to End performances is better.

We can conclude that the choice of the routing algorithm will be made according to the metric that network want optimize, and this; depending on the service demand.

Our future work will focus on implementing other reinforcement learning algorithms based on difference temporal for both local and global approach. Also, testing the proposal in different network conditions (high/low mobility, high/low density…)

References

1. Giordano, S.: Mobile Ad Hoc Networks. Handbook of Wireless Networks and Mobile Computing, pp. 325–346 (2002)
2. Sutton, R.S., Barto, A.G.: Reinforcement Learning, 2nd edn. MIT Press, Cambridge (2014)
3. Perkins, C., Belding-Royer, E., Das, S.: Ad Hoc On-Demand Distance Vector (AODV) Routing. Network Working Group, July 2003. ftp://ftp.nordu.net/rfc/rfc3561.txt
4. Watkins, C.J.C.H., Dayan, P.: Q-learning. Mach. Learn. **8**, 279–292 (1992)
5. Chettibi, S., Chikhi, S.: A survey of reinforcement learning based routing protocols for mobile ad-hoc networks. In: Özcan, A., Zizka, J., Nagamalai, D. (eds.) CoNeCo/WiMo - 2011. CCIS, vol. 162, pp. 1–13. Springer, Heidelberg (2011). https://doi.org/10.1007/978-3-642-21937-5_1
6. Vassileva, N., Barcelo-Arroyo, F.: A survey of routing protocols for maximizing the lifetime of ad hoc wireless networks. Int. J. Softw. Eng. Appl. **2**(3), 77–79 (2008)
7. Cho, W., Kim, S.L.: A fully distributed routing algorithm for maximizing life time of a wireless ad hoc network. In: Proceedings of IEEE 4th International Conference on Workshop-Mobile & Wireless Communication Network, pp. 670–674, September 2002
8. Naruephiphat, W., Usaha, W.: Balancing tradeoffs for energy-efficient routing in MANETs based on reinforcement learning. In: The IEEE 67th Vehicular Technology Conference (2008)
9. Nurmi, P.: Reinforcement learning for routing in ad-hoc networks. In: Proceedings of the Fifth International Symposium on Modeling and Optimization in Mobile, Ad-Hoc, and Wireless Networks (WiOpt) (2007)
10. Ravi, G., Kashwan, K.R.: A new routing protocol for energy efficient mobile applications for ad hoc networks. Comput. Electr. Eng. **48**, 77–85 (2015)
11. Xu, Y., Heidemann, J., Estrin, D.: Geography informed energy conservation for ad-hoc routing. In: Proceedings of 7th Annual International Conference on Mobile Computing and Networking, pp. 70–84 (2001)
12. Srinivasan, V., Nuggehalli, P., Chiasserini, C.F., Rao, R.R.: Cooperation in wireless ad hoc networks. In: Proceedings of the 22nd Annual Joint Conference of the IEEE Computer and Communications Societies (INFOCOM), pp. 808–817. IEEE Computer Society (2003)

13. Altman, E., Kherani, A.A., Michiardi, P., Molva, R.: Non-cooperative forwarding in ad-hoc networks. In Proceedings of the 15th IEEE International Symposium On Personal, Indoor and Mobile Radio Communications (2004)
14. Chettibi, S., Chikhi, S.: An adaptive energy-aware routing protocol for MANETs using the SARSA reinforcement learning algorithm. In: IEEE Conference on Evolving and Adaptive Intelligent Systems (EAIS), pp. 84–89 (2012)
15. Chettibi, S., Chikhi, S.: Adaptive maximum-lifetime routing in mobile ad-hoc networks using temporal difference reinforcement learning. Evol. Syst. **5**, 89–108 (2014)
16. Chettibi, S., Chikhi, S.: Dynamic fuzzy (local routing) logic and reinforcement learning for adaptive energy efficient routing in mobile ad-hoc networks. Appl. Soft Comput. **38**, 321–328 (2016)
17. Chettibi, S., Chikhi, S.: FEA-OLSR: an adaptive energy aware routing protocol for manets using zero-order sugeno fuzzy system. Int. J. Comput. Sci. Issues (IJCSI) **10**(2), 136–141 (2013)
18. Das, S.K., Tripathi, S.: Intelligent energy-aware efficient routing for MANET. Wirel. Netw. **24**(4), 1139–1159 (2018)
19. Sutton, R., Barto, A.: Reinforcement Learning. MIT Press, Cambridge (1998)
20. Kim, D., Garcia-Luna-Aceves, J.J., Obraczka, K., Cano, J.C., Manzoni, P.: Power-aware routing based on the energy drain rate for mobile ad-hoc networks. In: 11th International Conference on Computer Communications and Networks (2002)
21. NS: The UCB/LBNL/VINT Network Simulator (NS) (2004). http://www.isi.edu/nsnam/ns/

Deep Neural Ranking for Crowdsourced Geopolitical Event Forecasting

Giuseppe Nebbione[1(✉)], Derek Doran[2], Srikanth Nadella[3], and Brandon Minnery[3]

[1] Department of Electrical and Computer Engineering, University of Pavia, Pavia, Italy
giuseppe.nebbione01@universitadipavia.it

[2] Department of Computer Science and Engineering, Wright State University, Dayton, OH, USA
derek.doran@wright.edu

[3] Wright State Research Institute, Dayton, OH, USA
{srikanth.nadella,brandon.minnery}@wright.edu

Abstract. There are many examples of "wisdom of the crowd" effects in which the large number of participants imparts confidence in the collective judgment of the crowd. But how do we form an aggregated judgment when the size of the crowd is limited? Whose judgments do we include, and whose do we accord the most weight? This paper considers this problem in the context of geopolitical event forecasting, where volunteer analysts are queried to give their expertise, confidence, and predictions about the outcome of an event. We develop a forecast aggregation model that integrates topical information about a question, meta-data about a pair of forecasters, and their predictions in a deep siamese neural network that decides which forecasters' predictions are more likely to be close to the correct response. A ranking of the forecasters is induced from a tournament of pair-wise forecaster comparisons, with the ranking used to create an aggregate forecast. Preliminary results find the aggregate prediction of the best forecasters ranked by our deep siamese network model consistently beats typical aggregation techniques by Brier score.

Keywords: Event forecasting · Deep learning · Crowdsourcing · Siamese networks

1 Introduction

The science (and art) of forecasting has been studied in domains such as computer and mobile network monitoring and evaluation [13,16,30], meteorology [7], economics [10], sports [27], finance [8], and geopolitics [11,14]. A closely related line of research involves leveraging the power of aggregation to improve forecast accuracy. In this approach, multiple forecasts from different sources (human or algorithm) are merged to achieve results that are, on average, superior to

E. Renault et al. (Eds.): MLN 2018, LNCS 11407, pp. 257–269, 2019.
https://doi.org/10.1007/978-3-030-19945-6_18

those of even the best individual forecaster. This "wisdom of crowds" effect [32] arises from the fact that individual forecasters possess different information and biases [28], leading to uncorrelated (or better yet, negatively correlated) prediction errors that cancel when combined (e.g., by averaging). Implicit in the above description is the assumption that forecasters are more than just proverbial "dart-throwing chimps": they possess some degree of relevant knowledge and expertise enabling them to perform above chance. By extension, not all forecasters have equal expertise, suggesting that it may be possible to identify smaller, "wiser" ensembles of the most skilled forecasters [19]—or alternatively, to assign different weights to different forecasters based on their expected contribution to the crowd's accuracy.

Indeed, a common approach to forecast aggregation is to use weighted models that favor the opinion of forecasters by their experience and past accuracy. Yet even weight aggregation methods are imperfect. This is because they tend to oversimplify the forecaster representation and do not take into account how skilled a forecaster may be within a specific context. This issue becomes even more apparent when considering the problem of *geopolitical forecasting* [34], where an analyst is asked to predict the outcome of an international social or political event. In geopolitics, a substantial number of factors, some of which may be highly unpredictable, must be considered. While a forecaster may possess expertise about a particular international region, political regime, or event type (e.g., Russian military operations in Ukraine), the countless factors and hidden information contributing to an event outcome guarantee that expertise will correlate imperfectly with prediction accuracy. Moreover, such expertise may not generalize to other forecasting questions involving different regions, regimes, or event types. We thus suggest a deep learning solution to the problem of crowd-sourced geopolitical forecaster aggregation. By the architecture of its network layers, and the number and size of such layers, deep neural networks can be designed to become arbitrarily expressive [26]. Such expressiveness is necessary in identifying good predictions from a crowd of forecasters because the factors that determine the accuracy of a prediction are numerous and latent. A challenge of applying expressive deep networks, however, is the expense of acquiring the substantial amount of data needed to train a generalizable model. The data that do exist come from past forecasting competitions [20], yet such data may still not be adequate to train a deep network with sufficient expressive power.

To overcome this challenge, we propose a method for geopolitical prediction aggregation from a crowd based on a *neural ranking* of forecasters. This refers to a deep neural network that defines a ranking of forecasters based on their *predicted relative accuracy* for a given forecasting question. Forecasters' rankings are based on their performance in previous questions, their self-reported confidence in their prediction, and on the latent topics present in the previous questions they performed well on. More formally, given a set of m geopolitical questions asked to n forecasters, where some subset of $k < n$ forecasters provide a prediction to each question, a neural ranker yields a *ranking* of the n forecasters based on how likely their response is to be "closest" to the correct outcome.

The ranking is derived from a tournament where all pairs of $\binom{k}{2}$ forecasters are compared by a deep siamese network to identify the one whose forecast is more likely to be correct. This scheme also yields a training dataset of $\sum_{i=1}^{m}\binom{k_i}{2}$ pairs of examples for training and testing the deep siamese network, which is far more likely to be of sufficient size for training a generalizable model. We evaluate our neural ranking approach against a standard weighted aggregation algorithm in which weights are based on forecasters' average past performance. Using data provided by the IARPA Good Judgement Project [2], we find that our neural ranking method produces substantially improved Brier scores (a measure of prediction accuracy) compared to the standard weighted aggregation model.

The layout of this paper is as follows: Sect. 2 gives details about the state of the art in the context of aggregation methods for forecasts. Section 3 details the design of our neural ranking system and its siamese network and ranking components. Section 4 describes evaluation results for our system compared to standard weighted aggregation models. Section 5 concludes the paper and offers directions for future research.

2 Related Work

Sir Francis Galton published the first modern scientific study of crowd wisdom more than a century ago in a paper that analyzed data from a contest in which hundreds of individuals independently attempted to estimate the weight of a prize-winning ox [18]. More recent research has similarly capitalized on data from large-scale public or private forecasting contests, where forecasters compete (often for financial award) individually or in teams to produce the most accurate predictions within a particular domain. Popular examples of such forecasting tournaments include fantasy sports [3,19] and, more recently, geopolitics [33]. These tournaments, which can include thousands of competitors, are a rich source of quality data because they naturally attract participants who are both knowledgeable and motivated.

With respect to forecast aggregation, recent work has focused on integrating forecaster features and past performance. Budescu *et al.* [9] proposes a method for measuring a single judge's (e.g., a single forecaster's) contribution to the crowd's performance and uses positive contributors to build a weighting model for aggregating forecasts. Forlines *et al.* [17] shows how heuristic rules based on the qualities of each forecaster can significantly improve aggregated predictions. Hosen *et al.* [23] applies a neural network to forecast a range of likely outcomes based on a weighted average aggregation of forecaster predictions. Further work by Ramos [29] uses a particular nonparametric density estimation technique called L2E, which aims at making the aggregation robust to clusters of opinions and dramatic changes. The Good Judgement project by Tetlock *et al.* [33] introduces an approach based on a novel cognitive-debiasing training design. Atanasov *et al.* [2] shows that team prediction polls outperform prediction markets once forecasts are statistically aggregated using various techniques such as temporal decay, differential weighting based on past performance, and recalibration.

In this work, we consider an innovative method to tap into the wisdom of the crowd for geopolitical forecasting. We reformulate the crowd forecasting problem as a neural ranking problem, in which a deep neural network is used to rank forecasters based on their expected relative accuracy for a given forecasting question. The neural network learns a complex representation of forecasters, their forecasts, and contextual information about the question of interest. The resulting ranking is then used to create a weighted aggregation of forecasts (i.e., a crowd forecast) for each unique forecasting question.

3 Methodology

We introduce a new crowd aggregation technique based on a novel *ranking* of forecasters who have provided a series of predictions for a particular question as illustrated in Fig. 1. We assume that forecasters can submit multiple predictions for a given question; i.e., forecasters are allowed to update their predictions as they receive new information over time. All pairs of forecasts made by unique forecasters are passed into a deep siamese network [25] that will evaluate which of the two forecasts are most likely to be closer to the true event outcome.

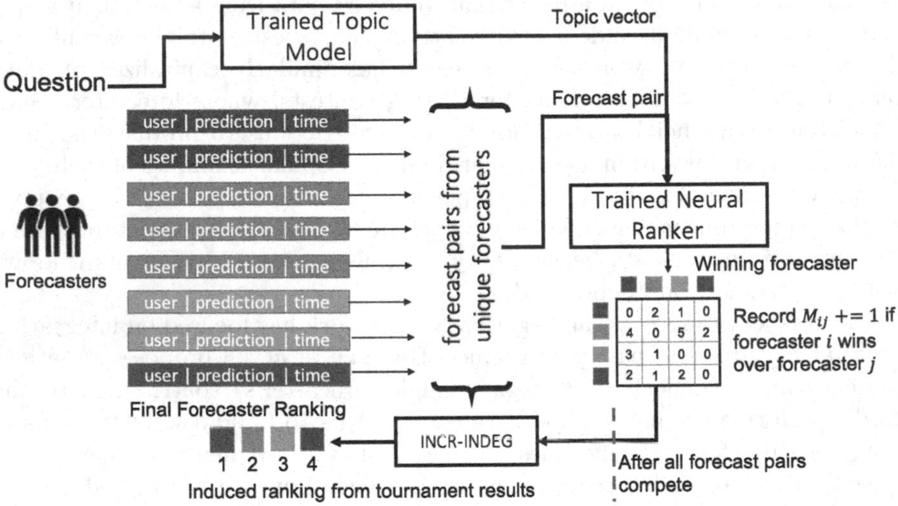

Fig. 1. Approach overview. Given multiple predictions on the outcome of a question from multiple forecasters, we apply a siamese neural network to identify, for all pairs of predictions submitted by unique forecasters, which of the two is more likely to be closer to the true outcome. The network additionally considers the latent topics within the question text to establish relationships between forecasters and the topics they make predictions over. Comparisons are recorded in a matrix that can be represented as a weighted tournament over the forecasters. The INCR-INDEG algorithm is applied to the tournament to infer a ranking. The predictions made by those forecasters in a top percentile of the ranking are aggregated to form the crowd's prediction.

The submitted forecasts include a forecaster-provided confidence score about their prediction. The siamese network further considers the kinds of topics featured in each question, which is derived by running the question text through a trained topic model. The outcome of all comparisons by the neural ranker is composed in a matrix M that defines a tournament graph over the forecasters. The tournament graph is processed through the INCR-INDEG algorithm [15] to induce a ranking of forecasters by their comparative ability to provide better predictions. An aggregation of the highest ranking forecasters (defined by some top percentile of the ranking) are used as the crowd's prediction. In the event that multiple forecasts are submitted by the same ranker, we compute a prediction conditioned on a time t by using the latest prediction submitted by each forecaster prior to t. We elaborate on the topic model, neural ranker, and ranking algorithm next.

3.1 Topic Modeling

The first step is to learn a topic model over a corpus of questions about geopolitical events that have been asked to a to a crowd of forecasters. The purpose of the topic model is to identify *a priori* the topics latent within the set of questions asked to forecasters. In doing so, we seek to provide the neural ranker with *topical* information about the question being asked, so that the ranker becomes able to learn associations between forecasters and the topics that they are (not) proficient in. We specifically apply Latent Dirichlet Allocation (LDA) [4] across an entire corpus of geopolitical questions that have been asked to forecasters in the past. LDA is a widely used unsupervised learning technique that represents a topic t_j $(1 \leq j \leq T)$ as a multinomial probability distribution $p(w_i|t_j)$ over M words taken from a corpus $D = \{d_1, d_2, \cdots, d_N\}$ of documents d_i where words are drawn from a vocabulary $W = \{w_1, w_2, \cdots, w_M\}$. The probability of observing a word in document is defined as $p(w_i|d) = \sum_{j=1}^{T} p(w_i|t_j)p(t_j|d)$. Gibbs sampling is used to estimate the word-topic distribution $p(w|t)$ and the topic-document distribution $p(t|d)$. While various metrics to identify the ideal number of topics T for a given corpus have been identified [5,21], no metric is a silver bullet, and manual inspection is often the preferred approach. Manual inspection ensures that words clustered into various topics are collectively meaningful, in the sense that a reasonable person may look at the words in a cluster to conclude that they are representative of a topic. For example, a cluster of words including *nuclear, peninsula, dictatorship, DMZ* likely represents a topic about North Korea. Admitting a larger number of topics in the model carries the risk of producing clusters of words that are not topically related. A learned model over T topics is then used to produce a T-dimensional topic vector for each question. The ith component of this vector simply represents the proportion of words in a question from topic i of the model.

3.2 Neural Ranker

A pair of forecasts made and topic vector for a question are then fed into a neural ranker. A neural ranker that simply takes as input all forecasts and the question to produce a ranking may be intuitively appealing, but may also not be possible in most geopolitical forecasting contexts. This is because such a network would require a large amount of training data to fit its myriad of parameters while ensuring generalization, and because the size of the input must be fixed. But for geopolitical forecasting, it is difficult to find a dataset consisting of thousands of questions with forecasts provided by a fixed set of forecasters; in fact there is often a variable number of forecasters who make predictions about a small number of questions [6].

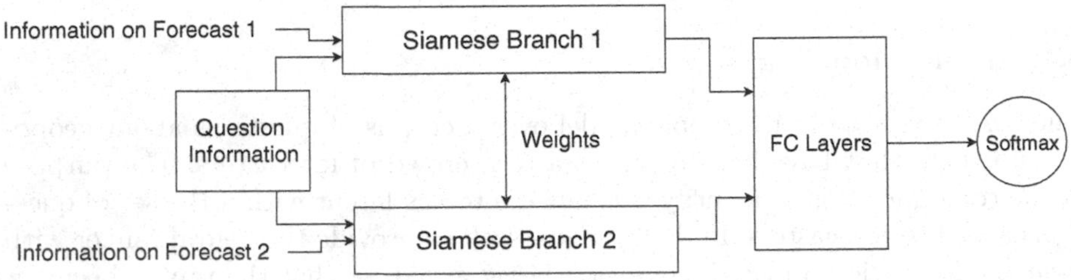

Fig. 2. Neural ranking architecture with two siamese branches. Both branches take the same topic vector of a question and share network weights. The output of a siamese branch can be thought of as a vector embedding of data about a forecaster, a question, and the forecaster's prediction for the question. A softmax output is interpreted as how likely the first forecast is closer to the true answer than the second.

We instead consider a neural ranker that evaluates which of a pair of forecasts submitted for a question is more likely to be closer to the correct response. By training such a ranker over forecasts submitted by the same set of forecasters across a set of questions with a variety of topics, associations between forecasters, question topics, and their prediction patterns can be learned and exploited for identifying the "better" of two predictions. The neural ranker is designed as a deep siamese network [22, 31] illustrated in Fig. 2. The siamese network is composed of two identical multi-layer perceptrons that share weights. Each branch takes as input a forecast concatenated with meta-data about the forecaster and the topic vector quantifying the information content of each question. The output of a siamese branch is a vector embedding representative of information about the forecaster, the question asked, and their prediction. The embeddings from each branch are concatenated and fed into a series of fully connected layers. A softmax output scores whether the prediction of forecaster 1 is superior. The network is trained by stochastic gradient descent with momentum using a binary cross-entropy loss function.

3.3 Ranking Process

The trained neural ranker evaluates all pairs of predictions for a given question submitted by two unique forecasters. For a question asked to n forecasters, we define an $n \times n$ matrix M where M_{ij} counts the number of times forecaster i's prediction was chosen over forecaster j's. We then define the matrix T where $T_{ij} = M_{ij}/(M_{ij} + M_{ji})$ as the adjacency matrix of a weighted tournament with probability constraints $T = (V, E, w)$. Here, V represents forecasters and $w : E \rightarrow [0,1]$ is a weight function such that $w((v_i, v_j)) + w((v_j, v_i)) = 1$ for any $(v_i, v_j) \in E$. A ranking of the $V \in T$ can be defined by an ordering $\sigma : V \rightarrow 1, 2, ..., |V|$ minimizing the sum of all $w((v_i, v_j))$ of all backedges induced from σ, where a backedge (v_a, v_b) has $\sigma(v_a) > \sigma(v_b)$. Finding σ is NP-hard [1,12], but Coopersmith *et al.* discovered a simple 5-approximation algorithm called INCR-INDEG [15] where σ orders vertices by their weighted in-degree $D(v_i) = \sum_{v_j \in V \setminus v_i} w(v_i, v_j)$ with ties broken randomly. We apply INCR-INDEG to T to produce a final ranking of forecasters for a question.

4 Evaluation

We evaluate our neural ranker using public data from the IARPA Good Judgment Project[1]. This data is a product of a four-year long (2011–2015) prediction tournament over geopolitical events. Questions from the project were published for a variable amount of time, during which a forecaster could submit a prediction along with a confidence score (an integer value between 1 and 5). A forecaster could update his or her forecast at any time, and the most recent forecast on any given day was taken as the forecast for that day. For each question, forecasters were presented with disjoint possible outcomes and were asked to submit the probability that each outcome would occur. Forecasters were free to determine which questions to attempt and how often to update their forecasts, resulting in a variable number of forecasters and forecasts per question. Questions from each year of the project were answered by a different set of forecasters, with a subset of forecasters participating across multiple years.

Questions greatly varied in terms of subject but were always related to determining if, when, or how a political, geographical, social, or economic event would occur in the future. Question also carried a brief but detailed description and links to news articles and on-line sources for a forecaster to begin investigations. Examples of questions from the IARPA forecasting tournament include:

- *By 1 January 2012 will the Iraqi government sign a security agreement that allows US troops to remain in Iraq?*
  1. Yes, by 15 October 2011
  2. Yes, between 16 October and 1 January
  3. No

[1] https://www.dni.gov/index.php/newsroom/press-releases/item/1751-iarpa-announces-publication-of-data-from-the-good-judgment-project.

- *Will the United Nations Security Council pass a new resolution concerning Iran by 1 April 2012?*
  1. Yes, a new resolution will be passed
  2. No, a new resolution will not be passed
- *Who will be inaugurated as President of Russia in 2012?*
  1. Medvedev
  2. Putin
  3. Neither

The responses to each question are constructed so that the actual outcome will always correspond to exactly one of the responses. We analyzed our approach using questions from the first year of the competition (extensive evaluations across all four years will be pursued in future work). The 101 questions in the first year had an average of 1,440.52 forecasts with standard deviation of 668.

4.1 Neural Model Evaluation

We ran LDA to extract the latent topics from a corpus of each question's text concatenated with their description. Manual inspection of words organized into clusters by LDA were used to determine that six topics would be appropriate. This was reached by a trial-and-error approach, starting from a large number of topics, where we continued to reduce the number of topics in the model until there was insignificant overlap of words within different topics.

The neural ranker is composed of two siamese branches of three fully connected layers. Each layer is composed of 32 ReLU activation functions. The 32 dimensional output of each branch is concatenated and passed into four fully connected layers, each composed of 64 ReLU activation functions. The final fully connected layer outputs a single logistic activation. We carried a preliminary evaluation of the neural ranker using all questions from the first year of the IARPA forecasting tournament featuring 101 questions or "IFIPs". The neural ranker was trained over a random sampling of 5 million response pairs across year 1[2]. Model selection was performed by a validation set composed of all pairs of responses from two questions (IFIP 1050 and 1051 in the dataset, respectively). The network was trained with mini-batches 512 prediction pairs. Model parameters were optimized to minimize cross-entropy loss by stochastic gradient descent with a learning rate of $\lambda = 0.01$ and with a momentum term $\rho = 0.9$ added. The network was trained over multiple epochs and instantiated an early stoppage procedure when the validation error exceeded training error for more

[2] We must mention that the 5M response pairs were meant to be sampled from the first 50 questions of the Year 1 data set, but due to a bug identified after submission of the paper, the sampling occurred across all Year 1 questions. This caused some testing set examples to have bled into our training data as well. However, this evaluation bug does **not** affect our evaluation of the crowd's performance – which is the ultimate aim of the model – to be discussed in Sect. 4.2. Our subsequent work will be carried out with this bug fixed.

than 3 consecutive epochs. After training, we ultimately select the model parameter settings from the end of the latest epoch whereby the training error rate did not fall below validation error over IFIP 1050 and 1051 questions. The training process was repeated for a small number of different networks where the number of layers and size of the layers varied. The layer sizes and counts listed above were selected based on these repeated experiments; a comprehensive sensitivity analysis of the rankings against different architectures will be the subject of future work.

The neural ranker was tested against all forecast pairs from the 54 questions asked in the Year 1 dataset, constituting the later half of questions asked to the forecast crowd. These questions were chosen to simulate a test set where questions asked in the future are evaluated on a model trained on examples from the past. Figure 3 shows the distribution of the number of questions for which the neural ranker is able to choose the better of two predictions at some accuracy. The performance is encouraging: despite the variety and complexity of the questions asked, the neural ranker is able to identify the better of two predictions over 80% of the time on average. We found performance to decrease slightly after one epoch, while training accuracy continued to improve. We thus fixed the model after training for one epoch to minimize overfitting.

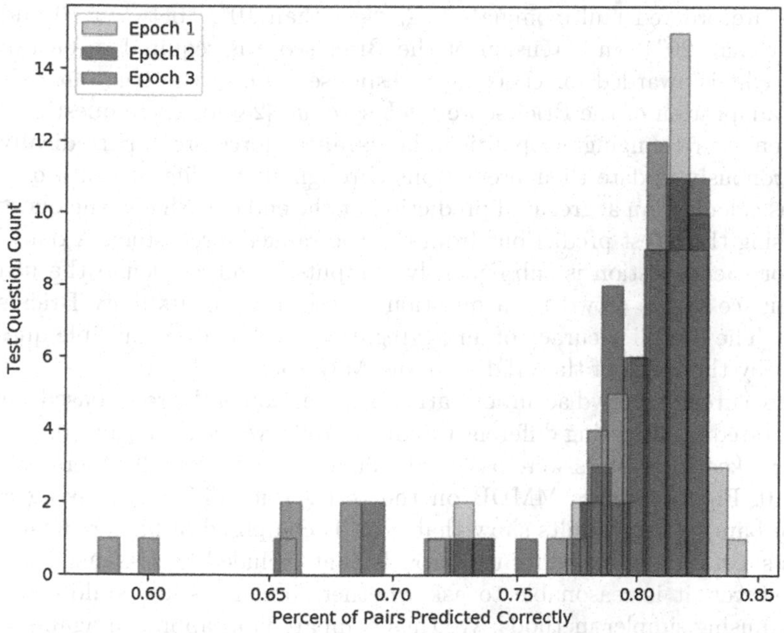

Fig. 3. Distribution of the proportion of forecast pairs where the neural ranker identifies the better prediction, per question in the validation set. Colors correspond to the distribution after one, two, and three training epochs. The distributions suggest decreasing validation set performance for later epochs, despite increasing accuracy over the training set. (Color figure online)

4.2 Crowd Performance

We defined a crowd's prediction as the weighted arithmetic mean prediction for each question, where weights were based on a forecaster's ranking for that question. For simplicity's sake, we used binary weights: that is, a forecaster c_i's prediction was either included ($w_i = 1$) or not ($w_i = 0$) in the aggregate based on his or her rank computed by the neural ranker and INCR-INDEG algorithm. We used a proper scoring rule called the Brier score [7], to measure the accuracy of a crowd's prediction for a given question. The Brier score, or Mean Quadratic Score, measures forecasting accuracy under discrete choice conditions such as the answer sets for the questions used in the IARPA tournament. The Brier score is proper as it encourages forecasters to report their true beliefs (i.e., their best estimate of the likelihood of a future event occurring). It is calculated as:

$$\frac{1}{n}\sum_{i=1}^{n}\sum_{j=1}^{r}(f_{ij} - o_j)^2$$

where n is the size of the crowd, r is the number of possible results, f_{ij} is the probability of result j predicted by forecaster i and $o_j \in \{0, 1\}$ is equal to 1 only if result j does occur. Brier scores can range from 0 to 2, with lower scores corresponding to greater accuracy. In the case of a question whose response options are ordered multinominals (e.g. "less than 10"; "between 10 and 20"; "greater than 20"), an extension of the Brier score is required to ensure that more credit is awarded for choosing a response close to the true outcome. We use an adaptation of the Brier score by Jose *et al.* [24] for these questions.

Given ever-changing geopolitical landscapes, forecasters periodically and asynchronously update their predictions throughout the lifespan of a question. We thus calculate an aggregated prediction at the end of each day the question is open using the latest predictions from the top-ranked forecasters. A daily Brier score for each question is subsequently computed, and we define the ultimate performance of the crowd for a question as the mean of its daily Brier scores (MDB). The overall accuracy of an aggregation method over multiple questions is given by the mean of the MDBs, or the MMDB.

To explore how crowd accuracy varies as a function of the rank-based weights, we generated results using different ranking cutoffs, where a varying top percentage of ranked forecasters were assigned weight $w_i = 1$ with all others receiving weight 0. Figure 4 shows MMDB on the test set for different ranking cutoffs (orange bars). These results show that crowds composed of higher-ranked forecasters substantially outperformed crowds that included lower-ranked forecasters. However, it is reasonable to ask whether similar results could have been achieved using simpler methods. We thus compared our approach against a typical aggregation strategy (Fig. 4, yellow bars) in which forecaster weights w_i are based on their MMDB scores computed at the end of each day while the question was open for predictions by forecasters. This MMDB thus considers forecasts for each day a question was open. As Fig. 4 shows, the benchmark weighting scheme also yields more accurate forecasts compared to an unweighted

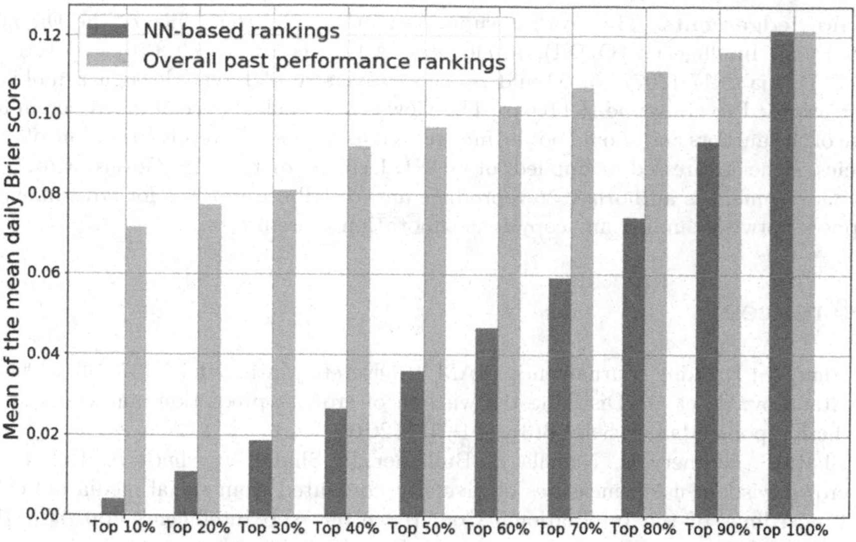

Fig. 4. Brier scores of crowd predictions based on the neural ranker and by past performance. (Color figure online)

aggregation (shown when using 100% of the crowd); however, our NN-based weightings produced consistently superior results, and this differential increased with greater selectivity.

5 Conclusion

This paper introduced a novel scheme to aggregate predictions from a crowd of forecasters for predicting geopolitical events. Whereas the current art bases the relevance of predictions by forecasters' past performance, we score relevance by a question-specific ranking of forecasters induced by a tournament where forecaster predictions are adjudicated by a siamese neural network. Preliminary results show that choosing the average prediction of forecasters in a top percentile of our neural ranking consistently yields a lower (superior) Brier score than using the same top percentile of forecasters based on past performance. Future experiments may include attempting the ranking phase over the following years and experimenting with different forecaster input representations, for example by including psychometric features included in the Good Judgment data set. Another interesting approach would be fine tuning the topic modeling for example by using more robust approaches for the determination of the number of topics.

Acknowledgements. This work is supported in part by the Office of the Director of National Intelligence (ODNI), Intelligence Advanced Research Projects Activity (IARPA), via 2017-17072100002 and by the University of Pavia through a mobility grant awarded to Giuseppe Nebbione. The views and conclusions contained herein are those of the authors and should not be interpreted as necessarily representing the official policies, either expressed or implied, of ODNI, IARPA, or the U.S. Government. The U.S. Government is authorized to reproduce and distribute reprints for governmental purposes notwithstanding any copyright annotation therein.

References

1. Alon, N.: Ranking tournaments. SIAM J. Discrete Math. **20**(1), 137–142 (2006)
2. Atanasov, P., et al.: Distilling the wisdom of crowds: prediction markets vs. prediction polls. Manag. Sci. **63**(3), 691–706 (2016)
3. Bhatt, S., Minnery, B., Nadella, S., Bullemer, B., Shalin, V., Sheth, A.: Enhancing crowd wisdom using measures of diversity computed from social media data. In: Proceedings of the International Conference on Web Intelligence, pp. 907–913. ACM (2017)
4. Blei, D.M., Ng, A.Y., Jordan, M.I.: Latent Dirichlet allocation. J. Mach. Learn. Res. **3**(Jan), 993–1022 (2003)
5. Bolelli, L., Ertekin, Ş., Giles, C.L.: Topic and trend detection in text collections using latent Dirichlet allocation. In: Boughanem, M., Berrut, C., Mothe, J., Soule-Dupuy, C. (eds.) ECIR 2009. LNCS, vol. 5478, pp. 776–780. Springer, Heidelberg (2009). https://doi.org/10.1007/978-3-642-00958-7_84
6. Bolger, F., Wright, G.: Use of expert knowledge to anticipate the future: issues, analysis and directions (2017)
7. Brier, G.W.: Verification of forecasts expressed in terms of probability. Mon. Weather Rev. **78**(1), 1–3 (1950)
8. Brock, W.A.: Causality, chaos, explanation and prediction in economics and finance. In: Beyond Belief, pp. 230–279. CRC Press (2018)
9. Budescu, D.V., Chen, E.: Identifying expertise to extract the wisdom of crowds. Manag. Sci. **61**(2), 267–280 (2014)
10. Bulligan, G., Marcellino, M., Venditti, F.: Forecasting economic activity with targeted predictors. Int. J. Forecast. **31**(1), 188–206 (2015)
11. Cahnman, W.J.: Methods of geopolitics. Soc. Forces **21**(2), 147–154 (1942)
12. Charbit, P., Thomassé, S., Yeo, A.: The minimum feedback arc set problem is NP-hard for tournaments. Comb. Probab. Comput. **16**(1), 1–4 (2007)
13. Chen, K.T., Chang, C.J., Wu, C.C., Chang, Y.C., Lei, C.L.: Quadrant of euphoria: a crowdsourcing platform for QoE assessment. IEEE Netw. **24**(2), 28–35 (2010)
14. Collins, R.: Prediction in macrosociology: the case of the soviet collapse. Am. J. Sociol. **100**(6), 1552–1593 (1995)
15. Coppersmith, D., Fleischer, L., Rudra, A.: Ordering by weighted number of wins gives a good ranking for weighted tournaments. In: Proceedings of the Seventeenth Annual ACM-SIAM Symposium on Discrete Algorithms, pp. 776–782. Society for Industrial and Applied Mathematics (2006)
16. Faggiani, A., Gregori, E., Lenzini, L., Luconi, V., Vecchio, A.: Smartphone-based crowdsourcing for network monitoring: opportunities, challenges, and a case study. IEEE Commun. Mag. **52**(1), 106–113 (2014)

17. Forlines, C., Miller, S., Prakash, S., Irvine, J.: Heuristics for improving forecast aggregation. In: AAAI Fall Symposium: Machine Aggregation of Human Judgment (2012)
18. Galton, F.: Vox populi (the wisdom of crowds). Nature **75**(7), 450–451 (1907)
19. Goldstein, D.G., McAfee, R.P., Suri, S.: The wisdom of smaller, smarter crowds. In: Proceedings of the Fifteenth ACM Conference on Economics and Computation, pp. 471–488. ACM (2014)
20. Grela, M., Kulesza, K., Zagórowska, M., Zioło, P.: Crowdsourcing and defence, in the age of big data. In: Institute of Mathematics and its Applications Conference on Mathematics in Defence (sic), Oxford, UK (2015)
21. Griffiths, T.L., Steyvers, M.: Finding scientific topics. Proc. Natl. Acad. Sci. **101**(suppl. 1), 5228–5235 (2004)
22. Hoffer, E., Ailon, N.: Deep metric learning using triplet network. In: Feragen, A., Pelillo, M., Loog, M. (eds.) SIMBAD 2015. LNCS, vol. 9370, pp. 84–92. Springer, Cham (2015). https://doi.org/10.1007/978-3-319-24261-3_7
23. Hosen, M.A., Khosravi, A., Nahavandi, S., Creighton, D.: Improving the quality of prediction intervals through optimal aggregation. IEEE Trans. Ind. Electron. **62**(7), 4420–4429 (2015)
24. Jose, V.R.R., Nau, R.F., Winkler, R.L.: Sensitivity to distance and baseline distributions in forecast evaluation. Manag. Sci. **55**(4), 582–590 (2009)
25. Koch, G., Zemel, R., Salakhutdinov, R.: Siamese neural networks for one-shot image recognition. In: ICML Deep Learning Workshop, vol. 2 (2015)
26. LeCun, Y., Bengio, Y., Hinton, G.: Deep learning. Nature **521**(7553), 436 (2015)
27. Leitner, C., Zeileis, A., Hornik, K.: Forecasting sports tournaments by ratings of (prob) abilities: a comparison for the euro 2008. Int. J. Forecast. **26**(3), 471–481 (2010)
28. Makridakis, S., Winkler, R.L.: Averages of forecasts: some empirical results. Manag. Sci. **29**(9), 987–996 (1983)
29. Ramos, J.J.: Robust methods for forecast aggregation. Ph.D. thesis, Rice University (2014)
30. Ren, J., Zhang, Y., Zhang, K., Shen, X.: Exploiting mobile crowdsourcing for pervasive cloud services: challenges and solutions. IEEE Commun. Mag. **53**(3), 98–105 (2015)
31. Sun, Y., Wang, X., Tang, X.: Deep learning face representation from predicting 10,000 classes. In: Proceedings of the IEEE Conference on Computer Vision and Pattern Recognition, pp. 1891–1898 (2014)
32. Surowiecki, J.: The Wisdom of Crowds. Anchor (2005)
33. Tetlock, P.E., Mellers, B.A., Rohrbaugh, N., Chen, E.: Forecasting tournaments: tools for increasing transparency and improving the quality of debate. Curr. Dir. Psychol. Sci. **23**(4), 290–295 (2014)
34. Warnaar, D.B., et al.: The aggregative contingent estimation system: selecting, rewarding, and training experts in a wisdom of crowds approach to forecasting. In: AAAI Spring Symposium: Wisdom of the Crowd (2012)

The Comment of BBS: How Investor Sentiment Affects a Share Market of China

Xuanlong Weng[1], Yin Luo[2(✉)], Jianbo Gao[3,4], Haishan Feng[1], and Ke Huang[1]

[1] School of Business, Guangxi University, Nanning 530004, Guangxi, China
[2] School of Computer, Electronics and Information, Guangxi University, Nanning 530004, Guangxi, China
13557229610@163.com
[3] State Key Laboratory of Earth Surface Processes and Resource Ecology, Beijing Normal University, Beijing 100875, China
[4] Center for Geodata and Analysis, Faculty of Geographical Science, Beijing Normal University, Beijing 100875, China

Abstract. This paper studies the influence of investor sentiment on the A share market of China. According to the behavioral financial theory, transformation of investor sentiment will trigger irrational transaction behavior and have an influence on the Chinese stock market. We study the effect of more than 23 million investor's comments posted on EastMoney.com, which is the biggest stock BBS in China. We utilize TextCNN to mine emotional tendency of investor comment stock comment, classify comments into positive, negative and neutral. The classified accuracy of validate set can reach 90%. And utilize such emotional tendency to define investor sentiment index. Based on our research, we find that the correlation between sentiment index and Shanghai Composite Index (SHCI) is positive, statistically significant and exponentially decays in period of time. Besides, with Hurst parameter H, it indicates that investor sentiment have long-range correlations, investor sentiment will develop as the current trend.

Keywords: Investor sentiment · Behavioral finance · Hurst parameter · TextCNN

1 Introduction

As social media gets more and more developed, investors can spread their feeling more quickly and widely through all kinds of social tools. Baker and Wurgler [1] shows that investor sentiment can have impact on stock return. The formation of stock price is a progress that is striking and being corrected continuously under sentiment and other economic factors. In traditional financial theory, investors are classified into rational and informed or irrational and sentiment-driven. Many financial theories are based on hypotheses-rational agents. For instances, Markowitz [2] proposes mean-variance optimizers and Fama [3] proposes efficient market hypothesis, which is based on rational agents, and the affection of the trade behavior of irrational investor is

É. Renault et al. (Eds.): MLN 2018, LNCS 11407, pp. 270–278, 2019.
https://doi.org/10.1007/978-3-030-19945-6_19

disseminated each other. In the end, the price stock is keeping stable. Compared with the traditional theories, behavior finance opinion takes sentiment-driven agents into account in the stock market. Blasco et al. [4] find that sentiment can lead investors to join the feedback type strategies of buying and selling with the people that exist in the stock market. Behavior of sentiment driven investor will affect stock price. Stambaugh et al. [5] show that investor sentiment contains a market-wide component with the potential to influence prices on many securities in the same direction at the same time.

As early as Keynes et al. [6], how sentiment and emotion drive stock price is puzzled by people. Every investor cares about this mystery. Nowadays how sentiment drives investor is a popular research field in the behavior finance. When the stock market is under extreme condition and out of control, efficient market hypothesis and traditional technical analysis is not enough to explain what happen (such as 2008 US sub-prime financial crisis). Therefore, the research of investor sentiment is very meaningful in the stock market.

2 Literature Review

Nowadays, there are many studies that make use of social media information to analyze American stock market. Antweiler et al. [9] collect 1.5 million messages posted on Yahoo Finance and Raging Bull, utilize the Naïve Bayes algorithm to classify these messages into three categories (buy, hold, sell) to construct bullishness index, and find that bullishness index has impact on Dow Jones index return and market volatility. Bollen et al. [10] utilize mode tracking tool to measures positive vs. negative mood, in terms of mood time series to predict the changes in DJIA closing values.

Chauet et al. [7] find there is also asymmetry in the role of sentiment refer to business conditions. Antoniou et al. [8] argues that sentiment can have impact on investor trade behavior and further influence on stock price. Blau [11] examines investor sentiment influences the degree of price clustering and show a contemporaneous correlation between price clustering and investor sentiment by the univariate and multivariate tests. Sun et al. [12] utilize GuBa comments to show the positive correlation between Guba-based sentiment and the stock market trends. Fang et al. [13] investigates the influence of the composite index of investor sentiment on the time varying long-term correlation of the U.S. stock and bond markets find the result that the composite index of investor sentiment has a significantly positive influence on the long-term stock-bond correlation.

Gao et al. [14] shows that an elegant generic multiscale theory about novels sentiment can be developed based on random fractal theory. So far, there is no one use fractal theory to depict investor sentiment in stock market. We show here that an elegant generic multiscale theory about investor sentiment can be developed based on random fractal theory.

The paper is organized as following: In Sect. 3, we raise our hypotheses. Section 4 describes data collection and preprocessing. In Sect. 5, we give the brief description of AFA, the method we used to analyze our data. In Sect. 6, we show the outcome of data analysis. Finally, in Sect. 7, our conclusion is proposed.

3 Testable Hypotheses

Hypotheses 1: the correlation between sentiment index and stock market return is highly positive.
Hypotheses 2: Sentiment index have long-range correlations (or long memories) and follow current trend.
Hypotheses 3: Investor Sentiment will keep persistent to affect stock market in short time and exponential decay.

4 Data Collection and Preprocessing

The collection and preprocessing of our data are shown as following:

1. We collect more than 23 million comments from EastMoney.com which covers 3065 stocks listed in China and date from 2017.01.01 to 2018.7.25. All the comments are collected by web crawler.
2. We randomly sample 56181 comments from the corpus we collected, in order to keep as much features as possible of investor comments. And we manually label these comment into positive, negative and neutral, in order to form the train set and validate set our CNN model.
3. After the labeling we find that the distribution of three categories (positive, negative and neutral) is unbalanced, which might be harmful to our classifying model. Considering the expensive time cost of manually labeling, we utilize data augmentation technology in natural language processing to enlarge our labeled corpus, in order to make the three categories balanced. We utilize seqGAN algorithm to generate text that can be added into the unbalanced categories.
4. We build a CNN model by utilizing python Tensorflow package, which structure is similar to TextCNN [15], including convolution layer, max-pooling layer and full-connected layer. We firstly utilize python jieba module to segment every comment into Chinese word, and then utilize Google Word2Vec to generate our word embedding, and we take the word embedding as the input of our CNN model. In our model we use batch normalization to accelerate the training process. After an hour training, the classified accuracy of validate set can reach 90%, which is high enough for our follow-up study, so we utilize early-stopping to prevent overfitting of our model.
5. After training we utilize our trained model to predict the remained comments. All the remained comments are automatically classified into three categories (positive, negative, neutral, the neutral might include unrelated corpus noise). Combined with our manually labeled data, we finally get the classified result of all comments.

After the collection and preprocessing the comments, we aggregate the sentiment classifications (positive, negative, neutral) in order to obtain a $Sentiment\_index_t$ for each of stock training day intervals t. our first measure is defined as

$$Sentiment\_index_t = \ln M_t^{total} * \ln \frac{M_t^{Positive} + 1}{M_t^{Negative} + 1} \quad (1)$$

Where $M_t^{Positive}$ is the number of positive comments in the given time interval. $M_t^{negative}$ is the number of negative comments in the given time interval. M_t^{total} is s the number of all kinds of comments in the given time interval.

In order to measure daily sentiment state, we define sentiment state factor $Positive_t$ where $Sentiment\_index_t > 1$. $Negative_t$ where $Sentiment\_index_t < 1$; $Neutral_t$ where $Sentiment\_index_t = 1$. We extract a time series of daily Shanghai Composite Index (SHCI) from 2017.1.1 to 2018.7.25 daily closing-values sh_t from WIND database. And compute log-return of Shanghai securities composite index r_t:

$$r_t = \ln \frac{sh_t}{sh_{t-1}} \quad (2)$$

5 Our Analyzing Method

Many fractal analyses concentrate explicitly on how to measure the variability scales with the size of a time window over which the measure is calculated [16]. Gao et al. [17] provided a succinct and comprehensive treatment of various fractal analysis methods.

A parameter called the Hurst exponent provides a way to quantify the 'memory' or serial correlation in a time series. Different H values have different meanings. In fact, H = 0.5 indicates the process is random. A finding of $0.5 < H < 1$ indicates that process will develop as the current trend, that is to say, the process has long-range correlations. In contrast, $0 < H < 0.5$ indicates an anti-persistent process, which means motion is likely to move in the opposite direction to the current trend.

AFA is based on a nonlinear adaptive multiscale decomposition algorithm [18]. The first step involves partitioning an arbitrary time series under study into overlapping segments of length $w = 2n + 1$, where neighboring segments overlap by $n + 1$ points. In each segment, the time series is fitted with the best polynomial of order M, obtained by using the standard least-squares regression; the fitted polynomials in overlapped regions are then combined to yield a single global smooth trend [14]. Denoting the fitted polynomials for the (i)th and (i + 1)th segments by $y^i(l_1)$ and $y(i + 1)(l_2)$, respectively, where $l_1, l_2 = 1,\ldots,2n + 1$, we define the fitting for the overlapped region as

$$y^{(c)}(l) = w_1 y^{(i)}(l+n) + w_2 y^{(i+1)}(l), l = 1, 2, \ldots, n+1 \quad (3)$$

where $w_1 = \left(1 - \frac{l-1}{n}\right)$ and $w2 = \frac{l-1}{n}$ can be written as $(1 - dj/n)$ for j = 1, 2, and where dj denotes the distances between the point and the centers of y(i) and y(i + 1), respectively. Note that the weights decrease linearly with the distance between the point and the center of the segment. Such a weighting is used to ensure symmetry and effectively eliminate any jumps or discontinuities around the boundaries of neighboring segments. As a result, the global trend is smooth at the non-boundary points, and has the right and left derivatives at the boundary [19]. The global trend thus determined can be used to maximally suppress the effect of complex nonlinear trends on the scaling analysis. The parameters of each local fit are determined by maximizing the goodness of fit in each segment. The different polynomials in overlapped part of each segment are combined using Eq. (3) so that the global fit will be the best (smoothest) fit of the overall time series. Note that, even if M = 1 is selected, i.e., the local fits are linear, the global trend signal will still be nonlinear. With the above procedure, AFA can be readily described. For an arbitrary window size w, we determine, for the random walk process u(i), a global trend v(i), i = 1,2,…,N, where N is the length of the walk. The residual of the fit, u(i) − v(i), characterizes fluctuations around the global trend, and its variance yields the Hurst parameter H according to the following scaling equation:

$$F(w) = \left[\frac{1}{N}\sum\nolimits_{i=1}^{N}(u(i) - v(i))^2\right]^{1/2} \sim w^H. \tag{4}$$

Thus, by computing the global fits, the residual, and the variance between original random walk process and the fitted trend for each window size w, we can plot log2 F(w) as a function of log2(w). The presence of fractal scaling amounts to a linear relation in the plot, with the slope of the relation providing an estimate of H. The above are the basic steps of applying AFA.

6 Data Analysis

As can be seen in Fig. 1, the performance of the r_t and $Sentiment\_index_t$ is almost the same. $Sentiment\_index_t$ is more sensitive and swift to the reaction of stock market. log-return of SHCI and sentiment index have large decline from January 29th 2018 to February 9th 2018. From the high point of SHCI in the several days has fallen by 521.55 to 3065.62. The first trough of log-return of SHCI and sentiment index is both at February 9th, 2018. The second trough of log-return of SHCI and sentiment index is on March 23rd, 2018. In the same day, President Trump signed a memorandum to impose tariffs on $60bn of Chinese goods.

As the Fig. 2 depicts, the green curve is the closing value of SHCI, and we use three kinds of asterisks to describe different sentiment state factors. The downtrend of SHCI is accompanied with negative state factor, while the uptrend of SHCI is accompanied with positive state factor. However, the state factor is not totally conformed to the trend of SHCI, it will appear alternation. In addition, the sentiment state factor also can seize the crest and the trough of SHCI.

Figure 3 (top) depicts the correlation of the log-return of SHCI and sentiment index. It is positively correlated in the scatter distribution, correlation coefficient is 0.61405. Because $p = 0.0 < 0.05$, it is statistically significant. As in the Fig. 3 (bottom), the correlation coefficient of the log-return of SHCI and lag sentiment index decay by time. The first-lagged correlation falls to 0.2724 quickly and then slowly

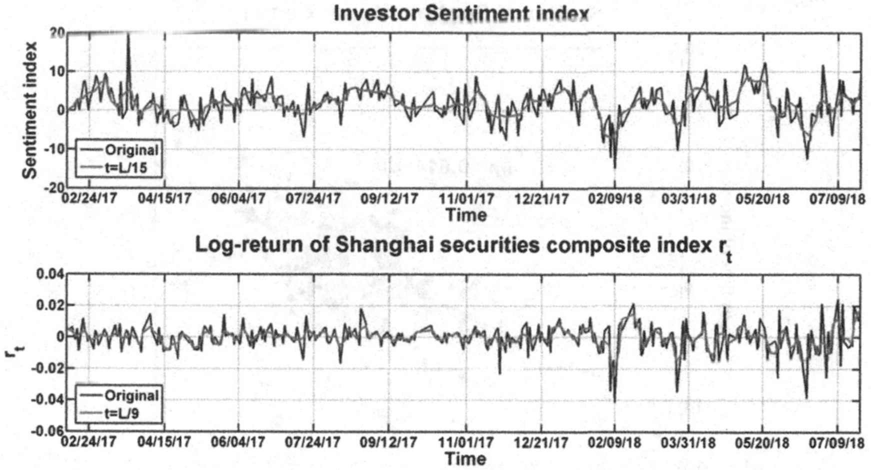

Fig. 1. Investor sentiment index and log-return of Shanghai Composite Index (SHCI).

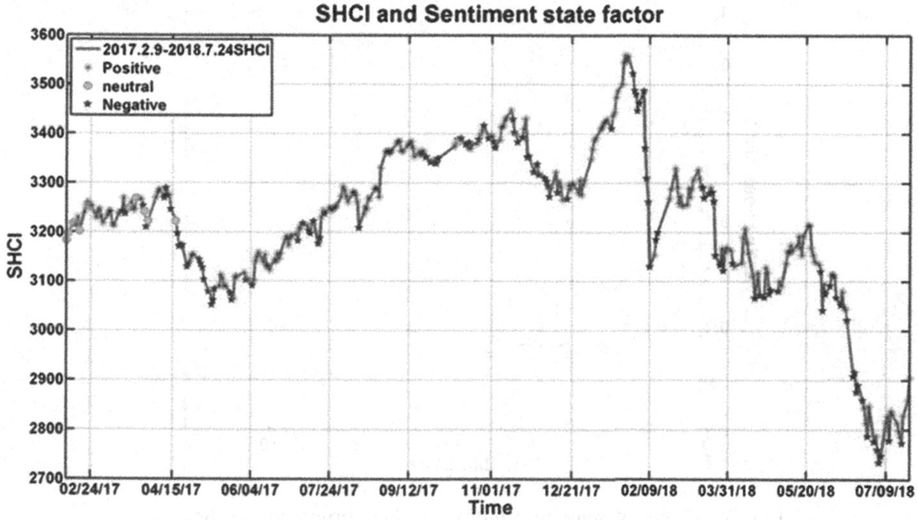

Fig. 2. SHCI and sentiment state factor.

decays to zero until the eighth lag. We know that sentiment effect on the stock market can last about 8 days and get weaker and weaker during these days. So we can extract the evidence of hypothesis 2. If the bad news cause investors scared excessively and make the irrational trade behavior, the negative sentiment will keep affecting the stock market. Therefore, our invest strategy is overselling stock. On the contrary, if investors are over positive, we can buy the stock. These evidences support hypotheses 3 (Fig. 3).

To better know how the investors sentiment change, we use the Hurst value to describe whether the investor sentiment exists long-range correlations. The Hurst value

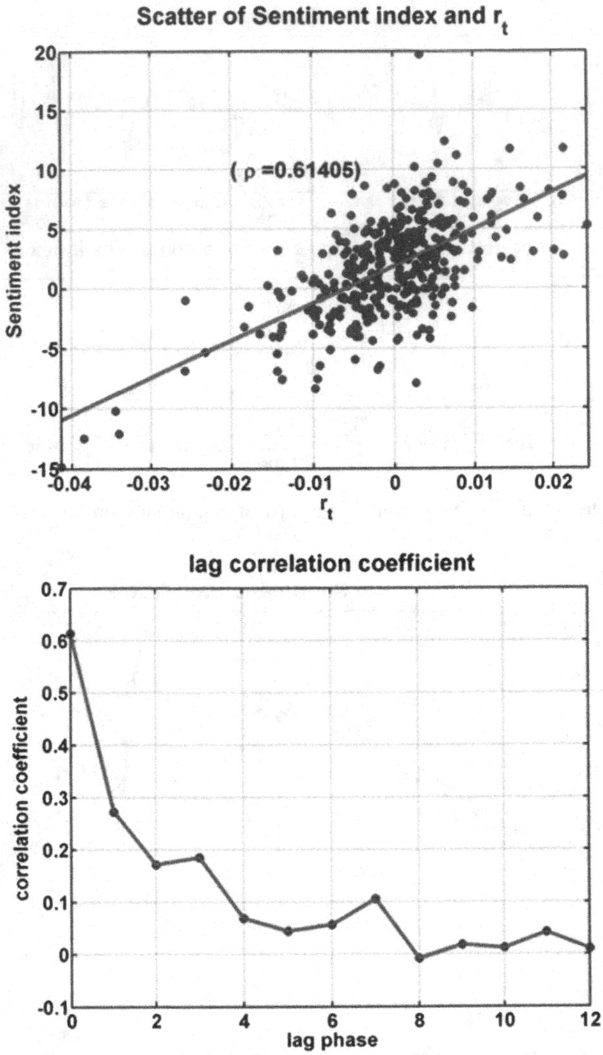

Fig. 3. Scatter of sentiment index and lag correlation coefficient.

is 0.85641 more than 0.5 shows that investor sentiment exists long-range correlations (Fig. 4). It means investor sentiment will keep over a period of time and explains why the sentiment state factor is not totally conformed to the trend of SHCI in Fig. 2. If the stock market is under bullish, the investor sentiment will keep positive. This is the reason why the investors are more and more confident to make the stock market overinflated in the bullish market. Investor can sell their position. On the contrary, if the stock market is under bearish, investors will be more and more negative in a period of time. The stock price will be underestimated, which is a good opportunity to hold a position. The result presented above suggests that the market panic will occur as burst,

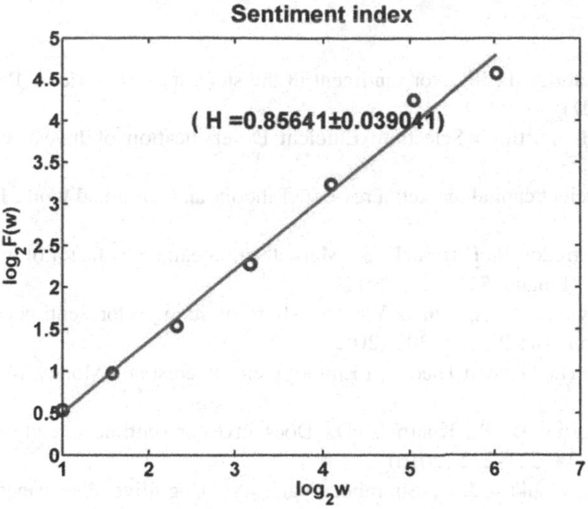

Fig. 4. Hurst parameter of sentiment index.

in a period of time, negative sentiment will lead to more serious market panic. When the negative sentiment is accumulated in the stock market, financial crisis will occur. These evidences support hypotheses 2.

7 Conclusion

Nowadays more and more financial researchers realize that investor sentiment plays an important role in the stock market. Researchers have attempted to utilize several ways to measure investor sentiment. Due to the complexity of Chinese language, it is hard to extract information from texts accurately in the past studies. In our paper, we design and improve the experiment process compared with other past studies. And the accuracy of our sentiment analyzing model can reach 90%, which is superior to majority of analyzed sentiment model used in past studies.

The classical financial theory is that sentiment-driven investors are irrational, their trade behaviors will be cancelled by each other. Thus, the stock market is not affected by sentiment-driven investors. However, it's quite possible that the investor sentiment make a great influence on stock market. This paper examines possibility and empirically test whether the investor sentiment influences the stock market. The key findings of our study can be summarized as follows. Firstly, we find that the correlation between sentiment index and stock market return is positive. Secondly, investor sentiment will persist for a period of time. Sentiment index have long-range correlations (or long memories), investor sentiment will keep current trend in future. Thirdly, the reaction between sentiment index and stock market is immediate and swift. Sentiment will remain impact on stock market for about the eighth day though its impact is decaying by time. According to our study, we can find the market state so as to change our investment strategy in time and predict market condition or crisis for reference.

References

1. Baker, M., Wurgler, J.: Investor sentiment in the stock market. J. Econ. Perspect. **2007**(21), 129–152 (2007)
2. Markowitz, H.: Portfolio Selection: Efficient Diversification of Investments. Wiley, New York (1959)
3. Fama, E.: Efficient capital market: a review of theory and empirical work. J. Finan. **25**, 382–417 (1970)
4. Blasco, N., Corredor, P., Ferreruela, S.: Market sentiment: a key factor of investors' imitative behavior. Acc. Financ. **51**, 1–27 (2011)
5. Stambaugh, R.F., Yu, J., Yuan, Y.: The short of it: investor sentiment and anomalies. J. Financ. Econ. **104**(2), 288–302 (2012)
6. Keynes, J.M.: The General Theory of Employment, Interest and Money. Macmillan, London (1936)
7. Chau, F., Deesomsak, R., Koutmos, D.: Does investor sentiment really matter? Int. Rev. Financ. Anal. **48**, 221–232 (2016)
8. Antoniou, C., Doukas, J.A., Subrahmanyam, A.: Cognitive dissonance, sentiment and momentum. J. Financ. Quant. Anal. **48**, 245–275 (2013)
9. Antweiler, W., Frank, M.Z.: Is all that talk just noise? The information content of internet stock message boards. J. Financ. **59**(3), 1259–1294 (2004)
10. Johan, B., Mao, H., Zeng, X.: Twitter mood predicts the stock market. J. Comput. Sci. **2**(1), 1–8 (2010)
11. Benjamin, M.B.: Price clustering and investor sentiment. J. Behav. Finan. **20**(1), 19–30 (2019)
12. Sun, Y., Fang, M., Wang, X.: A novel stock recommendation system using Guba sentiment analysis. Pers. Ubiquit. Comput. **6**, 1–13 (2018)
13. Fang, L., Yu, H., Huang, Y.: The role of investor sentiment in the long-term correlation between U.S. stock and bond markets. Int. Rev. Econ. Finan. **11**(58), 127–139 (2018)
14. Gao, J., Jockers, M.L., Laudun, J., Tangherlini, T.: A multiscale theory for the dynamical evolution of sentiment in novels. In: International Conference on Behavioral, pp.1–4 (2017)
15. Kim, Y.: Convolutional neural networks for sentence classification. In: EMNLP, pp. 1746–1751 (2014)
16. Riley, M.A., Bonnette, S., Kuznetsov, N.: A tutorial introduction to adaptive fractal analysis. Front. Physiol. **3**, 371 (2012)
17. Gao, J.B., Cao, Y.H., Tung, W.W., Hu, J.: Multiscale Analysis of Complex Time Series: Integration of Chaos and Random Fractal Theory, and Beyond. Wiley Interscience, Hoboken (2007)
18. Gao, J., Hu, J., Tung, W.: Facilitating joint chaos and fractal analysis of biosignals through nonlinear adaptive filtering. PLoS One **6**(9), e24331 (2011)
19. Gao, J., Fang, P., Liu, F.: Empirical scaling law connecting persistence and severity of global terrorism. Phys. A Stat. Mech. Appl. **482**, 74–86 (2017)

A Hybrid Neural Network Approach for Lung Cancer Classification with Gene Expression Dataset and Prior Biological Knowledge

Hasseeb Azzawi(✉), Jingyu Hou, Russul Alanni, and Yong Xiang

School of Information Technology, Deakin University, Burwood, VIC, Australia
{hazzawi, jingyu.hou, ralanni, yong.xiang}@deakin.edu.au

Abstract. Lung cancer has continued to be the leading cause of related mortality and its frequency is rising daily worldwide. A reliable and accurate classification is essential for successful lung cancer diagnosis and treatment. Gene expression microarray, which is a high-throughput platform, makes it possible to discover genomic biomarkers for cancer diagnosis and prognosis. This study proposes a new approach of using improved Particle Swarm Optimization (IMPSO) technique to improve the Multi-Layer Perceptrons (MLP) neural network prediction accuracy. The MLP weights and biases are computed by the IMPSO for more accurate lung cancer prediction. The proposed discriminant method (MLP-IMPSO) integrates the prior knowledge of lung cancer classification on the basis of gene expression data to enhance the classification accuracy. Evaluations and comparisons of prediction performance were thoroughly carried out between the proposed model and the representative machine learning methods (support vector machine, MLP, radial basis function neural network, C4.5, and Naive Bayes) on real microarray lung cancer datasets. The cross-data set validations made the assessment reliable. The performance of the proposed approach was better upon the incorporation of prior knowledge. We succeeded in demonstrating that our method improves lung cancer diagnosis accuracy with prior biological knowledge. The evaluation results also showed the effectiveness the proposed approach for lung cancer diagnosis.

Keywords: Lung cancer · Prior biological knowledge · Multilayer Perceptron · Particle Swarm Optimization · Classification

1 Introduction

Nowadays the death ratio of lung cancer is increasing according to the reports of different medical institutions [1–3]. Compared to the beginning of the century, the death ratio of cancer has increased greatly [4, 5]. Although some treatments, like chemotherapy, radiation, and surgery have been used widely, there is still a long way to go before achieving satisfactory results. For lung cancer, accurately classifying different kinds of lung cancer is remarkably critical to the treatment efficiency and toxic quality minimization on patients [6].

Millions of genes can be investigated with the help of microarray technology for retrieving important information about cell functionalities. This useful information can

É. Renault et al. (Eds.): MLN 2018, LNCS 11407, pp. 279–293, 2019.
https://doi.org/10.1007/978-3-030-19945-6_20

be applied to cancer diagnosis and prognosis [7–11]. However, due to the features of gene expression data (i.e., small sample size, high noise rate, and high dimension), it is necessary to have a high-quality method for selecting the important subsets of genes that can be used for better cancer classification. Such method will not only reduce the computational costs, but enable doctors to classify a small subset of biologically related gene-specified cancers and focus on some specific genes to design less expensive experiments [11]. Furthermore, this kind of classification technique with high accuracy can assist in early diagnosis of cancer patients and drug detection [12].

Recently, new approaches have - shown that the prior knowledge of a biological system can be applied to guide the statistical examination and improve the detection of new biological features while dropping the exposure of spurious relationships. Furthermore, the prior knowledge can be used for consistency checking for the available knowledge, as well as for experimental data checking to fill in possible gaps or add more details. In recent years, different gene expression signatures have been identified for precise classification between subtypes of tumors [13–19]. Researches have shown that rational utilization of the accessible biological information could not only effectively remove or abolish noise in gene chips, but avoid one-sided results of individual experimentation. However, the significance of prior information in cancer classification has not been well recognized [19–22].

Artificial Neural Network (ANN) systems can help to handle aforementioned issues [23–25]. The work efficiency of ANN is dependent on how well the data are trained and how the learning process occurs. Multi-Layer Perceptron (MLP) is one of the most popular neural networks. It is a feed-forward ANN model based on supervised training. Gradient-based and stochastic methods are two common methods of supervised training for MLP. Many researchers are using the backpropagation algorithm and its alternatives [26, 27] as a default example of gradient-based methods. However, the gradient-based approaches have some drawbacks, such as inclination to be confined in local minima, slow junction, and high dependence on the initial parameters [28, 29]. Metaheuristic algorithms are often nature-inspired, and becoming increasingly popular in optimization and applications.

Particle Swarm Optimization (PSO), introduced by Kennedy and Eberhart in 1995 [30], is a simple, but still dominant optimization algorithm that reproduces the flying and searching behavior of birds. PSO has been widely utilized in different areas to deal with numerous optimization difficulties, such as gene selection [12, 31–39]. Hybrid algorithms based on Particle Swarm Optimization and neural networks have been proposed [40–44]. However, to our knowledge, there is no research on using the prior biological knowledge in these hybrid algorithms to classify lung cancers from microarray data. This research propose a new method to improve lung cancer classification accuracy by using improved PSO on MLP. The method is denoted as MLP-IMPSO. In our research, we used PSO for its outstanding search abilities, and improved it, then combined it with MLP for its knowledge interpretation advantage. Furthermore, to achieve better performance, the prior biological knowledge was incorporated. The performance evaluation for our proposed method was conducted based on standard evaluation metrics: accuracy, sensitivity, specificity, and F1-score. We compared MLP-IMPSO (with prior knowledge), in terms of performance results, with MLP-MPSO1 the original without prior knowledge. In addition, another comparison was conducted between our proposed model and other well-known classifier algorithms, i.e., Support Vector Machine (SVM), Multilayer Perceptron (MLP), radial basis function neural network (RBFNN), C4.5, and Naive Bayes (NB). Furthermore, K-fold cross validation was used to assess the reliability of the models. The outcomes of our study showed that our proposed model outperformed current models in lung cancer diagnosis.

The paper is organized as follows: Sect. 2 presents the Multi-Layer Perceptron; Sect. 3 presents the proposed classification/prediction framework; Sect. 4 discusses the experimental outcomes of our model in lung cancer diagnosis, and the evaluation results; finally Sect. 5 presents our conclusions.

2 Multi-Layer Perceptron

Multi-Layer Perceptron (MLP) is the most common multilayer feed-forward neural network because of its characteristics, such as nonlinearity, robustness, adaptability, and easy to use. Moreover, MLP has been proven as a superior algorithm for different applications. Rosenblatt's effort produced much excitement, disagreement, and attention in neural network models for pattern classification in that era and directed to significant models abstracted from his effort in later years. Presently, the names (single-layer) Perceptron and Multilayer Perceptron are used to refer to explicit artificial neural network arrangements created on Rosenblatt's Perceptrons [45]. The structure of an Artificial Neural Network is a design of neurons assembled into layers. ANN has parameters, which are the layers' number, neurons' number in each layer, and connectivity level and kind of neuron interconnections [46]. ANN is similar to neurobiology synapses by connection, each connection is associated with a weight that can be varied in strength. A classic ANN construction is recognized as the multilayer perceptron where the neural network works quite easily. The input layer takes the input vector, each neuron in the input layer obtains a significance (a component of the input vector), and yields a new value (production) that directs to all neurons of the subsequent layer [45].

The training procedure of ANN is typically complex and highly dimensional. So far, countless investigators choose practical Backpropagation (BP) algorithms to train ANNs. The output error is measured by BP, and then the algorithm works on computing the gradient of this error and modifying the weight with biases of ANN in the descending gradient direction. These gradient approaches approximate the error in the network's judgment as related to a supervisor and spread the error to the weights all over the network. So one of the central problems is that searching of optimal weights is powerfully reliant on initial weights, and if they are located near local minimum, the algorithm would be stuck at a sub-optimal solution. Hence, the conventional gradient search method is prone to be converged at local optima. The neural network researchers have suggested many solutions to overcome the problem of slow convergence rate and being trapped in a local minimum.

Most of these solutions focused on adopting some parameters during learning or increasing the number of hiding layers in order to overcome the limitation in adjusting the weight. However, these studies get a limit enhancement while the complexities of the models were very notable [47–57].

The evolutionary approaches (EA), such as Genetic algorithm, Ant Colony Optimization, and PSO are successfully used in avoiding local minimum and improving convergence rate of training algorithm. These advantages give EA more robustness and attractiveness than many other search algorithms [58]. PSO is one of EA algorithms, it is used for different optimization problems and performs well. PSO could be used in neural network training like other evolutionary algorithms. In this research, we will replace BP in MLP with PSO to adjust the weight to overcome the limitation of the BP algorithm in MLP.

3 Proposed Prediction Framework

In this study, prior knowledge was used as a means pointing the classifier by using known lung adenocarcinoma genes. In our research, we combine two sources of genes (one is from attribute evaluator Chisquare and another is from published studies), there is no overlap between these two sources of genes. Then the proposed classification method was used to diagnose the lung cancer. Improved Particle Swarm Optimization (IMPSO) and multilayer perceptron are combined (MLP-IMPSO) as the classification model. Figure 1 shows the architecture of the proposed framework.

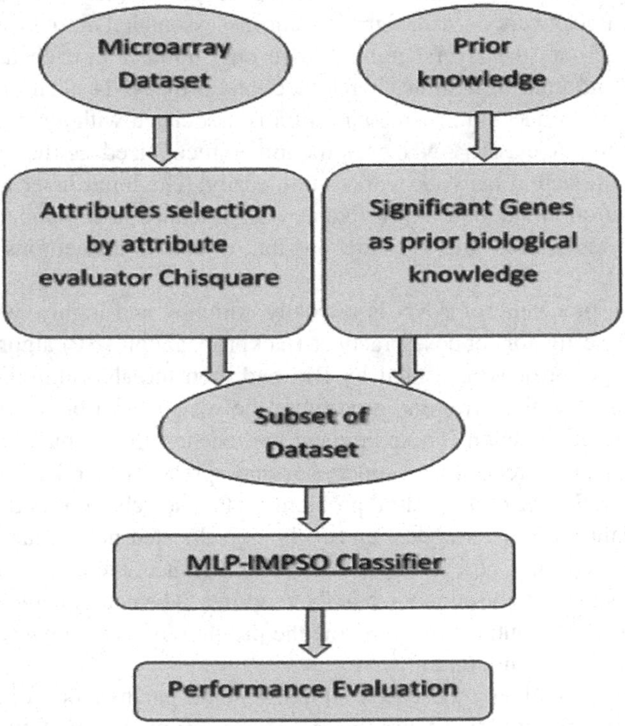

Fig. 1. Architecture of the proposed framework

3.1 Lung Cancer Dataset

In our study, we made use of the well-known and publicly accessible microarray dataset from Brigham and Women's Hospital, Harvard Medical School. This dataset had 12533 gene expression profiles and 181 samples of malignant tissue which include31 malignant pleural mesothelioma and 150 adenocarcinoma. With this Affymetrix Human GeneAtlas U95Av2 microarray dataset, we aim at conducting the expression ratio-based analysis to make a binary classification of malignant pleural mesothelioma and adenocarcinoma.

3.2 Microarray Attributes Selection

Attributes selection is a procedure of choosing relevant features for model construction. Feature selection is also known as other names such as variable selection and variable subset selection, which is to improve the classification accuracy by selecting minimized applicable features. In our experiments, we used the attribute evaluator Chisquare. Chisquare test is a method used for feature ranking and it was introduced by British statistician Karl Pearson initially for measuring the goodness of fit. The Chisquare is a commonly used feature selection method [59]. The Chi-square works on assessing the merit of each gene separately by the chi-squared statistical measure with respect to the classes. More details of Chi-square can be seen in [60]. In this study, we used ChiSquareTest.java package (http://home.apache.org/~luc/commons-math-3.6-RC2-site/jacoco/org.apache.commons.math3.stat.inference/ChiSquareTest.java.html) to implement Chi-Square.

Based on our previous study [9], we select the best ranked informative genes e.g. attributes to be used for creating a sub dataset from the original lung cancer microarray dataset. The number of patient (samples) of this sub dataset is the same as the original dataset but the number of genes is smaller.

3.3 Prior Biological Knowledge

Prior knowledge is used in this study as a means to guide the classifier by using the genes known to be related to lung adenocarcinoma. These genes were approved by other researchers. In this study, we select the genes for lung cancer diagnosis [21]. There are twenty-three lung adenocarcinoma-related genes gathered from the previous research. These genes have passed multiple testing procedures to come up with eight genes [61]. The details of eight of these genes are presented in Table 1. To create a sub dataset of the original lung cancer microarray dataset, a combination of these eight genes and attributes selected by attribute evaluator Chisquare test is used. There is no overlap between these two sources of lung cancer genes. This sub dataset of the original microarray dataset had the same number of samples (patients) as the original dataset, but a much smaller number of genes. This sub dataset will be used to direct our classification methods.

Table 1. The eight genes as prior biological knowledge [21]

Gene name	GenBank access No.	Location at HG_U95Av2
CXCL1	J03561	408_at
IL-18	U90434	1165_at
AKAP12	X97335	37680_at
KLF6	U51869	37026_at
AXL	M76125	38433_at
MMP-12	L23808	1482_g_at
PKP3	Z98265	41359_at
CYP2A13	U22028	1553_r_at

3.4 Improved Particle Swarm Optimization (IMPSO)

PSO is described as a population-based stochastic optimization technique that is motivated by the social behavior of swarm, like birds' flocking or schooling of fish. With this technique, a promising position could be obtained in accomplishing certain objectives [30]. Every particle, which is a "bird" in the flock, in PSOs represents a candidate solution for a given optimization problem [30]. Particles work cooperatively to find the position that corresponds to a best solution to the problem. The particles fly through the search space, constantly adjusting their velocity according to their own experience and their neighbor particle's experience. The performance of each particle position (i.e., each solution) is evaluated by a fitness function. The algorithm progressively stores and replaces the two best values, called the personal best (pbest) and global best (gbest). The pbest is the best previous experience for each particle during specific time/iterations, and gbest is the best previous position among all particles.

At each iteration (*t*) particles are searching in a j-diminutions space. The particle (*p*) denoted by a set of possible solutions $\{x1, \ldots.., xd\}$ represents the location of the particle in the proposed space. Each location will be modified by using the velocity set $\{v1, \ldots.., vd\}$ of the particle. We use *t, vij, xij* to denote the current iteration counter, current velocity of the *i*-th particle in the *j*-th dimension of the space, and new position of the *i*-th particle in the *j*-th dimension of the space, respectively. In this paper, the number of space dimensions refers to the number of weights in a MLP. At iteration *t* + 1, the values of *vij* and *xij* are modified according to Eqs. (1) and (2), respectively.

$$v_{ij}(t+1) = w(t)\ v_{ij}(t) + c_1 r_1 \left(pbest_{ij}(t) - x_{ij}(t)\right) + c_2 r_2 \left(gbest_j(t) - x_{ij}(t)\right) \quad (1)$$

$$x_{ij}(t+1) = v_{ij}(t+1) + x_{ij}(t) \quad (2)$$

Where $w(t)$ represents the inertia weight, c_1 and c_2 are cognitive and social acceleration coefficients, r_1 and r_2 are random variables, $pbest_{ij}(t)$ is the personal best position and $gbest_j(t)$ is the global best position, $x_{ij}(t)$ is the previous personal position.

In this study, in order to fulfil the problem requirement, we will use Mean Absolute Error (MAE) as a fitness function to examine the performance of each particle. MAE is effective in measuring the accuracy for continuous variables, which is suitable for microarray datasets. MAE simply calculates the average of the absolute differences between the prediction value (o_i) and the real value ($\hat{o}_i$) in the dataset for all samples (p). The prediction value is represented by the classification accuracy shown in Eq. (3), while the real value is the accuracy of the correctly classified cases on a particular dataset.

$$\text{Accuracy} = (\text{TP} + \text{TN})/(\text{TP} + \text{FP} + \text{FN} + \text{TN}) \tag{3}$$

where TP, TN, FP and FN are the true positive, true negative, false positive and false negative respectively.

The following MAE equation is used to assess the fitness of the ith particle at the tth iteration

$$\text{f}_\text{i}(\text{t}) = (1/\text{P}) \sum\nolimits_{\text{i}=1}^{\text{p}} |o_i - \hat{o}_i| \tag{4}$$

where P is the number of particles.

A personal best position *pbest* and a *global* best position *gbest* of the ith particles is adapted in the tth iteration by using (5) and (6).

$$Pbest_{ij}(t) = \begin{cases} Pbest_{ij}(t) & f_{ij}(t) > f_{ij}(t-1) \\ P_{ij}(t) & f_{ij}(t) \leq f_{ij}(t-1) \end{cases} \tag{5}$$

where $P_{ij}(t)$ is the practical i in the dimension j at iteration t

$$\text{gbest}_\text{j}(\text{t}) = \text{best } \{\text{pbest}_i(t)\}_{i=1}^{P} \tag{6}$$

3.5 IMPSO in Multilayer Perceptron

PSO is different from BP. It is a global and population-based algorithm that has been used for training neural networks, computing neural network constructions, regulating network learning constraints, and optimizing network weights. PSO is not based on gradient information, which makes it be able to avoid local minimum [63].

The purpose of using IMPSO in MLP is to get a best set of weights. The particles are representing the candidate solutions, i.e., candidate sets of weight. The dimensions of the space is the number of weights in MLP. In each iteration, all positions of the particles are evaluated by using the fitness function to find the pbest and gbest. The new velocity is calculated for each particle using Eq. (1), and the new location (the modified set of weights) is calculated accordingly using Eq. (2).

By repeating this process iteratively, a particle with fewer errors (best fitness value) will be recognized as the global best particle. This training process will stop when one of the following conditions are met: fitness result is zero, or exceeding the predefined number of iterations. After the training process is completed, the best IMPSO-MLP model will be the outcome of the model training phase. Then test of new patterns will be conducted by using the same weights obtained from the training phase. Learning process of IMPSO-MLP is shown in Fig. 2.

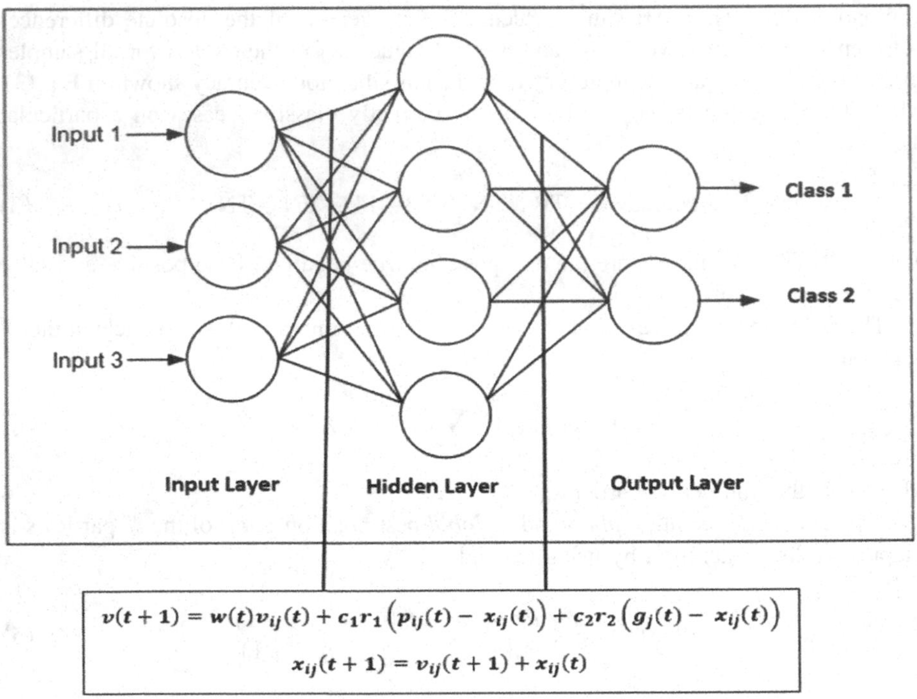

Fig. 2. Learning process of IMPSO-MLP

3.6 Building the Binary Classification Model

In our model, the input values are the gene expression values of the selected genes. PSO is used to train MLP. In this study we replace BP in MLP with PSO to search for the best set of weights for MLP. Each particle represents an individual solution of the problem, i.e., a set of weights. Each solution is evaluated by the fitness function defined in Eq. (3), where prediction results come from MLP. In the first iteration, all particles are saved as the best individual solution (pbest) and the best solution among all the solutions/particles in the iteration serves as the global best solution (gbest). Change the velocity of a particle according to Eq. (1) and update the particle position (modify the set of weights) according to Eq. (2). Using the new sets of positions to generate the new learning error (calculate the accuracy then the fitness function). Update the pbest (t) if the fitness function is better than the pbest(t-1). Find the best fitness results in the swarm/generation, compare it with the gbest, then update the gbest(t) if the minimal fitness result is better than gbest(t-1).

The optimization output is based on the gbest value. The iteration continues until one of the termination conditions is met. The gbest weights are used as the training results when the iteration is finished.

The parameters of the proposed IMPSO model are set as follows: Inertia weight = 1, c1 = c2 = 2, $v_{max} = 6.0$, $v_{min} = -6.0$, population size is 50, and the maximum number of iterations is 100.

4 Results and Discussion

The performance evaluation of our proposed modules (IMPSO) for cancer diagnosis is presented in this section. For simplicity in the following text without triggering confusion, we still use the notation MLP-IMPSO to represent the classifier created from the sub dataset with prior knowledge in Sect. 3.3, and MLP-IMPSO1 to represent the classifier created from the sub dataset alone without prior knowledge. Comparisons were done with other machine learning methods SVM, MLP, RBFNN, C4.5, and NB with prior knowledge. Quite a lot research employ these classification techniques in classifying and predicting objectives [64, 65]. Weka platform with the software default settings was used for implementations when doing comparisons.

To assess the reliability, we used the ten-fold cross validation with randomly dividing the dataset into ten equal sub datasets. Nine of these sub datasets were utilized as training datasets for the construction of the model, while the remaining one sub dataset was utilised as a testing dataset in each prediction run. The average result of the ten iterations was used as the final measurement.

The setting of the two models was for the investigation of their performance in diagnosis. Analysis of their performance in lung cancer subtype diagnosis was conducted and the model that has the highest prediction accuracy rate was picked as the optimal model. This optimal model was then compared with other machine learning methods (SVM, MLP, RBFNN, C4.5, and NB). Figures 3, 4 and 5 show the results of the comparison between the proposed MLP-IMPSO method and other methods.

4.1 MLP-IMPSO Model Results (With and Without Prior Knowledge)

The results of Fig. 3 illustrate that MLP-IMPSO method achieved the highest classification accuracy compared to MLP-IMPSO1 method on the dataset. In fact, MLP-IMPSO achieved 99.45% accuracy while MLP-IMPSO1 achieved 97.79%. Additionally, in terms of sensitivity and specificity, MLP-IMPSO method also outperformed MLP-IMPSO1. Regarding F-Score, MLP-IMPSO method achieved 1, which was better than 0.99 achieved by MLP-IMPSO1, as shows in Fig. 4. Overall, MLP-IMPSO obtained better results compared to the method MLP-IMPSO1, and IMPSO enables the MLP to dynamically evolve weights and biases effectively.

4.2 Comparisons of MLP-IMPSO with Representative Classifiers

From Fig. 5 it can be observed that the classification accuracy of our proposed module MLP-IMPSO is superior to that of the representative methods when microarray dataset with prior knowledge are used. It can also be observed from Figs. 4 and 5 that the classification accuracy, sensitivity, specificity and F-Score of the MLP-IMPSO classifier are higher than those of the SVM, MLP, RBFNN, C4.5, and NB methods.

It can be concluded from Figs. 3, 4 and 5 that MLP-IMPSO with prior knowledge is effective for lung cancer diagnosis, and can achieve better performance over conventional representative methods (i.e., SVM, MLP, RBFNN, C.4., and NB).

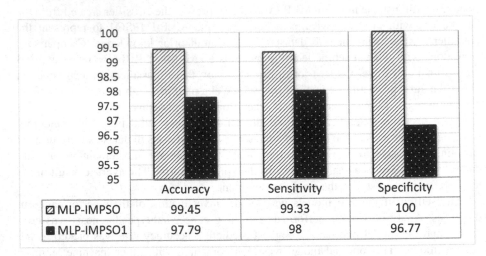

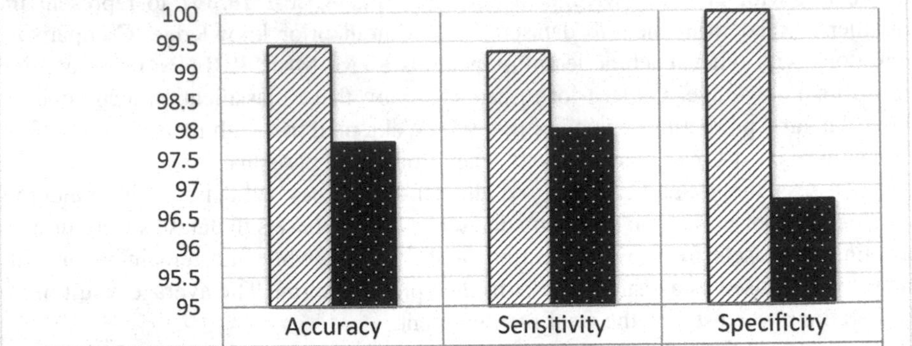

	Accuracy	Sensitivity	Specificity
MLP-IMPSO	99.45	99.33	100
MLP-IMPSO1	97.79	98	96.77

Fig. 3. Accuracy, Sensitivity, Specificity results of MLP-IMPSO and MLP-IMPSO1

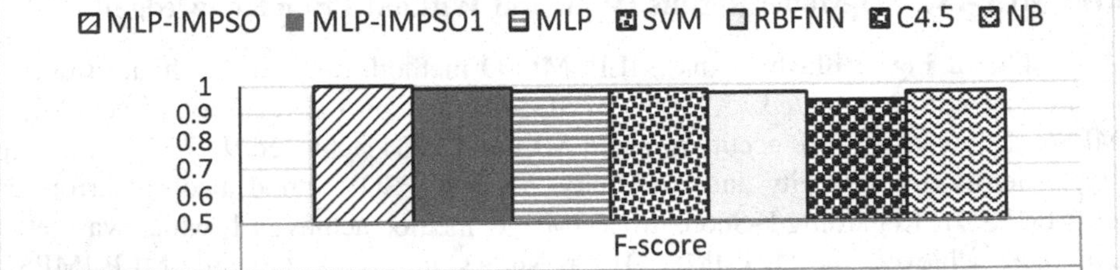

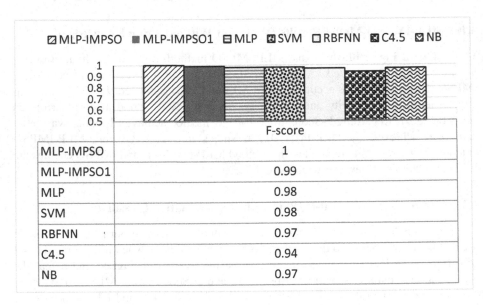

	F-score
MLP-IMPSO	1
MLP-IMPSO1	0.99
MLP	0.98
SVM	0.98
RBFNN	0.97
C4.5	0.94
NB	0.97

Fig. 4. F-Score results obtained of all classifiers

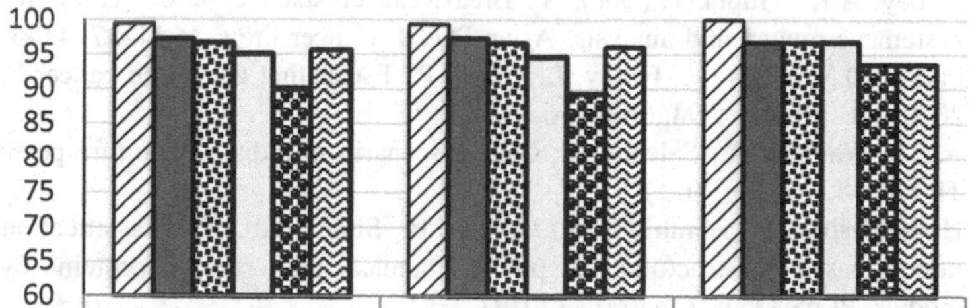

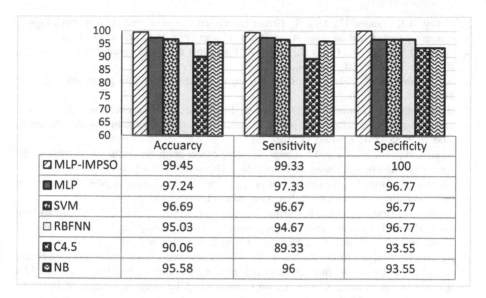

	Accuarcy	Sensitivity	Specificity
MLP-IMPSO	99.45	99.33	100
MLP	97.24	97.33	96.77
SVM	96.69	96.67	96.77
RBFNN	95.03	94.67	96.77
C4.5	90.06	89.33	93.55
NB	95.58	96	93.55

Fig. 5. Accuracy, Sensitivity, Specificity results of all classifiers

5 Conclusion

In this paper, for the diagnosis of lung cancer, a new hybrid method is proposed. We incorporate prior biological knowledge into the classification and combine PSO algorithm with MLP to improve the classifier. We used the microarray dataset from Brigham and Women's Hospital, Harvard Medical School in our study because it is a commonly used dataset among researchers who applied machine learning methods to lung cancer problems. Classification accuracy, sensitivity, specificity, and F-score of the proposed method MLP-IMPSO (with prior knowledge) were compared with the original method MLP-IMPSO1 (without prior knowledge). Also, the comparisons with other well-known classifiers SVM, MLP, RBFNN, C4.5, and NB were conducted. We used the ten-fold cross-validation method to justify the method performance. The evaluation results showed that our proposed method outperformed all the other methods, and was effective in lung cancer diagnosis. However, in this paper, our concentration was not on the physiological plausibility of the obtained results. This will be one of our goals in the future work. Besides, in the future research we plan to investigate the effectiveness of the proposed method in solving more complex multi-class lung cancer classification problems, as well as the problems of other cancer types.

References

1. Fan, Z., et al.: Smoking and risk of meningioma: a meta-analysis. Cancer Epidemiol. **37**, 39–45 (2013)
2. Ferlay, J., et al.: Cancer incidence and mortality worldwide: sources, methods and major patterns in GLOBOCAN 2012. Int. J. Cancer **136**, E359–E386 (2015)

3. Dubey, A.K., Gupta, U., Jain, S.: Breast cancer statistics and prediction methodology: a systematic review and analysis. Asian Pac. J. Cancer Prev. **16**, 4237–4245 (2015)
4. Parkin, D.M., Bray, F., Ferlay, J., Pisani, P.: Estimating the world cancer burden: globocan 2000. Int. J. Cancer **94**, 153–156 (2001)
5. Ali, I., Wani, W.A., Saleem, K.: Cancer scenario in India with future perspectives. Cancer Therapy **8**, 56–70 (2011)
6. Hosseinzadeh, F., Ebrahimi, M., Goliaei, B., Shamabadi, N.: Classification of lung cancer tumors based on structural and physicochemical properties of proteins by bioinformatics models. PLoS ONE **7**, e40017 (2012)
7. Azzawi, H., Hou, J., Xiang, Y., Alanni, R.: Lung cancer prediction from microarray data by gene expression programming. IET Syst. Biol. **10**(5), 168–178 (2016)
8. Al-Anni, R., Hou, J., Abdu-aljabar, R.D.A., Xiang, Y.: Prediction of NSCLC recurrence from microarray data with GEP. IET Syst. Biol. **11**, 77–85 (2017)
9. Azzawi, H., Hou, J., Alanni, R., Xiang, Y., Abdu-Aljabar, R., Azzawi, A.: Multiclass lung cancer diagnosis by gene expression programming and microarray datasets. In: International Conference on Advanced Data Mining and Applications, pp. 541–553 (2017)
10. Azzawi, H., Hou, J., Alnni, R., Xiang, Y.: SBC: a new strategy for multiclass lung cancer classification based on tumour structural information and microarray data. In: 17th IEEE/ACIS International Conference on Computer and Information Science (ICIS 2018), Singapore, pp. 68–73 (2018)
11. Alanni, R., Hou, J., Azzawi, H., Xiang, Y.: New gene selection method using gene expression programing approach on microarray data sets. In: Lee, R. (ed.) ICIS 2018. SCI, vol. 791, pp. 17–31. Springer, Cham (2019). https://doi.org/10.1007/978-3-319-98693-7_2
12. Alba, E., Garcia-Nieto, J., Jourdan, L., Talbi, E.-G.: Gene selection in cancer classification using PSO/SVM and GA/SVM hybrid algorithms. In: IEEE Congress on Evolutionary Computation. CEC 2007, pp. 284–290 (2007)
13. Botting, S.K., Trzeciakowski, J.P., Benoit, M.F., Salama, S.A., Diaz-Arrastia, C.R.: Sample entropy analysis of cervical neoplasia gene-expression signatures. BMC Bioinform. **10**, 66 (2009)
14. Abba, M.C., et al.: Breast cancer molecular signatures as determined by SAGE: correlation with lymph node status. Mol. Cancer Res. **5**, 881–890 (2007)
15. Xu, L., Geman, D., Winslow, R.L.: Large-scale integration of cancer microarray data identifies a robust common cancer signature. BMC Bioinform. **8**, 275 (2007)
16. Guinney, J., et al.: The consensus molecular subtypes of colorectal cancer. Nat. Med. **21**, 1350 (2015)
17. Bailey, P., et al.: Genomic analyses identify molecular subtypes of pancreatic cancer. Nature **531**, 47 (2016)
18. Ceccarelli, M., et al.: Molecular profiling reveals biologically discrete subsets and pathways of progression in diffuse glioma. Cell **164**, 550–563 (2016)
19. Dienstmann, R., Vermeulen, L., Guinney, J., Kopetz, S., Tejpar, S., Tabernero, J.: Consensus molecular subtypes and the evolution of precision medicine in colorectal cancer. Nat. Rev. Cancer **17**, 79 (2017)
20. Tai, F., Pan, W.: Incorporating prior knowledge of gene functional groups into regularized discriminant analysis of microarray data. Bioinformatics **23**, 3170–3177 (2007)
21. Guan, P., Huang, D., He, M., Zhou, B.: Lung cancer gene expression database analysis incorporating prior knowledge with support vector machine-based classification method. J. Exp. Clin. Cancer Res. **28**, 103 (2009)
22. Le, P.P., Bahl, A., Ungar, L.H.: Using prior knowledge to improve genetic network reconstruction from microarray data. In Silico Biol. **4**, 335–353 (2004)

23. Santoso, S., Powers, E.J., Grady, W.M., Hofmann, P.: Power quality assessment via wavelet transform analysis. IEEE Trans. Power Deliv. **11**, 924–930 (1996)
24. Ribeiro, P.F.: Wavelet transform: an advanced tool for analyzing non-stationary harmonic distortions in power systems. In: Proceedings IEEE ICHPS VI, pp. 365–369 (1994)
25. Gaouda, A., Salama, M., Sultan, M., Chikhani, A.: Power quality detection and classification using wavelet-multiresolution signal decomposition. IEEE Trans. Power Deliv. **14**, 1469–1476 (1999)
26. Wang, L., Zeng, Y., Chen, T.: Back propagation neural network with adaptive differential evolution algorithm for time series forecasting. Expert Syst. Appl. **42**, 855–863 (2015)
27. Kim, J., Jung, S.: Implementation of the RBF neural chip with the back-propagation algorithm for on-line learning. Appl. Soft Comput. **29**, 233–244 (2015)
28. Faris, H., Aljarah, I., Mirjalili, S.: Training feedforward neural networks using multi-verse optimizer for binary classification problems. Appl. Intell. **45**, 322–332 (2016)
29. Mirjalili, S., Hashim, S.Z.M., Sardroudi, H.M.: Training feedforward neural networks using hybrid particle swarm optimization and gravitational search algorithm. Appl. Math. Comput. **218**, 11125–11137 (2012)
30. Kennedy, J., Eberhart, R.: Particle swarm optimization. vol. 4, pp. 1942–1948 (1995)
31. Chen, L.-F., Su, C.-T., Chen, K.-H., Wang, P.-C.: Particle swarm optimization for feature selection with application in obstructive sleep apnea diagnosis. Neural Comput. Appl. **21**, 2087–2096 (2012)
32. Mohamad, M.S., Omatu, S., Deris, S., Yoshioka, M.: Particle swarm optimization for gene selection in classifying cancer classes. Artif. Life Robot. **14**, 16–19 (2009)
33. Shen, Q., Shi, W.-M., Kong, W.: Hybrid particle swarm optimization and tabu search approach for selecting genes for tumor classification using gene expression data. Comput. Biol. Chem. **32**, 53–60 (2008)
34. Banka, H., Dara, S.: A hamming distance based binary particle swarm optimization (HDBPSO) algorithm for high dimensional feature selection, classification and validation. Pattern Recogn. Lett. **52**, 94–100 (2015)
35. Sheikhpour, R., Sarram, M.A., Sheikhpour, R.: Particle swarm optimization for bandwidth determination and feature selection of kernel density estimation based classifiers in diagnosis of breast cancer. Appl. Soft Comput. **40**, 113–131 (2016)
36. Kar, S., Sharma, K.D., Maitra, M.: Gene selection from microarray gene expression data for classification of cancer subgroups employing PSO and adaptive K-nearest neighborhood technique. Expert Syst. Appl. **42**, 612–627 (2015)
37. Xi, M., Sun, J., Liu, L., Fan, F., Wu, X.: Cancer feature selection and classification using a binary quantum-behaved particle swarm optimization and support vector machine. Comput. Math. Methods Med. **2016**, 1–9 (2016)
38. Moradi, P., Gholampour, M.: A hybrid particle swarm optimization for feature subset selection by integrating a novel local search strategy. Appl. Soft Comput. **43**, 117–130 (2016)
39. Jain, I., Jain, V.K., Jain, R.: Correlation feature selection based improved-binary particle swarm optimization for gene selection and cancer classification. Appl. Soft Comput. **62**, 203–215 (2018)
40. Geethanjali, M., Slochanal, S.M.R., Bhavani, R.: PSO trained ANN-based differential protection scheme for power transformers. Neurocomputing **71**, 904–918 (2008)
41. Dudoit, S., Fridlyand, J., Speed, T.P.: Comparison of discrimination methods for the classification of tumors using gene expression data. J. Am. Stat. Assoc. **97**, 77–87 (2002)
42. Riget, J., Vesterstrøm, J.S.: A diversity-guided particle swarm optimizer-the ARPSO. Dept. Comput. Sci., Univ. of Aarhus, Aarhus, Denmark, Technical Report, vol. 2, p. 2002 (2002)

43. Shi, Y., Eberhart, R.C.: Parameter selection in particle swarm optimization. In: International Conference on Evolutionary Programming, pp. 591–600 (1998)
44. Almeida, L.M., Ludermir, T.B.: A multi-objective memetic and hybrid methodology for optimizing the parameters and performance of artificial neural networks. Neurocomputing **73**, 1438–1450 (2010)
45. Rosenblatt, F.: Perceptions and the theory of brain mechanisms (1962)
46. Liang, X.: Removal of hidden neurons in multilayer perceptrons by orthogonal projection and weight crosswise propagation. Neural Comput. Appl. **16**, 57–68 (2007)
47. Singhal, S., Wu, L.: Training multilayer perceptrons with the extended Kalman algorithm. In: Advances in Neural Information Processing Systems, pp. 133–140 (1989)
48. Sarkaleh, M.K., Shahbahrami, A.: Classification of ECG arrhythmias using discrete wavelet transform and neural networks. Int. J. Comput. Sci. Eng. Appl. **2**, 1 (2012)
49. Suykens, J.A., Vandewalle, J.: Training multilayer perceptron classifiers based on a modified support vector method. IEEE Trans. Neural Networks **10**, 907–911 (1999)
50. Tzikas, D., Likas, A.: An incremental bayesian approach for training multilayer perceptrons. In: International Conference on Artificial Neural Networks, pp. 87–96 (2010)
51. Ni, J., Song, Q.: Dynamic pruning algorithm for multilayer perceptron based neural control systems. Neurocomputing **69**, 2097–2111 (2006)
52. Battiti, R.: First-and second-order methods for learning: between steepest descent and Newton's method. Neural Comput. **4**, 141–166 (1992)
53. Riedmiller, M., Braun, H.: A direct adaptive method for faster backpropagation learning: the RPROP algorithm. In: IEEE International Conference on Neural Networks, pp. 586–591 (1993)
54. Møller, M.F.: A scaled conjugate gradient algorithm for fast supervised learning. Neural Networks **6**, 525–533 (1993)
55. Nasir, A.A., Mashor, M.Y., Hassan, R.: Classification of acute leukaemia cells using multilayer perceptron and simplified fuzzy ARTMAP neural networks. Int. Arab J. Inform. Technol. **10**, 356–364 (2013)
56. Süt, N., Çelik, Y.: Prediction of mortality in stroke patients using multilayer perceptron neural networks. Turkish J. Med. Sci. **42**, 886–893 (2012)
57. Abid, S., Fnaiech, F., Jervis, B., Cheriet, M.: Fast training of multilayer perceptrons with a mixed norm algorithm. In: Proceedings of the 2005 IEEE International Joint Conference on Neural Networks. IJCNN 2005, pp. 1018–1022 (2005)
58. Fogel, D.B.: An introduction to simulated evolutionary optimization. IEEE Trans. Neural Networks **5**, 3–14 (1994)
59. Liu, H., Setiono, R.: Chi2: feature selection and discretization of numeric attributes. In: Proceedings of the Seventh International Conference on Tools with Artificial Intelligence, pp. 388–391 (1995)
60. Chitsaz, E., Taheri, M., Katebi, S.D., Jahromi, M.Z.: An improved fuzzy feature clustering and selection based on chi-squared-test. In: Proceedings of the International Multi Conference of Engineers and Computer Scientists, pp. 18–20 (2009)
61. Gordon, G.J., et al.: Translation of microarray data into clinically relevant cancer diagnostic tests using gene expression ratios in lung cancer and mesothelioma. Can. Res. **62**, 4963–4967 (2002)
62. Abdullah, M., Bakar, A., Rahim, N., Mokhlis, H., Illias, H., Jamian, J.: Modified particle swarm optimization with time varying acceleration coefficients for economic load dispatch with generator constraints. J. Electr. Eng. Technol. **9**, 15–26 (2014)
63. Abbass, H.A., Sarker, R., Newton, C.: PDE: a Pareto-frontier differential evolution approach for multi-objective optimization problems. In: Proceedings of the 2001 Congress on Evolutionary Computation, pp. 971–978 (2001)

64. Kourou, K., Exarchos, T.P., Exarchos, K.P., Karamouzis, M.V., Fotiadis, D.I.: Machine learning applications in cancer prognosis and prediction. Comput. Struct. Biotechnol. J. **13**, 8–17 (2015)
65. Joseph, A.C., David, S.W.: Applications of machine learning in cancer prediction and prognosis. Cancer Inform. **2**, 59–77 (2006)

Plant Leaf Disease Detection and Classification Using Particle Swarm Optimization

Rishabh Yadav[1(✉)], Yogesh Kumar Rana[2], and Sushama Nagpal[3]

[1] Mobikwik, Gurugram, India
rishabhydv@gmail.com
[2] HSBC, Pune, India
1314.yogesh@gmail.com
[3] Netaji Subhas University of Technology, New Delhi, India
sushmapriyadarshi@yahoo.com

Abstract. The loss of crops due to diseases is a major danger to food security. It is important to develop the requisite infrastructure and tools for the detection of diseases in crops. The opportunity to detect diseases in crops has increased manifolds with the rise in the number of smartphone users and improved network connectivity. In this paper, we provide an approach to detect and classify plant leaf diseases. The methodology involves image acquisition, pre-processing of the images, feature extraction followed by feature selection and finally the classification of plant diseases. A deep convolutional neural network was trained to extract features from the input image. An optimal set of features is selected using Particle Swarm Optimization (PSO) and are classified into 23 different classes, including both healthy and diseased categories. Apropos, by employing this technique, the plant leaf images are classified with an accuracy of 97.39%.

Keywords: Deep convolutional neural network · Particle Swarm Optimization · AlexNet · SVM

1 Introduction

Humans are capable to produce sufficient food to feed the whole world by using modern technologies. In spite of producing enough food, its reach is restricted by different factors such as pollinators decline, diseases in plants, climate change, and others. Plant diseases are responsible for the food shortage at global scale which poses a threat to small farmers, who are largely dependent on the healthy yield of crops. Food security is a major threat to mankind. Globally more than 800 million people do not have adequate food. Moreover, in India itself annually 17.5% of crops are lost due to pests and disease infestation [14]. The problem of disease identification in developing countries is exacerbated due to lack of adequate resource allocation by the respective governments.

É. Renault et al. (Eds.): MLN 2018, LNCS 11407, pp. 294–306, 2019.
https://doi.org/10.1007/978-3-030-19945-6_21

Agricultural production is largely dependent on small farmers, who are responsible for 80% of the total crop yield, but 50% of this yield is lost due to pests and diseases in crops. More than 50% of food deprived people live in small farmer households, which makes small farmers the worst affected group in food supply due to diseases in plants and crops.

Earlier, identification of diseases in crops has been carried out by plant clinics, agriculture based organizations or institutions. The main approach that is adopted by experts for detection of diseases in plants is through naked eye observation. Visual identification of diseases by experts requires constant examination which is unaffordable for most of the farmers. Moreover, in some developing countries, to contact expert's, farmers have to travel long distances. This makes the whole process of consultation time consuming and at the same time expensive. Most of the times farmers, are not aware of non-native diseases which forces them consult experts.

Recently, mobile phones based tools have increased in numbers, due to more number of mobile devices and availability of the internet in remote locations. It is envisaged that in 2018, 66% of the population in 52 countries, including India would have access to smartphones which has witnessed an increase from 58% in 2016 to 62% in 2017 [7]. The number of smartphone users is expected to reach 6 billion by the end of 2020. The combination of large smartphone users, high definition cameras and increased performance of mobile device processors facilitates the usage of automated disease identification in plants using mobile phones. Mobile broadband subscribers have increased by more than 20% annually since the last 5 years, and had predicted it to reach 4.3 billion by the end of 2017 [4]. This has led to disease identification to be supported by online diagnosis by providing more information about the disease.

The advancement in the field of object recognition provides an opportunity for a more robust and accurate disease recognition application employing smartphones. Recent enhancements in the field of deep neural networks has provided exceptional results and outperformed the previous state of the art techniques in the field of object recognition. Significant advances have been made with the use of deep neural networks in various fields such as Speech recognition, Natural language processing and Healthcare. Here we demonstrate the use of deep convolutional neural network architecture AlexNet [6] for feature extraction and PSO for feature selection, and later classification using support vector machine as this approach is more technically feasible for smartphone users. The deep neural network was trained using open source dataset PlantVillage [3]. Dataset consists of 54,306 images of both diseased and healthy leaves incorporating 14 crop species and 26 diseases. Our paper utilizes 8750 images of 7 crop species with 23 classes for classification, including both diseased and healthy classes, due to computational limitations.

2 Related Works

There are a lot of procedures that employ computer vision for the detecting the disease in plants. In [1] authors have presented an approach to detect disease using image processing techniques for segmentation of the diseased spot. For the process of disease spot detection, they have compared CIELB, YcbCr and HSI colour models and used a median filter for image smoothing. This provided an algorithm that successfully detected diseases and remained independent of background noise.

Patil and Kumar in [9] describe a method for extracting colour and texture features of diseased leaf. The texture features such as correlation, energy, inertia and homogeneity are extracted by calculating the gray level co-occurrence matrix of an image. Also, colour features were extracted by obtaining the HSV of an image, together texture and colour features were used to detect disease in maize leaves.

The primary aim of feature selection is to optimize the maximum number of irrelevant features, as also maintaining classification accuracy [18]. It is statistically proven that PSO is computationally more efficient as compared to genetic algorithm. PSO when applied to solve unconstrained nonlinear problems with continuous design variables performs better than genetic algorithm by greater computational efficiency differential. At the same time it outperformed genetic algorithm by less efficiency differential when they were applied to constrained nonlinear problems which involved continuous or discrete design variables [2].

In [10] the authors extracted colour, edge and texture features from the diseased leaf image. PSO was used for selection of features which would be used to train deep forward neural network. The proposed system was able to identify cotton diseases with an accuracy of 95%.

In the past few years, there has been paradigm shift in the fields of computer vision and object recognition. Large Scale Visual Recognition Challenge (ILSVRC) [11], which is based on the ImageNet database [5] is considered as benchmark for numerous computer vision related problems. AlexNet [6] a large, deep convolutional neural network for classification of 1.2 million images into 1000 possible categories in ILSVRC, achieved a top-1 and top-5 error rates of 37.5% and 17% which was better than the previous state of art.

The authors in [15] employed deep learning for classification of plant leaves diseases. They had used 30880 images for classification of fifteen different types of classes. The images were used to train a convolutional neural network based on CaffeNet architecture which contains eight learning layers, five convolutional and three fully connected layers. The network was able to achieve classification accuracy of 96.3%.

In another work, the authors [8] used different types of deep learning architecture for plant leaf disease classification. The authors compared AlexNet [6] and GoogLeNet architecture [17] by training them from scratch as well as using the transfer learning approach. The network was trained with PlantVillage dataset [3] which consisted of 54306 images containing 38 classes of 14 crop species and

26 diseases. The overall accuracy varied from 85.53% to 99.34% wherein in this case GoogLeNet architecture was trained using the transfer learning approach.

Training neural networks is a time-consuming process as it involves large dataset. In our study, we aim to employ deep learning for the purpose of feature extraction. All the above-mentioned approaches used an end to end neural networks, including for the purpose of classification. We present a computationally optimal approach which can easily be used on smartphones. We employ deep neural network for feature extraction and PSO for the feature selection. The optimal set of features is used to train a classifier for the classifying images into their respective classes.

3 Materials and Methodology Used

3.1 Dataset Description

We have analyzed 23 classes of plant leaves with a total of 8750 images. Each class is assigned as a crop-disease pair and after training the model, it tries to predict the correct pair using image of the diseased plant leaf. In our model to predict the correct class, we resized all images to 256×256 pixels, and the model training and predictions are performed on these resized images. Across all our process, we have used coloured resized images of the PlantVillage dataset [3] (Fig. 1).

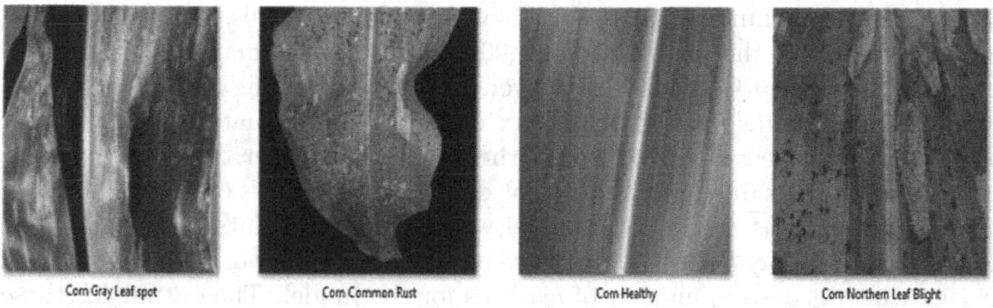

Fig. 1. Some images from the PlantVillage dataset (Color figure online)

3.2 Approach

Our approach is divided into three parts viz feature extraction, feature selection and classification. A deep neural network is quite useful in extracting features out of colored raw images. We trained one such architecture AlexNet [6] using transfer learning for extracting features out of input image. The softmax layer of AlexNet is removed and features are extracted by considering the output of 100 nodes. These 100 features extracted from AlexNet are fed as input to PSO for feature selection. In PSO for feature selection, we considered representing each feature with a binary selection. Thus, it provides us with sample space ranging from 1 to 2100, on which we select a population size of 70. By using the multi objective fitness function, the velocity was updated and parameters were updated using global best and local best of each particle. For the velocity function, we set self-confidence range as 0.75 and swarm confidence range as 0.75 [13]. These values were chosen for best results after experimentation with different values. This provides us with an optimal set of features for training our SVM classifier (Fig. 2).

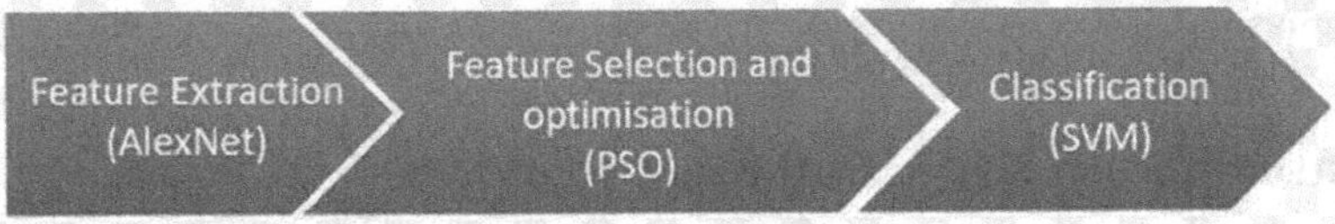

Fig. 2. Flowchart depicting main process of proposed method

3.3 Feature Extraction

The performance of any detection or recognition system is highly dependent on the features which are extracted. A deep neural network was trained to extract features. As the size of dataset was not sufficient to train the neural network to the desired level from scratch, therefore transfer learning was used. Transfer learning optimizes the training process with some tweaks in the existing model. The classification function in the original AlexNet model is a softmax classifier that computes the probability of 1,000 classes of the ImageNet dataset. In AlexNet there are 5 Convolutional Layers with Max Pooling and 3 Fully Connected Layers giving 1000-way SoftMax output. The ultimate fully connected layers predicts based on the output of the penultimate layer. Since the transfer learning model is trained with a different dataset, the weights of the penultimate layer are reset so as to train the model with the current dataset. As AlexNet, results in a 1000-way SoftMax output we reduce the number of classes from 1000 to 100, which is desired number of features for the model. The output for these 100 classes were taken as features.

3.4 Particle Swarm Optimization (PSO)

A total of 100 features were extracted for each image in the preceding stage using deep neural networks. Deep neural networks can be trained for extracting more than 100 features. As the above step is computationally extensive, it is necessary to use an optimization technique that can specify only the dominant or important features. PSO which is one such heuristic technique, is used for the selection of optimal set of features. PSO is motivated by social behaviors such as bird flocking and fish schooling. The underlying phenomenon of PSO is that knowledge is optimized by social interaction with the population where thinking is not only personal but also social [19]. In Particle Swarm Optimization, each solution can be represented as a particle in the swarm. The position of the particle in the search space is represented by a vector $P_i = (P_{i1}, P_{i2}, P_{iN})$ where N is the dimensionality of the vector. Also each particle has velocity associated with it, which is represented by $V_i = (V_{i1}, V_{i2}, V_{iN})$, as they move in the search space to find the optimal solution. During this phase each particle updates its position and velocity. The variable 'pbest' represents the best previous position of the particle and 'gbest' represents the best position of the entire population. According to the values of the parameter's pbest and gbest, PSO searches for optimal solutions by updating the position and the velocity of each particle according to the following equations:-

$$Vi(t+1) = w * Vi(t) + c_1 * ud * [pbesti(t) - Pi(t)] + c_2 * ud * [gbest(t) - Pi(t)] \quad (1)$$

$$Pi(t+1) = Pi(t) + Vi(t+1) \quad (2)$$

3.5 Feature Selection

The task of PSO is to search for the most descriptive feature subset among the extracted features. Each particle in the population represents a possible candidate solution. Fitness function drives evolution which provides an indication of the expected fitness on future trials. The initial coding of the particle was randomly generated and each particle was coded to a binary alphabetic string P = $F_1F_2F_3F_4 \ldots F_n$, n = 1, 2, 3, ..., m; where 'm' is the length of the feature vector extracted which is 100 in this case. Each index of the binary string B represents which feature is to be selected, '1' depicts that particular feature which is to be selected and correspondingly '0' denotes rejection.

The algorithm is used to search for the optimal solution in the search space which is denoted by 2^m. For example, when 10 dimensional dataset (n = 10) P = $F_1F_2F_3F_4F_5F_6F_7F_8F_9F_10$ is analyzed using PSO to select features, we can select any subset of features smaller than n. i.e. PSO can choose a random 5 features $F_1F_4F_6F_7F_9$ by setting bits 1, 4, 6, 7 and 9 [18]. For each particle, the effectiveness of the selected feature subset in retaining the maximum accuracy in representing the original feature set is evaluated based on the fitness value. Indices denoted by m in the particle represent the parameters to be iteratively evolved by the PSO. The fitness function F, in each iteration evaluates the fitness or goodness value for each particle. This evolution is driven by the fitness function that evaluates the quality of evolved particles in terms of their ability to maximize the classification accuracy. The fitness function is a k-fold cross validation function; where k = 10, which trains the classifier with the features represented by the particle. The accuracy of the classifier is used as fitness value and used for evolution.

3.6 Classification

The last stage is to classify the image into various classes of diseases. The optimal set of features selected by PSO, from the features extracted by AlexNet are used

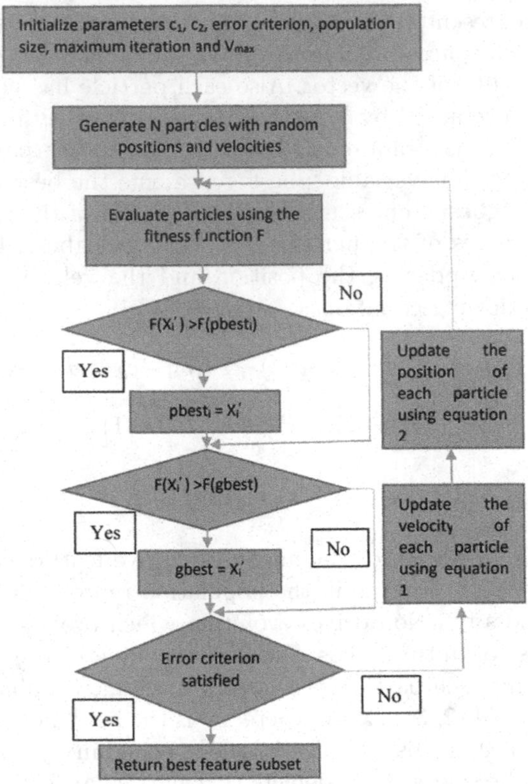

Fig. 3. Flowchart representing the process of feature selection using PSO

to train the classifier. For this, a number of classifiers XGBoost [12], Support Vector Machine (SVM), Random Forest and K-Nearest Neighbor (KNN) were tested on their cross validation accuracy based on which SVM was selected for the classification of the images (Fig. 3).

4 Experimental Setup and Results

4.1 Dataset Description

The dataset includes 8750 images comprising of 23 classes consisting of both diseased as well as healthy leaves. Table 1 provides a description of 23 classes and the number of images which are used for each class.

4.2 Measurement of Performance

To gauge the performance of the proposed approach on unseen data, and to keep a track of overfitting in our approach, we trained our model with approximately 300 images of each class which had a 70%–30% train-cross validation split. The testing dataset comprised of 10 images for each of the 23 classes totaling to 230 images.

Table 1. Dataset description

Disease	Number of images
Apple Black Rot	270
Apple Healthy	703
Apple Rust	114
Apple Scab	260
Corn Gray Leaf Spot	180
Corn Common Rust	508
Corn Northern Leaf Blight	363
Corn Healthy	434
Grape Black Rot	466
Grape Spanish Measles	541
Grape Leaf Spot	410
Grape Healthy	161
Peach Bacterial Spot	952
Peach Healthy	98
Pepper Bell Healthy	567
Pepper Bell Bacterial, Spot	418
Potato Early Blight	402
Potato Late Blight	370
Potato Healthy	54
Squash Healthy	150
Squash Powdery Mildew	739
Strawberry Leaf Scorch	380
Strawberry Healthy	204

4.3 Results

After training the model for 23 classes, 100 features were extracted. By applying PSO, 34 optimal features were extracted, which were used to train our classifier. The AlexNet model for feature extraction was trained for 23 classes with approximately 6700 number of images for training and 2050 number of images for cross-validation. The ratio for training to cross validation was 70%–30%. A total of 8750 images were used for training the model.

Table 2 denotes the accuracy which was achieved for the stated number of iterations. 100 features were extracted for each image. PSO was used to select an optimal feature subset from the 100 features that were extracted using the neural network. Table 3 provides the values of the parameters which are required by PSO.

Table 2. AlexNet training accuracy

Epochs	Training accuracy
500	97.33
1000	97.33

Table 3. PSO parameter setting

Population size	50
Error criterion	0.01
c_1	0.75
c_2	0.75

Table 4. PSO results

gbest fitness	0.987
gbest parameter	55212075364990419967829606400O
Iterations	101

Table 4 depicts the results of PSO. The 'gbest' fitness specifies the global best fitness value that was obtained and subsequently the global best parameter which represents the optimal feature subset. The Global best parameter when converted to binary represents which column needs to be selected. 552120753649904199678296064000 when converted to binary is 01101111011111-1111110010101010011010001001111111000010000000000000000000000000000000 -0000000000000000. Out of which 34 features with indices's: '3', '4', '6', '7', '8', '9', '11', '12', '13', '14', '15', '16', '17', '18', '19', '20', '21', '22', '25', '27', '29', '31', '34', '35', '37', '41', '44', '45', '46', '47', '48', '49', '50', '55' were selected for feature selection. These optimal set of features were used to train the classifier.

Table 5 describes the cross validation results for different classifiers which were scored on Accuracy, Precision, Recall and F_1-score. The following classifiers: XGBoost, SVM, Random Forest and KNN are evaluated on their cross validation

Table 5. Classifier cross-validation results

Classifier	Accuracy	Precision	F_1-score	Recall
XGboost	0.9886	0.9849	0.9830	0.9823
SVM	0.9907	0.9886	0.9872	0.9866
Random forest	0.9804	0.9774	0.9739	0.9719
KNN	0.9851	0.9882	0.9803	0.9742

Table 6. Detailed final result

Class name	No of images	Precision	Recall	F_1-score
Apple Black Rot	10	1.0000	0.9000	0.9474
Apple Healthy	10	0.9091	1.0000	0.9524
Apple Rust	10	1.0000	1.0000	1.0000
Apple Scab	10	1.0000	0.9000	0.9474
Corn Gray Leaf Spot	10	1.0000	0.7000	0.8235
Corn Common Rust	10	1.0000	1.0000	1.0000
Corn Northern Leaf Blight	10	0.8333	1.0000	0.9091
Corn Healthy	10	1.0000	1.0000	1.0000
Grape Black Rot	10	1.0000	1.0000	1.0000
Grape Spanish Measles	10	1.0000	1.0000	1.0000
Grape Leaf Spot	10	1.0000	1.0000	1.0000
Grape Healthy	10	1.0000	1.0000	1.0000
Peach Bacterial Spot	10	0.9091	1.0000	0.9524
Peach Healthy	10	1.0000	0.9000	0.9474
Pepper Bell Healthy	10	1.0000	1.0000	1.0000
Pepper Bell Bacterial Spot	10	1.0000	1.0000	1.0000
Potato Early Blight	10	1.0000	1.0000	1.0000
Potato Late Blight	10	1.0000	1.0000	1.0000
Potato Healthy	10	1.0000	1.0000	1.0000
Squash Healthy	10	1.0000	1.0000	1.0000
Squash Powdery Mildew	10	0.8333	1.0000	0.9091
Strawberry Leaf Scorch	10	1.0000	1.0000	1.0000
Strawberry Healthy	10	1.0000	1.0000	1.0000
Micro Avg	230	0.9739	0.9739	0.9739
Macro Avg	230	0.9776	0.9739	0.9734

F_1-score. SVM outperformed the other classifiers with a F_1-score of 0.9866 and hence was selected for classification. SVM classifier was trained on the optimal feature subset consisting of 34 features. A total of 230 images, 10 images of each class were used for testing, of which only 6 images were incorrectly classified. Thus, an overall accuracy of 97.39% (mean F_1-Score of 0.9739) was achieved.

Table 6 provides a detailed result of the classification. For each class its corresponding Precision, Recall and F_1-score is calculated. The macro and micro average [16] of Precision, Recall and F_1-score is calculated, which results in a mean F_1-score of 0.9739. Figure 4 depicts the Precision-Recall curve with an area of 0.9989 which is derived from the results mentioned in Table 6.

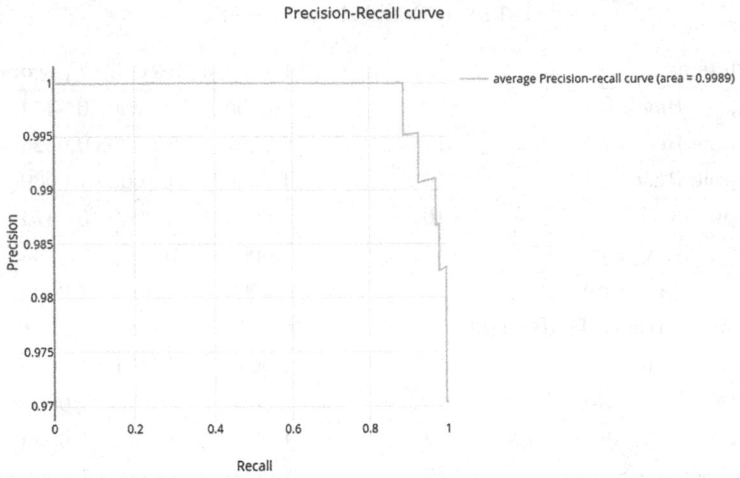

Fig. 4. Precision-Recall curve for the test data using SVM

5 Conclusion and Future Work

In previous works, neural networks have been used in the field of disease detection in leaves. In [10] the authors trained a deep forward neural network by selecting features using PSO and were able to identify cotton diseases with an accuracy of 95%. The system proposed in [15] which used an end to end deep convolutional network and trained using 30880 images, was able to classify images with an accuracy of 96.3% among 15 different classes. Our proposed approach outperformed above-mentioned methods by achieving an accuracy of 97.39% (mean F_1-Score of 0.9739). Although the dataset used was different, it cannot be directly compared. The authors in [8] compared various deep convolutional neural network architecture AlexNet [6] and GoogleNet [17]. The various neural network in different configurations were trained using 54306 images provided by PlantVillage dataset [3]. The classification accuracy ranged from 85.53% to 99.34% for GoogLeNet architecture trained by employing transfer learning. In comparison to the approach mentioned in [8], an accuracy of 99.34% was achieved by using GoogLeNet architecture, where as the proposed model employing AlexNet achieved an accuracy of 97.39%. As compared to various approaches, the proposed model outperformed majority of the approaches mentioned in [1, 10, 11, 16], but still lagged behind the approach mentioned in [8]. Therefore, it is imperative to explore areas in which the proposed model can be improved upon.

The proposed AlexNet model is trained using the 8750 images as compared to 54306 images available in PlantVillage dataset [3]. Due to computational limitations, only 8750 images were used to train the model. The increase in the number of training images provides an opportunity to improve the overall accuracy of the system. Since this study, considers only single leave images in clean environment, it also provides another area of improvement. This model can be extended to real life scenarios wherein leaves are in a cluttered environment and chances for detecting false positives are high. In the proposed model, only AlexNet is employed for feature extraction, different neural network architectures needs to be analyzed for better performance. To make this model more suitable for use in the real world, it is necessary to extend the number of classes of plants and diseases thereby enriching the dataset.

In essence, it is imperative to develop a smartphone application in order to make it user friendly for the public in general and specifically for farmers, students, researchers and enthusiasts in the fields of agriculture. The application would be able to integrate the proposed model for disease detection. The scope could be further enhanced by incorporating more functionality such as suitable climate conditions, fertilizers for plants and remedies for plant diseases, thus providing wholesome information regarding crops under a single roof.

References

1. Chaudhary, P., et al.: Color transform based approach for disease spot detection on plant leaf. Int. J. Comput. Sci. Telecommun. **3**(6), 65–70 (2012)
2. Hassan, R., et al.: A comparison of particle swarm optimization and the genetic algorithm. In: 46th AIAA/ASME/ASCE/AHS/ASC Structures, Structural Dynamics and Materials Conference, Structures, Structural Dynamics, and Materials and Co-located Conferences (2005)
3. Hughes, D.P., Salathé, M.: An open access repository of images on plant health to enable the development of mobile disease diagnostics. arXiv:1511.08060 (2015)
4. ICT Facts and Figures 2017. www.itu.int/en/ITU-D/Statistics/Pages/facts/default.aspx
5. Deng, J., et al.: ImageNet: a large-scale hierarchical image database. In: 2009 IEEE Conference on Computer Vision and Pattern Recognition (2009). ieeexplore.ieee.org/document/5206848
6. Krizhevsky, A., et al.: ImageNet classification with deep convolutional neural networks. In: Proceedings of the 25th International Conference on Neural Information Processing Systems, vol. 1 (2012)
7. Mobile Advertising Forecasts 2017 - Zenith. Zenith. www.zenithmedia.com/product/mobile-advertising-forecasts-2017/
8. Mohanty, S.P., Hughes, D.P., Salathé, M.: Using deep learning for image-based plant disease detection. Front. Plant Sci. **7**, 1419 (2016). PMC. Web. 14 Jan 2018
9. Patil, S., Shrikant, Bodhe, K.: Leaf disease severity measurement using image processing. Int. J. Eng. Technol. **3** (2011). https://www.researchgate.net/publication/267430095_Leaf_disease_severity_measurement_using_image_processing
10. Revathi, P., Hemalatha, M.: Identification of cotton diseases based on cross information gain deep forward neural network classifier with PSO feature selection. Int. J. Eng. Technol. **5**(6), 4637–4642 (2013)
11. Russakovsky, O., et al.: ImageNet large scale visual recognition challenge. arXiv:1409.0575 (2015)
12. Schapire, R.E.: The boosting approach to machine learning: an overview. In: Denison, D.D., Hansen, M.H., Holmes, C.C., Mallick, B., Yu, B. (eds.) Nonlinear Estimation and Classification. LNS, vol. 171, pp. 149–171. Springer, New York (2003). https://doi.org/10.1007/978-0-387-21579-2_9

13. Seal, A., Ganguly, S., Bhattacharjee, D., Nasipuri, M., Gonzalo-Martin, C.: Feature selection using particle swarm optimization for thermal face recognition. In: Chaki, R., Saeed, K., Choudhury, S., Chaki, N. (eds.) Applied Computation and Security Systems. AISC, vol. 304, pp. 25–35. Springer, New Delhi (2015). https://doi.org/10.1007/978-81-322-1985-9_2
14. Singh, T., Satyanarayana, J., Peshin, R.: Crop loss assessment in India-past experiences and future strategies. In: Pimentel, D., Peshin, R. (eds.) Integrated Pest Management. Springer, Dordrecht (2014). https://doi.org/10.1007/978-94-007-7796-5_9
15. Sladojevic, S., Arsenovic, M., Anderla, A., Culibrk, D., Stefanovic, D.: Deep neural networks based recognition of plant diseases by leaf image classification. Comput. Intell. Neurosci. **2016**, 11 (2016). https://doi.org/10.1155/2016/3289801. Article ID 3289801
16. Stąpor, K.: Evaluation of classifiers: current methods and future research directions. In: Position Papers of the 2017 Federated Conference on Computer Science and Information Systems, vol. 12, pp. 37–40 (2017). https://doi.org/10.15439/2017F530
17. Szegedy, C., et al.: Going deeper with convolutions. In: 2015 IEEE Conference on Computer Vision and Pattern Recognition (CVPR), Boston, MA, pp. 1–9 (2015). https://doi.org/10.1109/CVPR.2015.7298594
18. Chung-Jui, T., Chuang, L.-Y., Jun-Yang, C., Yang, C.-H.: Feature selection using PSO-SVM. IAENG Int. J. Comput. Sci. **33** (2007). https://www.researchgate.net/publication/26587209_Feature_selection_using_PSO-SVM
19. Xue, B., et al.: Particle swarm optimization for feature selection in classification: a multi-objective approach. IEEE Trans. Cybern. **43**(6), 1656–1671 (2013). ieeexplore.ieee.org/document/6381531

Game Theory Model for Intrusion Prevention in Computer Networks

Julián Francisco Mojica Sánchez[1], Octavio José Salcedo Parra[1,2](✉), and Miguel J. Espitia R.[3]

[1] Department of Systems and Industrial Engineering, Faculty of Engineering, Universidad Nacional de Colombia, Bogotá D.C., Colombia
{jfmojicas,ojsalcedop}@unal.edu.co

[2] Faculty of Engineering, Intelligent Internet Research Group, Universidad Distrital "Francisco José de Caldas", Bogotá D.C., Colombia
osalcedo@udistrital.edu.co

[3] Faculty of Engineering, GEFEM Group, Universidad Distrital "Francisco José de Caldas", Bogotá D.C., Colombia
mespitiar@udistrital.edu.co

Abstract. No system is completely secure, but it's possible to analyze system vulnerabilities, take security measures and thus reduce the likelihood of system intrusions. One tool to reduce this probability of intrusion is to implement intrusion prevention systems, which analyze network traffic to determine anomalous behavior and take security measures with the agents that exhibit these behaviors. This document proposes a game theory model based in an evaluation of works related in the area that could be the basis of an algorithm for the prevention of intrusions in networks.

Keywords: Intrusion prevention system · Game theory · Vulnerabilities

1 Introduction

WIFI technology has been imposed in recent years over other types of connection in local networks due to its versatility and low cost compared to wired networks, this has allowed the creation of different network configurations from small home and office networks up to business networks, with a considerable amount of devices sharing information at all time. This versatility has been achieved leaving behind security, therefore it is necessary to work on tools and protocols that increase the security of WIFI networks [1]. Currently, the security protocols designed by IEEE for WIFI networks are the 802.11 family, where the most used form of encryption due to its efficiency is WPA2. In spite of the efforts made by IEEE, the advance in new devices has been much greater than the advance in the creation, implementation and updating of security protocols so that each time new attacks are created taking advantage of this gap. Intrusion prevention systems are used in large companies and it is not feasible to implement them in public networks with a large number of connected devices sharing information at all times and in different locations. As for office and home networks, router security mechanisms are obsolete against various types of attacks [2].

É. Renault et al. (Eds.): MLN 2018, LNCS 11407, pp. 307–320, 2019.
https://doi.org/10.1007/978-3-030-19945-6_22

The internet of things every day is being introduced more into our daily lives due to the access that people have both to internet and new technology, with the increase of everyday devices connected to the internet also increases the amount of sensitive information transmitted wirelessly [3].

The increment of devices connected to Internet and the sharing of a big amount of sensitive information and the intrinsic vulnerabilities of wireless communications, made the infrastructure of the Internet of Things networks a point of interest for criminals. For that reason it is necessary to ensure the privacy, available and integrity of this information through security adaptive mechanisms to different types of attacks [4]. The simplicity required in the Internet devices of the thing makes impossible the implementation of internally security systems in these. Because the creation of devices has been increasing to a greater extent than the development in network security with IOT devices, these devices have become a gateway to networks for attackers. Among the different attacks suffered by Wi-Fi networks connected to IOT devices we have: lowpassword exploitation, reverse engineering hardware, remote code execution, man in the middle and hidden monitoring functions. The most common use of remotely controlling IOT devices by cybercriminals is DoS, due it is easy to achieve if the devices have not been configured with basic security measures and allows them to obtain money by preventing access to a server or website. [5] The sensitivity of information and the failures in the security of wireless networks have led the research community to undertake research and development efforts simultaneously in new IOT devices and security mechanisms for wireless networks. An important point is the need for real-time processing and reaction to attacks because IOT devices transmit information constantly. [6] In this document, a model based on game theory will be created to establish the intrusion detection criteria of an IPS (Intrusion Prevention System) for networks, based on the review of previously developed related works.

2 Related Works

In 2016 P. Sharma, S. Moon, D. Moon and J. Park designed a DFA-AD (distributed framework architecture for the detection of advanced persistent threats), in which one of the 4 traffic classification modules was Dynamic bayesian game model Based, in this case the game model is dynamic since each player selects his behavior depending on the current state of the system and the information he possesses. The attackers identify users and special targets of the system in an exhaustive way, therefore, the attackers have more data about the module than the protectors, which creates a system in incompleteness and asymmetry [5]. In 2016 K. Wang, M. Du, D. Yang, C. Zhu, J. Shen and Y. Zhang propose an attack-defense game model for detecting malicious nodes in Embedded Sensor Networks (ESN) using a repeated game approach, where they define the function of rewards that attackers and defenders will receive for their actions [7]. To solve errors and absences in detection use a game tree model. They show that the game model does not have a pure Nash equilibrium but mixed strategy, where the nodes are changing due to the strategies of attackers and defenders so that they are in dynamic equilibrium, where limited resources are used and provided Security protection at the same. Finally, they perform simulations where they show that with the proposed model they can reduce energy consumption by 50% compared to the existing model All Monitor (AM) and improve the detection rate from 10% to 15% compared to the existing model Cluster Head (CH). In 2013 M. Manshaei, Q. Zhu, T. Alpcan, T. Bacar and J. Hubaux carry out a review of the investigations in privacy and security in communication and computer networks that have a game theory approach [8]. In their content they have a section of Intrusion Detection Systems where they present the different works found in the review of the literature; the way in which the IDS are configured; Networked IDS, where different IDSs operate in the network independently and the security of each subsystem that they individually protect depends on the performance of the other IDS; Collaborative Intrusion Detection System Networks, in this case in the network operate different IDS in collaborative way, that is, they share the knowledge of the new attacks they detect, but the system can be compromised if the control of an IDS is taken by an attacker and finally the response to intrusions, where they expose an intrusion response system based on Stackelberg stochastic game called Response and Recovery Engine (RRE).

3 Game Theory Models

The prevention of intrusions can be understood as an attack-defense scenario, in which the person in charge of the security of the network decides if it is necessary or not to put the intrusion prevention system into operation, since such operation has a cost that would not be necessary if the network is not being attacked. The game consists of a defender (in charge of starting or not the IPS) and an attacker (which seek to enter the network and take advantage of this intrusion), this was taken from the model proposed by Wang in 2016 [7] but limited to a single attacker and defender, since it took multiple attackers and defenders at multiple nodes and time periods. The defender has two strategies (U_D): defend or not defend and in the case of the attacker (U_A): attack or not attack. The realization of these strategies has rewards and costs that will determine the way the two actors act. These costs and rewards are defined below:

- Cost of starting the IPS Cm
- Average loss when the system is attacked Ci
- Cost to attack by the attacker Ca
- Cost of not attacking by the attacker Cw
- Payment to the defender for taking an action strategy defensive Ui
- Payment to the attacker for taking an action strategy offensive Ua

Now you can understand that the reward of the attackers Pa is equal to average losses when the system is attacked, that is:

$$Pa = Ci \tag{1}$$

It is necessary to define when it is profitable for the attacker to carry out the attack:

$$Cw < Pa - Ca \tag{2}$$

The above equation means that the attacker will perform an attack when its reward minus the cost of attacking greater than the cost of not attacking.

On the other hand, the attacker will not make an attack when the cost of starting the IPS is much lower than the loss average when the system is attacked, because in this case surely the defender would have started the IPS, therefore this will be in operation and the attack will be detected and the attacker isolated from the network.

From this it is possible to define the reward matrix as:

$$\begin{bmatrix} P_a - C_a, U_i - C_i & -U_a, U_i - C_m \\ C_w, U_i & C_w, U_i - C_m \end{bmatrix}$$

Where the columns correspond to the strategies of the defender, that is, not defend and defend; and the rows do reference to the attacker's strategies, that is, attack and not attack.

They determine that there is no pure Nash equilibrium, therefore they analyze if the game model is in mixed Nash equilibrium.

Analyzing the mixed Nash equilibrium for the game they found the probability that the attacker attacks σ and the probability that the defender defends δ.

From Eq. (10) it is possible to find δ:

$$\delta = \frac{P_a - C_a - C_w}{P_a - C_a + U_a} \tag{3}$$

From Eq. (11) it is possible to find σ:

$$\sigma = \frac{C_m}{C_i} \tag{4}$$

With these rewards depending on the probability of take the strategy of attacking and defending, we can find the rewards of not attacking and not defending like $(1 - \delta)$ and $(1 - \sigma)$ respectively. Therefore, the strategies of the Attackers under a mixed Nash equilibrium model are:

$$(\delta, 1 - \delta) = \left(\frac{P - C_a - C_w}{P_a - C_a + U_a}, \frac{U_a + C_w}{P_a - C_a + U_a} \right) \tag{5}$$

$$(\sigma, 1 - \sigma) = \left(\frac{C_m}{C_i}, \frac{C_i - C_m}{C_i} \right) \tag{6}$$

To analyze the Nash equilibrium by mixed strategy, you can start assuming that the probability of attacking δ be high, for this to be $c_m \gg c_i$, that is, the attack occurs when it is not profitable to start the IPS, which makes the probability of defense is low.

In case the defense probability is high, he wants say that IPS has probably been put in place due to that the losses from being attacked are greater than the cost to have the IPS in motion, that is, $c_m \ll c_i$ which indicates that the probability of attack must be low.

In conclusion, the chances of attack and defense are inversely proportional and the system will be found in Mixed Nash equilibrium when:

$$\delta = \sigma \tag{7}$$

Now Manshaei [8] explains a Bayesian game of two players, a defense node and a malicious or regular one. He malicious node can choose between attacking and not attacking, while that the defense node can choose between monitoring and not monitor The defender's security is quantifiable from according to the property that protects w, therefore, when there is a safety failure the damage is represented by $-w$. Then the payoff matrix is presented:

$$\begin{bmatrix} (1-\alpha)w - c_a, (2\alpha - 1)w - c_m & w - c_a, -w \\ 0, -\beta w - c_m & 0,0 \end{bmatrix}$$

In this matrix the columns represent the behaviors of the defender (monitor and not monitor) and the rows attacker behaviors (attack and not attack), *Ca* and *Cm* do they refer to the costs of attacking and monitoring, α and β are the detection rate and the false alarm rate of the IDS respectively and μ_0 the probability that a player is malicious.

Finally they show that when $\mu_0 < \frac{(1+\beta)w + Cm}{(2\alpha+\beta-1)w}$ the game supports a strategy of pure balance (attack if it is malicious, do not attack if it is regular), do not monitor, μ_0 and when $\mu_0 > \frac{(1+\beta)w+Cm}{(2\alpha+\beta-1)w}$ the game does not have a pure strategy.

4 Proposed Model

From the model described by Manshaei and establishing that the two players are intruder and defender, since the intruder is ready to carry out the attack because has done a vulnerability study and has planned the different strategies to follow in order to enter authorized to the network, the time when no attack represents a *Cw* cost (waiting cost) because the network can change and the investment mentioned above both of time and of resources can be lost. Therefore, the payment matrix is:

$$\begin{bmatrix} (1-\alpha)w - c_a, (2\alpha - 1)w - c_m & w - c_a, -w \\ -c_w, -\beta w - c_m & -c_w, 0 \end{bmatrix}$$

The following explains each of the possible scenarios and the respective payments for the intruder and the defender:

- When the intruder attacks and the defender monitors: the attacker's reward, the times the detection system fails for the good that protects less the cost of attacking; the defender's reward, the times the system works less the times it fails for the good it protects, all this less the cost of monitoring.
- When the intruder attacks and the defender does not monitor: reward the attacker, the good to be achieved less the cost of attack; The reward of the defender, in this case is the loss of good.

- When the intruder does not attack and the defender monitors: the attacker's reward, in this case it is the loss of wait to make the attack; the defender's reward, false alarm rate degrades the good and in turn also has the cost of monitoring.
- When the intruder does not attack and the defender does not monitor: the attacker's reward, in this case it is the loss of wait to make the attack; The defender's reward, in this case, is zero since he does not spend on monitoring and is not attacked.

Depending on the strategy that the other actor takes and the respective payments they obtain, it is possible to determine if there is a Nash equilibrium. When the defender does not monitor, the attacker has two possible strategies: Attack, with gain $w - c_a$; and do not attack, where he gets $-c_w$, therefore you will always choose to attack.

When the defender also monitors the attacker can choose between attacking and not attacking, assuming a detection rate greater than 90%, attacking would lose the cost of attacking, while not attacking would lose the cost of waiting.

Generally the deployment of an attack to take control of the network or the information it contains is more expensive than carrying out a recognition and learning of the network and its vulnerabilities, therefore the attacker will choose not to attack.

$$-c_w > -c_a$$

Now it is necessary to fix the behavior of the attacker and analyze the possible strategies that the defender will perform.

First, when the attacker decides to attack and assuming a detection rate greater than 90%.

$$(2\alpha - 1)w - c_w > -w$$

This means that the defender will choose to monitor.

Second, when the attacker decides not to attack.

$$-\beta w - c_m < 0$$

Then, the defender will always choose not to monitor.

After analyzing the different strategies, it is clear that in neither scenario will both actors be satisfied with their reward. Which prevents that pure Nash equilibrium exists in the proposed game.

5 Evaluation of the Model

Because there is no point in the rewards matrix in which both the defender and the attacker feel comfortable with the situation, it is necessary to determine if the model is in mixed Nash equilibrium, for this the probability that the attacker attack σ and the probability that the defender defends δ.

The mixed strategy of the attackers is:

$$U_A = [(1-\alpha)w - ca]\delta\sigma + [w - c_a](1-\delta)\sigma + (-c_w)\delta(1-\sigma) + (-c_w)(1-\delta)(1-\sigma) \quad (8)$$

The mixed strategy of the defender is:

$$U_I = [(2\alpha - 1)w - c_m]\delta\sigma + (-w)(1-\delta)\sigma + [-\beta w - c_m]\delta(1-\sigma) \quad (9)$$

Using the extreme value method to solve the Nash mixed model strategy, Eqs. (8) and (9) are derived from σ and δ respectively and equal to zero.

$$\frac{\partial U_A}{\partial \sigma} = -\delta\alpha w + w - c_a + c_w = 0 \quad (10)$$

$$\frac{\partial U_I}{\partial \delta} = 2\sigma\alpha w - \beta w - c_m + \sigma\beta w = 0 \quad (11)$$

From the Eq. (10) its possible find δ:

$$\delta = \frac{w - c_a + c_w}{w\alpha} \quad (12)$$

From the Eq. (11) its possible find σ:

$$\sigma = \frac{\beta w + c_m}{2\alpha w + \beta w} \quad (13)$$

To analyze the Nash equilibrium by mixed strategy, we can start assuming that the probability of attacking σ is high, so that $c_m \gg 2\alpha w$, this means that the attacker can attack comfortably when goods that the defender protects are not so valuable to him, for which he will not have activated the IPS. In the case where the goods that are protected are valuable, the defense probability will increase and the probability of attack will decrease.

Therefore it is found again that the probabilities of attack and defense are inversely proportional and the system will be in mixed Nash equilibrium when:

$$\sigma = \delta \quad (14)$$

Additionally, in case the defense probability is high, this situation occurs when the cost of waiting for the attacker is greater than the cost of attacking, that is, when it is more profitable for the attacker to make his attack than to continue waiting for the right moment.

It is known that:

$$0 \leq P(a) \leq 1 \quad (15)$$

Then:

$$\delta = \frac{w - c_a + c_w}{w\alpha} \leq 1 \tag{16}$$

$$\sigma = \frac{\beta w + c_m}{2\alpha w + \beta w} \leq 1 \tag{17}$$

From Eq. (17)

$$c_m \leq 2\alpha w \tag{18}$$

Equation 18 indicates that when the detection rate or value of the good to be protected decreases in the same way, the detection cost used to protect the good should decrease.

From Eq. (16)

$$c_w \leq c_a - (1 - \alpha)w \tag{19}$$

Equation 19 indicates that the attacker's maximum wait cost is directly linked to the undetected rate of the intrusion prevention system, as that rate increases the maximum wait cost should decrease.

From Eq. (14)

$$w = \frac{-(c_a(2\alpha + \beta) + c_m\alpha - c_w(2\alpha + \beta)}{\alpha(\beta - 2) - \beta} \tag{20}$$

The Eq. 20 describes the condition to be in Mixed Nash Equilibrium, it means that each player has the same probability in their options of behavior. Below is presented a graphical function analysis that defines the probability that the attacker attacks vs the cost of monitoring under certain parameters. Initially, the good that seeks to protect (w) was set up as 100 and the cost of monitoring varied from 0 to 100 per 1, this was established because it is illogical to use more resources protecting a good than the value for the defender. The alpha value was also set at 95% and different beta values were used to analyze the behavior of the probability that the attacker attacks based on the false alarm rate, this behavior is presented in Fig. 1.

When analyzing Fig. 1 it can be observed that by decreasing the false alarm rate (beta) the probability of the attacker attacking is diminished, which is because if the IPS is more efficient the attacker will tend not to carry out an attack until you are sure that it will not be detected.

Second, w = 100 and beta = 2% were set to observe the behavior of the probability of the attacker attacking with different values of detection rate, this can be seen in Fig. 2.

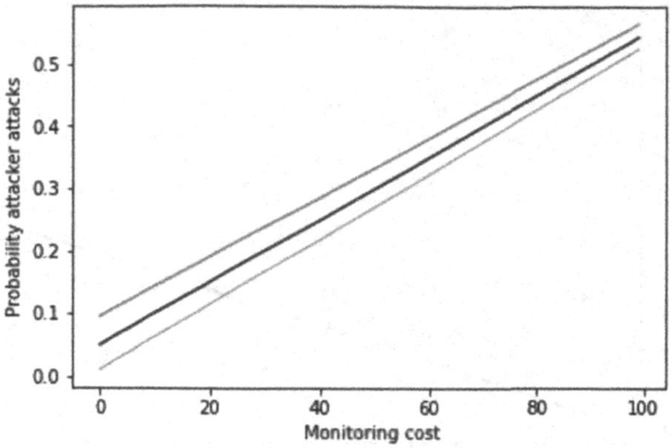

Fig. 1. Probability attacker attacks vs monitoring cost with: w = 100, alpha = 95%, beta = 20% (red), beta = 10% (blue) and beta = 2% (green) (Color figure online)

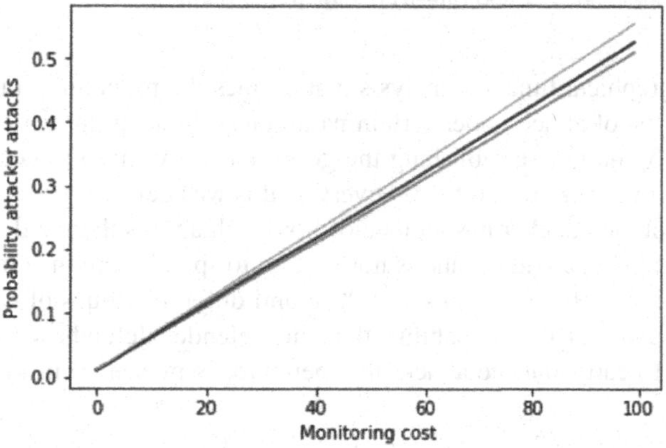

Fig. 2. Probability attacker attacks vs monitoring cost with: w = 100, beta = 2%, alpha = 98% (red), alpha = 95% (blue) and alpha = 90% (green) (Color figure online)

Figure 2 shows that by decreasing the detection rate the probability that the attacker will attack will be greater, this happens in the same way as in the previous case, because if the system becomes less efficient it will be more profitable for the attacker to make an attack, therefore, the probability that the attacker will attack will increase.

Third, alpha = 98%, beta = 2% was set and the value of w was varied to analyze how the probability that the attacker attacks with respect to the variation of the cost of the desired good is affected. opt to make the attack. Said analysis was carried out starting from Fig. 3.

Figure 3 shows that increasing the cost of the good that the attacker wants to obtain, the likelihood that the attacker attacks is reduced, this seems to go against the logic but it is because if the good is much more valuable than the cost of monitoring, the defender will always want to protect said good by starting the IPS and therefore it will be more complex for the attacker to violate the security of the system.

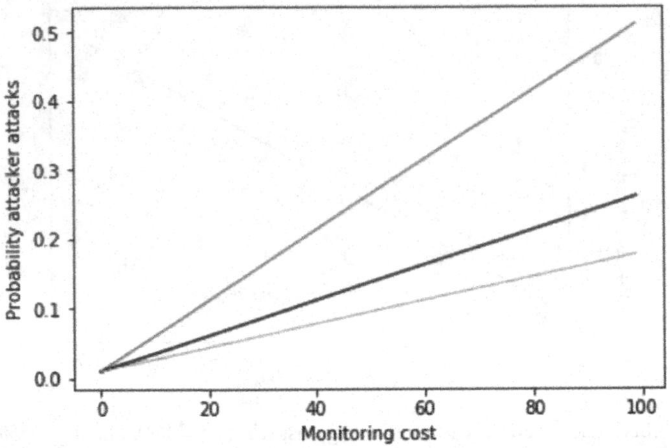

Fig. 3. Probability attacker attacks vs monitoring cost with: alpha = 98%, beta = 2%, w = 100 (red), w = 200 (blue) and w = 300 (green) (Color figure online)

Below is a graphical function analysis that defines the probability that the defender defends vs the cost of attack under certain parameters. In a similar way to the previous case we started by initially establishing the good that we want to protect w as 100 and varying the cost of attack from 0 to 100 every 1, this was defined in that way because it was assumed that the attacker has an idea of how Valuable is the good that you expect to obtain when making an attack and is not willing to spend more than the value of said good. The value of alpha was also set at 95% and different values of C_w were used to analyze the behavior of the probability that the defender defends with respect to the cost of waiting to carry out the attack, this behavior is presented in Fig. 4.

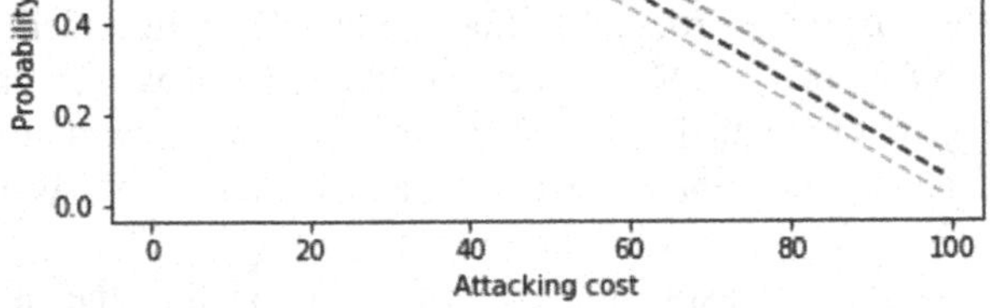

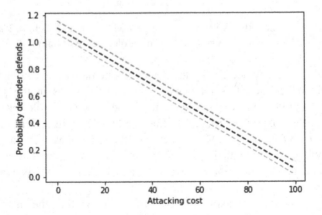

Fig. 4. Probability defender defends vs attacking cost with: w = 100, alpha = 95%, C_w = 10 (red), C_w = 5 (blue) and C_w = 1 (green) (Color figure online)

Figure 4 shows that by decreasing the value of the waiting cost to carry out the attack, the probability that the defender defends will also decrease. This is because the attacker is so clear about the vulnerabilities of the network that it is very cheap for him to wait for the attack, therefore the defender must activate the IPS as soon as possible to correct the failures in network security.

Second, $w = 100$ and $C_w = 10$ were set to observe the behavior of the probability that the defender defends with different values of detection rate, this can be seen in Fig. 5.

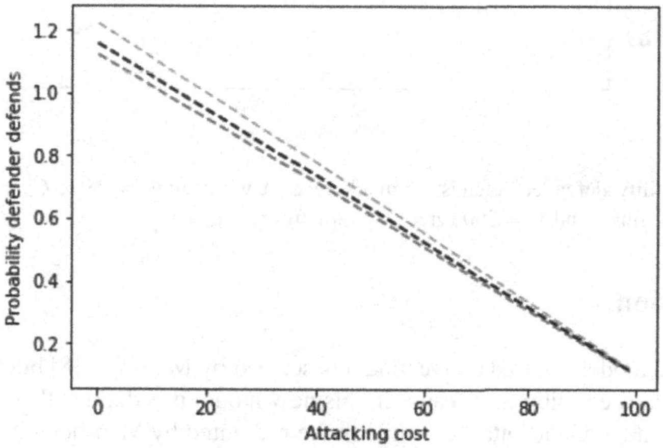

Fig. 5. Probability defender defends vs attacking cost with: w = 100, C_w = 10, alpha = 98% (red), alpha = 95% (blue) and alpha = 90% (green) (Color figure online)

When looking at Fig. 5 it is understood that when the detection rate of the IPS decreases the probability that the defender defends increases, this behavior is due to the fact that when the IPS is less efficient the defender must have it active for a longer time in order to detect the intrusions and therefore the probability that the defender defends will be greater.

Finally alpha = 98% and C_w = 10 were set to see how the cost of the good that protects the defender impacts the probability that the defender defends, this was done by variations in w and can be seen in Fig. 6.

It is possible to observe in Fig. 6 that when the cost of the good that the defender protects increases, the probability that the defender defends also increases. This is logical, since the more valuable the asset that is protected the defender will allocate more time and resources for the protection of the system.

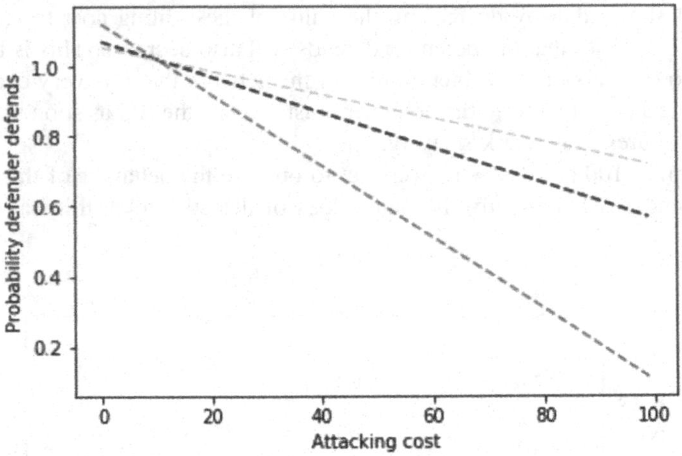

Fig. 6. Probability defender defends vs attacking cost with: alpha = 98%, C_w = 10, w = 100 (red), w = 200 (blue) and w = 300 (green) (Color figure online)

6 Discussion

The presented model is based on the model described by Manshaei [8] but this does not have a pure Nash equilibrium because in this new model it is defined that there are two players: a defender and an attacker, while in the presented by Manshei it is possible that the attacker is not malicious in which there is a pure balance and is not to attack and not monitor.

Regarding the model presented by Wang [7] the same conclusion was reached that the system will be in equilibrium when the probability of attacking:

$$\delta = \frac{\beta w + c_m}{2\alpha w + \beta w} \tag{21}$$

Equal to the probability of defending

$$\delta = \frac{(w - c_a + c_w)}{w\alpha} \tag{22}$$

And in case these are not equal, the model is regulated over time until reaching this equilibrium, although in this case the detection rate and false alarms of the IDS are taken into account, which is raised in the model described by Manshaei [8] and makes the model closer to reality. The analysis of the probability functions that the attacker attacks and that the defender defends allows to obtain a clearer view of the decisions that the players of this game can take and the reasons to act in a certain way.

7 Conclusions

This model of game theory adds a new feature to intrusion prevention systems, where you can evaluate how complex it is to perform a vulnerability analysis on the network that is defending itself. Because if it is much less expensive to perform the analysis than the attack, it will be more profitable for the attackers to launch their attack plan and therefore the IPS must be in operation to prevent such attacks and avoid loss of information, money and reputation.

The proposed model is regulated over time until reaching the equilibrium point at which the probability of attacking σ is equal to the probability of defending δ.

$$\sigma = \frac{\beta w + c_m}{2\alpha w + \beta w} = \frac{w - c_a + c_w}{w\alpha} = \delta \tag{23}$$

Decreasing the false alarm rate (beta) the probability of the attacker attacking is diminished; decreasing the detection rate the probability that the attacker will attack will be greater; increasing the cost of the good that the attacker wants to obtain, the likelihood that the attacker attacks is reduced; and decreasing the value of the waiting cost to carry out the attack, the probability that the defender defends will also decrease.

When the detection rate of the IPS decreases the probability that the defender defends increases, additionally When the cost of the good that the defender protects increases, the probability that the defender defends also increases.

$$w = \frac{-(c_a(2\alpha + \beta) + c_m\alpha - c_w(2\alpha + \beta))}{\alpha(\beta - 2) - \beta} \tag{24}$$

Describes the condition to be in Mixed Nash Equilibrium, it means that each player has the same probability in their options of behavior.

$$c_m \leq 2\alpha w$$

Indicates that when the detection rate or value of the good to be protected decreases in the same way, the detection cost used to protect the good should decrease. From Eq. (24).

$$c_w \leq c_a - (1 - \alpha)w$$

Indicates that the attacker's maximum wait cost is directly linked to the undetected rate of the intrusion prevention system, as that rate increases the maximum wait cost should decrease.

References

1. Kolias, C., Kambourakis, G., Stavrou, A., Gritzalis, S.: Intrusion detection in 802.11 networks: empirical evaluation of threats and a public dataset. IEEE Commun. Surv. Tutorials **18**(1), 184208 (2016). https://doi.org/10.1109/COMST.2015.2402161
2. Huang, H., Hu, Y., Ja, Y., Ao, S.: A whole-process WiFi security perception software system. In: 2017 International Conference on Circuits, System and Simulation, ICCSS 2017, 151156 (2017). https://doi.org/10.1109/CIRSYSSIM.2017.8023201
3. Michaels, S., Akkaya, K., Selcuk Uluagac, A.: Inducing data loss in Zigbee networks via join/association handshake spoofing. In: 2016 IEEE Conference on Communications and Network Security (CNS), Philadelphia, PA, pp. 401—405 (2016). https://doi.org/10.1109/cns.2016.7860527
4. Sforzin, A., Marmol, F.G., Conti, M., Bohli, J.M.: RPiDS: raspberry Pi IDS - a fruitful intrusion detection system for IoT. In: Proceedings - 13th IEEE International Conference on Ubiquitous Intelligence and Computing, 13th IEEE International Conference on Advanced and Trusted Computing, 16th IEEE International Conference on Scalable Computing and Communications, IEEE Internationa, pp. 440–448 (2017). https://doi.org/10.1109/UIC-ATC-ScalCom-CBDCom-IoPSmartWorld.2016.0080
5. Sharma, P.K., Moon, S.Y., Moon, D., Park, J.H.: DFA-AD: a distributed framework architecture for the detection of advanced persistent threats. Cluster Comput. **20**(1), 597609 (2017). https://doi.org/10.1007/s10586-016-0716-0
6. Chen, J., Chen, C.: Design of complex event-processing IDS in internet of things. In: Proceedings - 2014 6th International Conference on Measuring Technology and Mechatronics Automation, ICMTMA 2014, pp. 226–229 (2014). https://doi.org/10.1109/ICMTMA.2014.57
7. Wang, K., Du, M., Yang, D., Zhu, C., Shen, J., Zhang, Y.: GameTheory-based active defense for intrusion detection in cyber-physical embedded systems. ACM Trans. Embed. Comput. Syst. **16**(1), 121 (2016). https://doi.org/10.1145/2886100
8. Manshaei, M.M.H., Zhu, Q., Alpcan, T., Bacar, T., Hubaux, J.-P.: Game theory meets network security and privacy. ACM Comput. **45** (2013). https://doi.org/10.1145/2480741.2480742
9. Rafsanjani, M.K., Aliahmadipour, L., Javidi, M.M.: A hybrid intrusion detection by game theory approaches in MANET. Indian J. Sci. Technol. **5**(2), 2123–2131 (2012)

A Survey: WSN Heterogeneous Architecture Platform for IoT

Naila Bouchemal[3(✉)], Sondes Kallel[4], and Nardjes Bouchemal[1,2]

[1] University Center of Mila, Mila, Algeria
n.bouchemal.dz@ieee.org
[2] LIRE Laboratory of Constantine2, Constantine, Algeria
[3] École Centrale d'Electronique Paris, Paris, France
naila.bouchemal@ece.com
[4] LiParad Laboratory, University of Versailles, Versailles, France
sondes.khemiri-kallel@uvsq.fr

Abstract. Internet of Things (IoT) is a novel paradigm that allows millions of smart devices to be connected to the Internet. Such devices can be sensors/actuators, which are able to operate and transmit data to other systems in an autonomous way. In fact, IoT is about autonomous and heterogeneous devices, data and connectivity. As millions of devices are connected, internet of things will require more improvement in terms of platform deployment. In this context, IoT must coexists with several other paradigms (Cloud-Computing, Big-data, SDN…) in other to satisfy its new features. Next to that, Wireless Sensor Networks (WSNs) are one of the most important components in IoT.

This paper is the result of a state of the art analysis and performance evaluation of several SDN architectures designed for the Internet of Things platforms. At the end of the paper, we present an overview of our architecture based on SDN and SDR methods for WSN as a part of the Internet of Things.

Keywords: IoT architecture · WSN · SDN

1 Introduction

Today, smart grid, smart homes, smart water networks, intelligent transportation, are infrastructure systems that connect our world more and more. The common vision of such systems is usually associated with one single concept, the internet of things (IoT). As IoT continues to grow, there is several combination with other technologies and concepts such as cloud computing, future internet, big data, robotics, and semantic technologies. Indeed, the idea is not new but since these concepts overlap in some parts (technical and service architecture, virtualization, interoperability, automation), we see more the aspect of complementarity rather than defending each area individually. Indeed, this growth has resulted in a number of extended deployments, where several heterogeneous wireless communication solutions must coexist: cellular networks, WiFi, ZigBee and Bluetooth, Ad-hoc and protocols. They all need to be effectively integrated to create a seamless communication platform (Fig. 2). Managing these open,

É. Renault et al. (Eds.): MLN 2018, LNCS 11407, pp. 321–332, 2019.
https://doi.org/10.1007/978-3-030-19945-6_23

geographically distributed and heterogeneous network infrastructures, especially in dynamic environments, is a major technical challenge (Fig. 1).

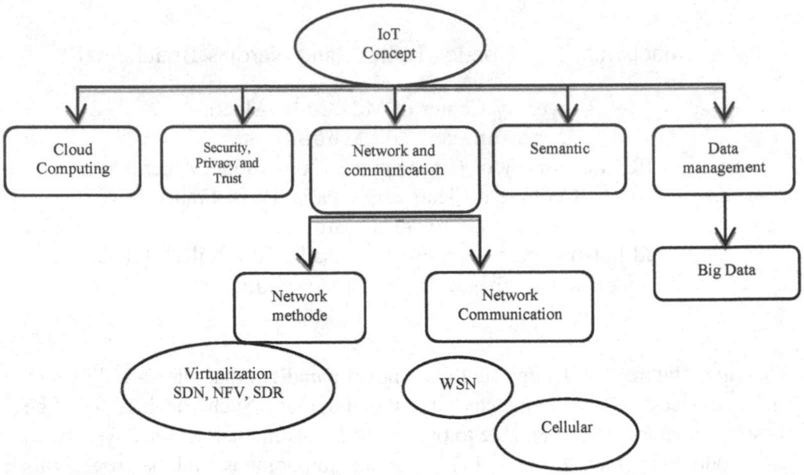

Fig. 1. Overall IoT concepts

Moreover, at present according to the studies, there are 9 billion connected devices and a number of 24 billon is expected for 2020. All devices use different ways to connect to the network including the traditional network infrastructure. In addition to that, in the traditional network architecture, the growth of the number of devices and communication protocols embedded on each device; generates the growth of the number of gateways between each communication technologies. For example for a number of technologies N we will need [(N-1) N]/2 gateways. However, the traditional equipment and network protocols are not designed to support the high level of scalability, high amount of traffic and mobility.

Indeed, the current architectures are inefficient and have significant limitations to satisfy these new requirements.

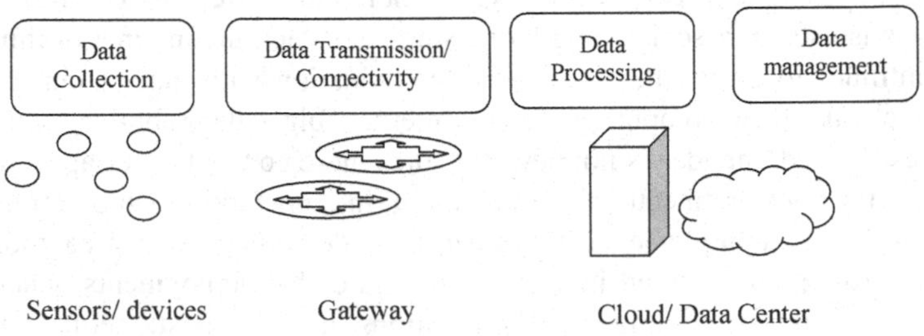

Fig. 2. A general model of IoT

To overcome the above issues, several IoT architectures have been provided [1–4], A general IoT architecture is shown in Fig. 2, which consists of the following different levels:

- Data collection: this layer is composed of devices. They collect data or connect with Internet gateways for data aggregation or data forwarding.
- Data Transmission/Connectivity: This layer consists of data communication and networking infrastructures for delivering data gathered from devices at the physical layer to higher layers.
- Data management and processing: This layer can be an application software, which provides access services for IoT users. The processing could be supported by physical hardware and platforms in data centers or services in the cloud with the aim of providing facility for the data access and storage.

In this paper, we focus on Connectivity and networking tier in Internet of Thing we address both:

- *Network communication:* Wireless Sensors Network.
- *Network methods/*virtualizations *methods:* (Software Defined Networking SDN, Network Function Virtualization NFV, and Software Defined Radio SDR).

The rest of the paper is organized as follow. In Sect. 2 we present the overall SDN architecture, then we analyze related works in Sect. 3, in Sect. 4 we introduce our proposal.

2 SDN Architecture

The SDN is a network optimization architecture [5] that eliminates the rigidity present in traditional networks. Its structure allows the behavior of the network to be more flexible and adaptable to the needs of each organization, campus, or group of users. Besides, its centralized design allows important information to be collected from the network and used to improve and adapt their policies dynamically. SDN has led to the design of models that integrate and finally achieve convergence of commonly separate architectures (Wifi-4G-LTE). Important issues such as convergence with existing networks, scalability, performance, and security are the challenges that should be overcome to be positioned in the market. The SDN architecture is divided into three distinct layers, the communication between layers is through Applications Programming Interface APIs. In the SDN, the control is decoupled from the infrastructure layer as shown in the following Fig. 3.

2.1 Infrastructure Layer

The data plan includes the physical elements of the network, in the SDN architecture the control is completely centralized, the data plan has the function of directing the traffic according to the routing policy defined by the controller.

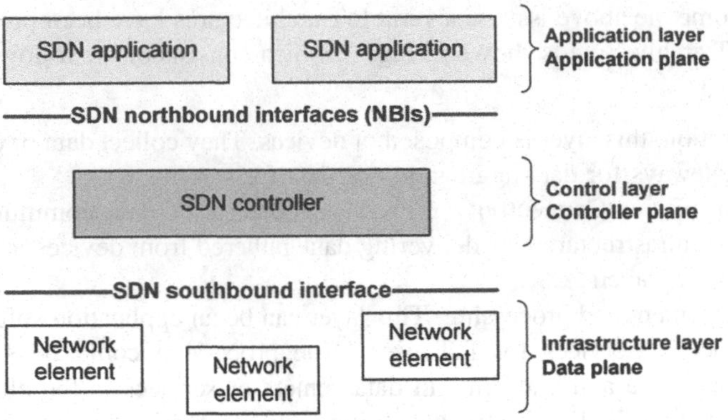

Fig. 3. SDN architecture

2.2 Control Layer

The control plane includes a set of SDN controllers, each controller has exclusive control over a set of radio resources, its main function is to send commands to the data plane, the number of controllers depends on the number of applications and the size of the network, the communication between the controllers is possible through the API East /Oust, it allows to better share the load of the network.

2.3 Application Layer

The application plan includes one or more applications, which gives all the programmability needed for the data plan in terms of applications, routing, security, QoS.

2.4 Applications Programming Interface

The southern API allows communication between the control plane and the data plane. The Northern API, for its part, connects and establishes the communication between the application layer and the control layer; finally, if several controllers need to exchange information, this is possible through the two East and West APIs.

3 SDN for WSN State of the Art

In such a sophisticated dynamic system, devices are interconnected to transmit useful measurement information and control instructions via distributed sensor networks. A wireless sensor network is a network formed by a large number of sensor nodes where each node is equipped with a sensor to detect physical phenomena such as light, heat, pressure, etc. With the rapid technological development of sensors, WSNs will become the key technology for IoT.

In this context, network virtualization has received special attention. Indeed this technology is considered as the key concept of the IoT. Various projects have emerged from numerous international initiatives such as PlanetLab [6], and GENI [7] in the United States, Akari [8] in Asia and 4ward [9] in Europe.

These initiatives look for providing a virtualized network infrastructure to help researchers deploy and evaluate new services over virtual networks. These projects are classified according to the network technology used or according to the level of virtualization; one can also mention the projects UCLP [10], VNET [11], AGAVE [12] and VIOLIN [13].

In parallel with these projects, several prototype network virtualization environments have been developed in order to exploit the characteristics of one or more network technologies. For example the project Tempes [14] X-Bone [15] and AGAVE [16], which focused only on IP networks. Projects such as VNRMS [13], NetScript [17], Genesis [18] and FEDERICA [19] emphasize the domain and management of architecture. As for the implementation in the CABO [20] and GENI [7] projects, it supports several heterogeneous networks. The VITRO project [21] focuses on the virtualization of WSN.

Most of these works use the SDN for sensor networks, but the three main architectures have attracted our attention. The architecture of Jacobsoon [22], OpenFlow [23], and SDNWISE [24]. In what follows, we will detail these three works.

3.1 The Architecture of Jacobson and al

Figure 4 shows the SDN-WSN architecture proposed by Jacobsson et al. [22], the infrastructure layer contains the sensor nodes where each node is equipped with a local controller, and the control layer contains one or more central controllers. Above the control layer is the application layer, which can contain automated applications such as routing and management of the network topology.

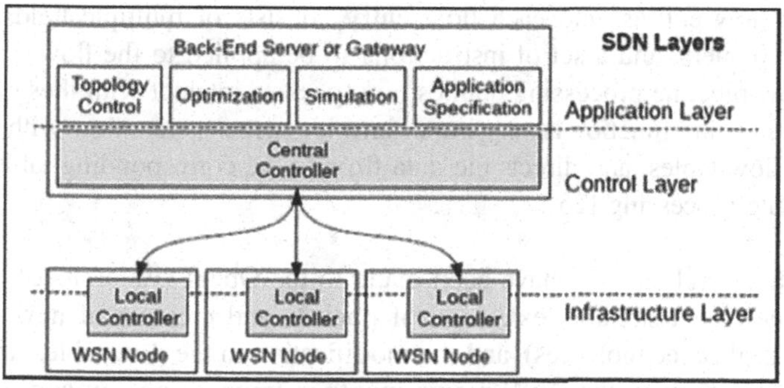

Fig. 4. The architecture of Jacobson and al

At each sensor, there is a local controller that runs. The role of the local controller is to receive and execute central controller commands such as configuration, reconfiguration, and necessary updates to applications. The local controller runs its programs by installing new code (script), which extends or modifies the functionality of a given application. Each application is implemented via a virtual machine (VM) built into the sensor, and the modification of this application requires a modification of the script already in place. The central controller must discover the actual topology of the network, but also know the quality of the links. The LinkQuality Estimation (LQE) method is based on packet trace information, a package trace contains detailed information about the behavior of packet loss on a link.

Critics:

- Using a local controller in a sensor node increases power consumption, and affects the node performance.
- Topology discovery is done through packets trace, which decreases the residual bandwidth.
- Estimating link quality requires a lot of recent data collection at regular intervals. However, in a sensor network, the collection of information is irregular, and depends on application needs.
- If the links are overloaded, the system cannot collect data to make link estimation.
- Quality of service depends on the link estimation, if the latter is imprecise, the QoS will be affected.
- The absence of simulation and real tests, does not validate the assumptions and assertions of the authors.

3.2 Sensor OpenFlow Architecture

Luo et al. in [23] uses the OpenFlow protocol as the communication protocol between the controller and the sensors, and they have adapted the SDN network flow tables to the requirements of the sensor network. We will first briefly explain the operation of the OpenFlow protocol in a conventional SDN network, in order to better understand the major changes that have been made by Luo et al.

Each OpenFlow switch consists of multiple flow tables, each flow table contains a multitude of flow entries, and each flow entry consists of multiple fields, mappings (matching), counters, and a set of instructions to be applied to the flow.

OpenFlow pipeline processing defines how packets interact with these flow tables. The packet matching function is very important; it matches a data flow with a particular input in the flow tables, and directs the data flow to the corresponding table to receive the appropriate processing [25].

Critics:

Although several changes have been made to the OpenFlow protocol, such as the creation of new data streams (extension of packets and creation of new fields compatible with Zigbee technologies), and the modification in the flow tables, this does not make it possible to meet the WSN requirements in terms of energy consumption and bandwidth.

Indeed, the OpenFlow protocol requires the reservation of a bandwidth dedicated to OpenFlow messages, but the links bandwidth between the sensors is limited and low, also the exchange messages frequency between the sensor and the controller is much too high for a WSN use this will result in high power consumption.

The OpenFlow protocol has been developed and thought for wired networks, several additional modifications must be considered to provide a satisfactory implementation, such as the reduction of messages exchanged.

3.3 SDN-WISE Architecture

In previous architecture, the implementation of OpenFlow protocol in the WSN generates high energy consumption, this is due to the large number of messages exchanged between the controller and sensors, it is from this observation that was born the SDN-WISE architecture [24, 25]. SDN-WISE is a stateful architecture with two objectives:

(i) To reduce the amount of information exchanged between the sensors and the controller.
(ii) To make programmable sensors as finite state machines in order to enable them to perform operations that cannot be supported by stateless solutions.

SDN-WISE is based on physical layers and MAC IEEE 802.15.4. Network elements can be distinguished into nodes (Node) and sink (Sink). The difference between nodes and sinks is that they are equipped with a network interface connected to a network infrastructure. Therefore, all control packets should find their paths to reach the controller (Fig. 5).

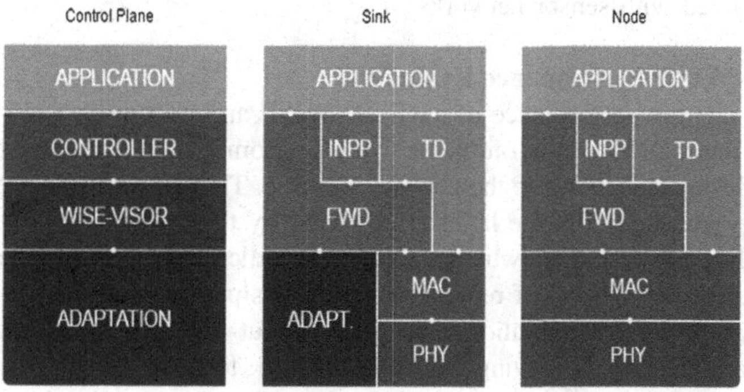

Fig. 5. SDNWISE protocol stack

The transfer layer (FWD) manages the incoming packets as specified in the WISE flow table, the tables are updated according to the configurations sent by the control plane.

The network packet processing layer (INPP) is responsible for operations such as aggregating data. Packets that must be sent in similar paths are grouped by the INPP layer [24].

If no entry in the WISE flow table matches the current packet, a request is sent to the control plane. In order to contact the controller, each node must know, its next best hop to a well node, this value is computed in a distributed manner using the topology discovery (TD) layer [25].

In the control plane, the behavior is managed by one to several controllers and a WISE-Visor, the latter summarizes the resources of the network, so that the different logical networks with different management strategies defined by different controllers can be run on the network.

The adaptation layer is responsible for formatting the messages exchanged between the well and the WISE-Visor and vice versa. The behavior of SDN-WISE nodes is organized according to the following three data structures:

- The WISE Status Table: SDN-WISE nodes are characterized by a current state for each active controller. The WISE States Array is the data structure that contains this information.
- Table of accepted identifiers: Allows a sensor to elect he packages to be processed.
- The WISE Flow Table: Each entry in the WISE Flow Table contains a match rule section that specifies the conditions under which the entry applies. If these conditions are met, the sensor will perform the action contained in the Action section of the input.

Critics:

Unlike previously treated architectures, reduces interactions between nodes and their controller, and introduces the notion of finite state machines into sensors, giving them the ability to define complex packet processing rules, in order to save more energy, as far as we are concerned, this architecture takes into account all the constraints associated with sensor networks.

3.3.1 SDN-WISE Performance Evaluation

We will evaluate the performance of the SDN WISE architecture in order to visualize the impact of the SDN solution on the network environment. A fundamental challenge in the WSN design is to enhance the network lifetime. The network lifetime is based on overall energy consumption; the latter is impacted by the duty cycle [26].

Duty cycling is a technique where a node is periodically placed into the sleep mode which is an effective method of reducing energy dissipation in WSN.

We will analyze two simulations, the first without SDN, and the second with the SDN-WISE architecture, following the same network topology.

Virtual machine (VMware), Operating System (Contiki), the simulator (Cooja) are the simulation environment that we used (Fig. 6).

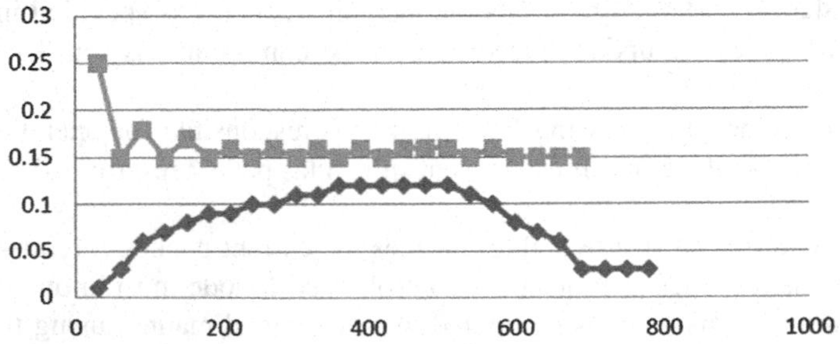

Fig. 6. Duty cycle performance evaluation VS time (ms)

For the same number of sensors (same topology), we can see that the duty cycle ratio for a conventional network is high, it reach 0, 15% against 0.04% for a SDN WISE network, the latter thus optimizes the use of radio resources, which has the effect of improving network autonomy. The more duty cycle improvement is, more the energy consumption will be, and the better network lifetime.

This solution shows real improvements in terms of energy consumption.

However, the architecture as it is proposed does not take into account the heterogeneity of the various sensors carried on different radio technologies. For this, we introduce our architecture proposal.

4 Proposal Architecture

Several architectures for sensor networks based on SDN have been proposed, previously we have studied and analyzed two architectures of sensor networks based on SDN. In [22] we saw that the control is not completely centralized, which generates large energy consumption. In [23] the adaptation of OpenFlow protocol in sensor networks is complex, the frequency of messages exchanged between the controller and the sensors are high, not to mention that the OpenFlow protocol was designed and designed for wired networks.

With this in mind, and taking into account all the constraints that sensor networks face (energy consumption, bandwidth), the research team at the University of Catania has defined its own architecture based on SDN [24].

In this paper we take into account WSN constraints and SDN advantages, we propose architecture based on SDN-WISE architecture in order to overcome heterogeneity problems that can we found in WSN environments.

(a) *First proposal architecture:*

Based on SDNWISE solution, we propose the following architecture, in fact in Sect. 3.3 we described SDNWISE protocols stack. The actual solution can be used only for homogeneous network. In order to overcome heterogeneity problems, we propose to duplicate SDNWISE architecture for each network Fig. 7.

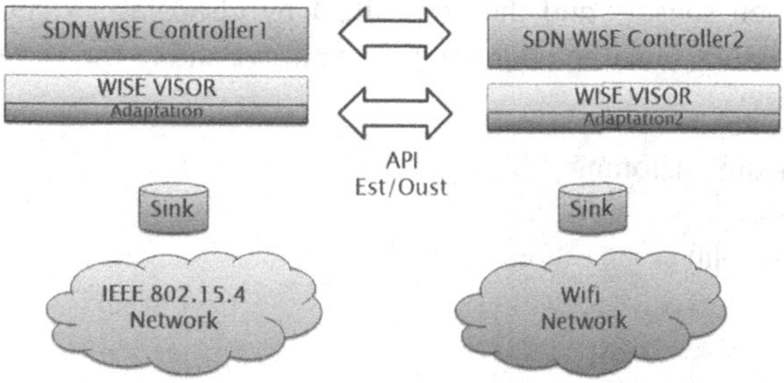

Fig. 7. First proposal SDN for heterogeneous WSN

However, this solution is "quick and dirty", which will generate several other issues:

- Problems generated by a multi domain network architecture.
- Need new development of API between controllers.
- Need coordination between distributed controllers.
- Need to have a coherent global view of the network to avoid anomalies between applications.
- Complexity network slicing, due to network heterogenity.
- Decreased WSN performance due to generated delay of interactions between different controllers.

(b) *Second proposal architecture:*

In this second proposal, the two wireless networks (WSN 1, WSN 2), have different transmission technologies, the software radio (SDR) [28] is implemented in the sink (Sink), this allows the two networks to communicate with a common antenna (Fig. 8).

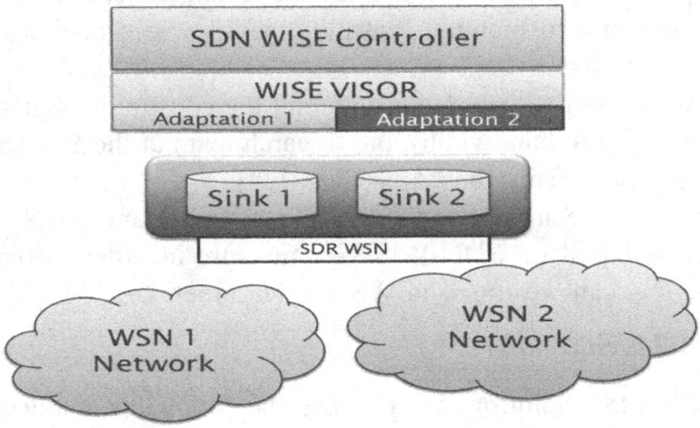

Fig. 8. Second Proposal SDN for heterogeneous WSN

Moreover, with this proposal, we conserve the SDNWISE first concept (reducing interactions between sensors and the controller) by choosing to centralize it to one controller. Other issues must be addressed as well; we quote the most important ones for our future work:

- The controller dimensioning
- Network slicing
- SDN, SDR cross-layer

5 Conclusion

In this paper, we discuss the topic of IoT platforms deployment that is present in last years. Indeed, because of the massive diversity of objects in terms of communication technologies as well as terms of use cases and data types, the proposals of one universal communicating platforms becomes a real challenge. Several works exist, but there is no standard to follow. For this, we have concentrated our efforts to propose a universal solution based on an open source virtualization methods. Moreover, our proposal does not depend on the wireless communication technology thanks to the uniformity of the exchanges offered by the SDNWISE solution. The heterogeneity issue and power consumption is then overcome. In future work, we want to test and put our proposal into practice.

References

1. Luong, N.C., Hoang, D.T., Wang, P., Niyato, D., Kim, D.I., Han, Z.: Data collection and wireless communication in Internet of Things (IoT) using economic analysis and pricing models: a survey. IEEE Commun. Surv. Tutorials **18**(4), 2546–2590 (2016)
2. Gubbi, J., Buyya, R., Marusic, S., Palaniswami, M.: Internet of things (iot): a vision, architectural elements, and future directions. Future Gener. Comput. Syst. **29**(7), 1645–1660 (2013)
3. Atzori, L., Iera, A., Morabito, G.: The internet of things: a survey. Comput. Netw. **54**(15), 2787–2805 (2010)
4. Uckelmann, D., Harrison, M., Michahelles, F.: An architectural approach towards the future internet of things. In: Uckelmann, D., Harrison, M., Michahelles, F. (eds.) Architecting the Internet of Things, pp. 1–24. Springer, Heidelberg (2011). https://doi.org/10.1007/978-3-642-19157-2_1
5. Akyildiz, I.F., Lee, A., Wang, P., Luo, M., Chou, W.: A roadmap for traffic engineering in SDN-OpenFlow networks. Comput. Netw. **71**, 1–30 (2014)
6. Spring, N., Peterson, L., Bavier, A., Pai, V.: Using PlanetLab for network research: myths, realities, and best practices. ACM SIGOPS Oper. Syst. Rev. **40**(1), 17–24 (2006)
7. Peterson, L., et al.: GENI design principles. IEEE Comput. **39**(9), 102–105 (2006)
8. Harai, H.: Designing new-generation network: overview of AKARI architecture design. In: Asia Communications and Photonics Conference and Exhibition (2009)
9. Achemlal, M., Almeida, T., et al.: "D-3.2.0 virtualisation approach: Concept," The FP7 4WARD Project, Technical report (2009)
10. Nandy, B., Bennett, D., Ahmad, I., Majumdar, S., Arnaud, B.S.: User controlled lightpath management system based on a service oriented architecture (2006). http://www.solananetworks.com/UCLP/files/UCLPv2-SOA.pdf
11. Sundararaj, A.I., Dinda, P.A.: Towards virtual networks for virtual machine grid computing. In: Proceedings Virtual Machine Research and Technical Symposium, San Jose, CA, May 2004, pp. 177–190 (2004)
12. Touch, J.: Dynamic internet overlay deployment and management using the X-Bone. Computer Net. **36**(2), 117–135 (2001)
13. Boucadair, M., et al.: "Parallel internets framework," AGAVE Consortium, Technical report, September 2006

14. van der Merwe, J.E., Rooney, S., Leslie, L., Crosby, S.: The tempest practical framework for network programmability. IEEE Network Mag. **12**(3), 20–28 (1998)
15. Grossman, D.: "New terminology and clarifications for DiffServ," Internet RFC 3260, Technical report, April 2002
16. Jun, A.-S., Leon-Garcia, A.: Virtual network resources management: a divide-and-conquer approach for the control of future networks. In: Proceedings IEEE Globecom, Sydney, NSW, November 1998
17. da Silva, S., Yemini, Y., Florissi, D.: The NetScript active network system. IEEE J. Sel. Areas Commun. **19**(3), 538–551 (2001)
18. Kounavis, M.E., Campbell, A.T., Chou, S., Modoux, F., Vicente, J., Zhuang, H.: The Genesis Kernel: a programming system for spawning network architectures. IEEE J. Sel. Areas Commun. **19**(3), 511–526 (2001)
19. Szegedi, P., Figuerola, S., Campanella, M., Maglaris, V., Cervelló-Pastor, C.: With evolution for revolution: managing FEDERICA for future internet research. IEEE Commun. Mag. **47** (7), 34–39 (2009)
20. Feamster, N., Gao, L., Rexford, J.: How to lease the internet in your spare time. ACM SIGCOMM Comp. Commun. Rev. (CCR) **37**(1), 61–64 (2007)
21. Sarakis, L., Zahariadis, T., Leligou, H.-C., Dohler, M.: A framework for service provisioning in virtual sensor networks. EURASIP J. Wireless Commun. Netw. **2012**(1), 1–19 (2012)
22. Jacobsson, M., Orfanidis, C.: Using software-defined networking principles for wireless sensor networks. In: 11th Swedish $National Computer Networking Workshop (SNCNW 2015) Karlstad, 28–29 May 2015
23. Luo, T., Tan, H.P., Quek, T.Q.: Sensor OpenFlow: enabling software-defined wireless sensor networks. IEEE Commun. Lett. **16**(11), 1896–1899 (2012)
24. Galluccio, L., Milardo, S., Morabito, G., Palazzo, S.: SDN-WISE: design, prototyping and experimentation of a stateful SDN solution for WIreless SEnsor networks. In: 2015 IEEE Conference on Computer Communications (INFOCOM), pp. 513–521. IEEE, April 2015
25. Le Protocole OpenFlow dans l'Architecture SDN (Software Defined Network) EFFORT, Copyright EFORT 2016
26. McKeown, N., et al.: OpenFlow: enabling innovation in campus networks. ACM SIGCOMM Comput. Commun. Rev. **38**(2), 69–74 (2008)
27. http://sdn-wise.dieei.unict.it/#Documentation
28. Chen, Z., Liu, A., Li, Z., Choi, Y.J., Li, J.: Distributed duty cycle control for delay improvement in wireless sensor networks. Peer-to-Peer Network. Appl. **10**(3), 559–578 (2017)
29. https://www.vmware.com/fr.html
30. Arslan, H. (ed.): Cognitive radio, software defined radio, and adaptive wireless systems. Springer Science & Business Media, Dordrecht (2007)

An IoT Framework for Detecting Movement Within Indoor Environments

Kevin Curran[1(✉)], Gary Mansell[1], and Jack Curran[2]

[1] School of Computing, Engineering and Intelligent Systems, Ulster University, Londonderry, Northern Ireland
Kj.curran@ulster.ac.uk

[2] Open University, Walton Hall, Milton Keynes, UK

Abstract. Tracking people indoors can be valuable in smart living scenarios such as tracking shoppers in a mall or in healthcare situations when tracking the movement of elderly patients can allow them to remain more independent. Determining accurate movement of people indoors is problematic however as there is no universal tracking system such as GPS which works indoors. Instead, a range of techniques are used based on technologies such as cameras, radio frequency identification, WiFi, Bluetooth, pressure pads and radar are used to track people and objects within indoor environments. The most common technologies for tracking are Bluetooth and WiFi. Many Internet of Things (IoT) devices support these protocols and can therefore act as beacons and hubs for movement detection indoor. We provide here an overview of an IoT focused framework which allows the plug and play of Bluetooth and WiFi devices in addition to integrating passive and active approaches to determining the movement of people indoors.

Keywords: Movement detection · IoT · Device Free Passive Localisation · Presence detection

1 Introduction

There is a substantial amount of work in determining the location or activity of an individual over time inside a building with wireless tracking [1, 2]. Movement detection is important for many smart living scenarios such as health care, games, manufacturing, logistics, shopping, security and tour guides. Indoor localisation systems can be classified into active and passive systems. Using Wireless signals is an attractive and reasonably affordable option to deal with the currently unsolved problem of widespread tracking in an indoor environment. A subset of tracking people is with context aware solutions. Context aware computing applications that adapts to their location of use, the collection of nearby people or objects, as well as changes to those objects over time. Location-aware services (LAS) are special context-aware application that recommend suitable services to a user based on the user's location. They enable location intelligence which provides many benefits such as personalisation of marketing communications, consumer analytics, locating a fireman in a burning building or concluding the optimal place for a group to meet for a common activity. Indoor location

É. Renault et al. (Eds.): MLN 2018, LNCS 11407, pp. 333–340, 2019.
https://doi.org/10.1007/978-3-030-19945-6_24

determination has become a crucial component in many applications and a standard for indoor localisation does not exist yet. LAS is therefore an important aspect for the Internet of Things (IoT). Current implementations of location intelligence (via LAS) in a mobile environment suffer from several issues and the choice of which technology to make use of is critical. Location tracking techniques can be classified into two broad categories; active localisation and passive localisation. The distinguishing factor is the participation of the tracked individual. In a passive system, the user is not required to participate, i.e. the system can track them without any need for an electronic device to be carried or attached which sends out signals to help deduce their location. In an active system, an electronic device is carried. Device Free Passive Localisation (DfPL) approaches can identify human presence by monitoring variances of the signal strength in wireless networks [3]. This is since human body contains about 70% water and it is known that waters resonance frequency is 2.4 GHz. An indoor wireless positioning system consists of at least two separate hardware components: a signal transmitter and a measuring unit. The latter usually carries the major part of the system intelligence. Mobile devices face many challenges that are intrinsic to mobility and are unlikely to be overcome easily. Wireless connectivity is highly variable and mobile devices must rely on limited energy sources. These issues often significantly hinder implementations of mobile LAS. The core concepts used to locate an object are explained in the next section which is then followed by a review of current location detection technologies which may be used to implement a Location based system [4, 5].

2 Detecting User Movement Indoors

A significant drawback of many indoor locating technologies is the requirement to deploy a costly and complex infrastructure composed of dedicated hardware. The existing IEEE 802.11 networks and support for wireless protocols by the clear majority of mobile devices makes Wi-Fi a logical choice for low-cost indoor location detection. Wireless networks are capable of tracking movement through the network using a technique known as radio mapping or more commonly fingerprinting, which most IEEE 802.11 based location detection approaches are based on. Fingerprinting requires a complex setup or training phase to construct a map of pre-recorded received signal strengths (RSS) from nearby access points (AP's) at every position of an interesting area. The results are stored in a fingerprint database which can be queried with any RSS to identify and map corresponding locations. This fingerprint database or radio map can be used to create a model such as the example shown in Fig. 1.

Fingerprinting provides good accuracy but is highly vulnerable to environmental changes such as rearranging furniture or moving the APs. One method of reducing this factor is collaborative feedback allowing a continually evolving radio map however the variation in RSS generated by different Wi-Fi chips could be a significant limitation in using a Wi-Fi based approach. An important consideration is that the decisions made when installing a Wi-Fi AP were typically to catch large congregations of users and primarily to provide the highest available throughput to those users. Indoor environments are also especially noisy with other wireless devices such as wireless headsets

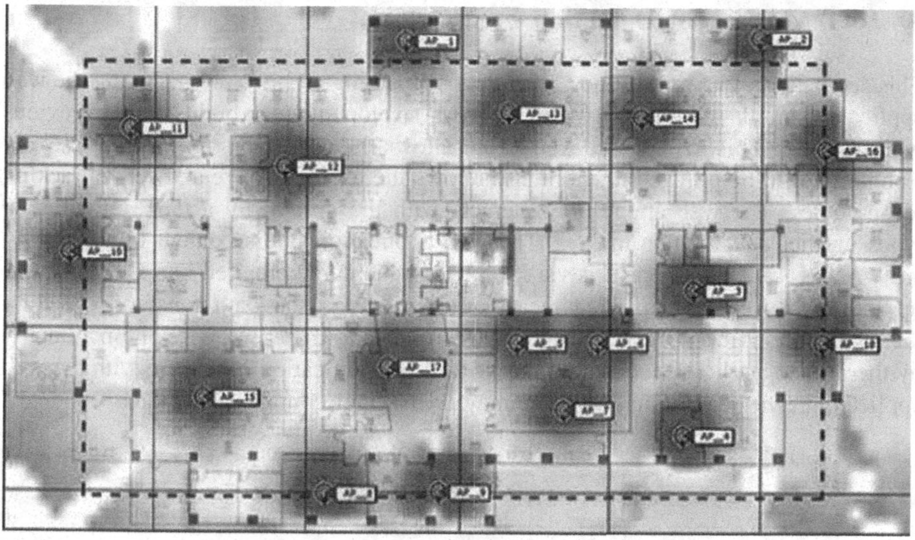

Fig. 1. Radio map

and cordless phones causing unpredictable interference. These factors result in a coverage area which is less than ideal for fingerprinting. Fingerprinting can be divided into deterministic and probabilistic approaches. There are many models within each category each with their own pros and cons, overall the probabilistic models are the most promising with the most notable being the Bayesian-Hidden Markov model. This model provides accuracy of 1.5 m but lacks reliability with only 70% of results being correct. It is also worth noting that depending on location, Wi-Fi based methods may be impossible or even dangerous to consider, for instance use in a hospital could be dangerous as the Wi-Fi signals may interfere with hospital equipment signals.

Received Signal Strength Indicator (RSSI) is the most crucial parameter in the localization of WLAN devices [6]. At the monitoring device e.g. laptop, phone, it shows the signal strength received from an access point, where the stronger signal received by the WLAN card, the closer the position of the card to the access point. One of the most important factors in the measurement of RSSI is the power attenuation due to distance; however, absorption gradient also affects the RSSI measurement. Sudden changes in signal absorption, due to walls for example, introduce discontinuities into the dependence between RSSI and distance which is normally considered a smooth function [7]. In addition to walls, the presence of humans, the thickness of walls and doors, the air temperature, nearby devices, the direction of the antenna, and the types of WLAN cards influence the absorption of the RF signal energy. The RSSI values can be reported by the device driver as a non-dimensional number or percentage and sometimes is converted to dBm through some nonlinear mapping process [8]. The error caused by different types of WLAN cards can be around 20% which is a considerable error value [9].

3 IOT Movement Detection Framework

This section provides an overview of an indoor tracking architecture including the web application and the mobile app which we have developed which builds on our previous work [10–13]. This IoT focused framework allows the use of different indoor tracking technologies to be plugged in and it also allows the use of passive and active tracking technologies. The web application interacts with the system by a range of different types of clients. It written in Java. The web application offers content in a RESTful way. Access to the web portal is via JSP pages which are compiled into HTML mark-up with Ajax used as the primary means of transferring data. Data transfer is based on the commonly used JSON data interchange format so as to permit reuse of functionality by the mobile application. Figure 2 shows the communication procedure with the web application. The request and response may or may not contain JSON data.

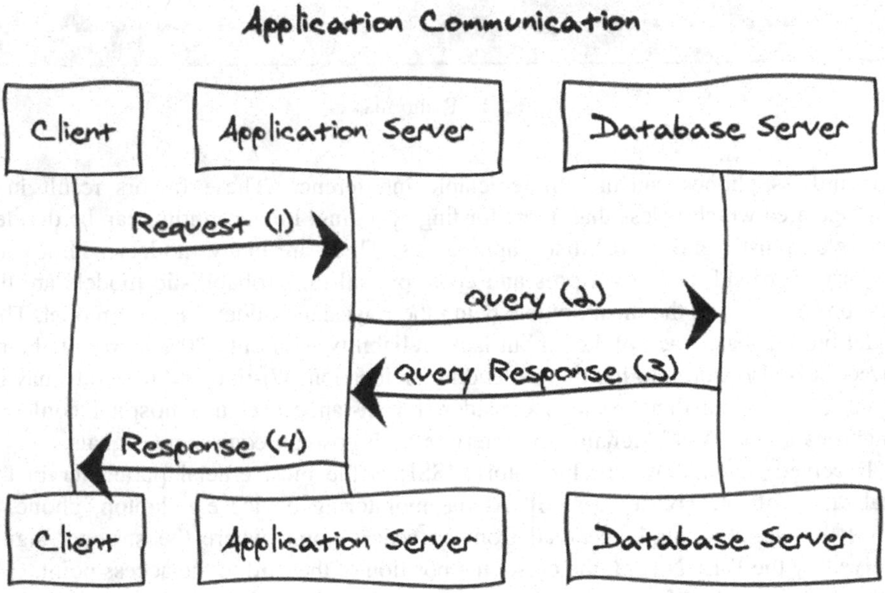

Fig. 2. Web app communication

The web application is modular and loosely coupled allowing functionality to easily be reused or replaced entirely if required. This allows us to create a framework where location sensors can be ibeacons or Eddystone beacons. They could also be another technology such as RFID or Zigbee. The framework makes it simple to slot in the location determination technology of choice. A three-tiered system is used to achieve this where three distinct tiers are made up from a data access layer, a business layer and a presentation layer. Each layer has clearly defined roles. The data access layer is used for communication with the database only. The business layer stores application specific logic. The presentation layer contains user interaction logic as well as offering an endpoint for client communication. It also communicates with the business layer and the business layer communicates with the data access layer which in turn communicates with the database. We took time to design the system like this so that future IoT location determination technologies can be more easily plugged in.

3.1 Mobile Application

The mobile application uses the Android beacon library to communicate with nearby beacons to retrieve their namespace id and instance ids. The Android beacon library is an open source library with a large support base and is regularly updated for bug fixes and support for new features is added regularly. Once retrieved, the Eddystone-UID communicates with the server and this determines the user's current position. The mobile app is used by most of the system users. The initial start page of the mobile application allows users to login with the username/password combination if they already registered. If no registration has been done previously, then new users can navigate to the registration page to create a new account.

Beacon scanning allows a user to enable scanning for beacons with a prompt shown if Bluetooth is not already enabled. Once scanning is enabled, the scanning functionality of the mobile application lets beacons be scanned for every ten seconds. If a beacon (closer than the ranging distance threshold) is found, then its Eddystone-UID details are sent to the web application which saves a new position history with the current timestamp. While remaining within range of the same beacon, new position histories are saved every minute. If multiple beacons are scanned, then the closest beacon is used. Once beacon scanning is enabled it will continue to run in the background even if the user closes the application completely. We choose this because beacon scanning is performed in a custom background service which the application communicates with. This becomes problematic however when attempting to get the current status of the background service after restarting the application as Android does not provide a clear way of checking if a service is running or not. Nevertheless, if the application does restart, the state of the background service is correctly shown by the state of the buttons. Although the normal scan cooldown time results in very little additional battery consumption, a power saving mode has been included which increases the scanning cooldown to thirty seconds. This shortens the time it takes to discover new beacons although user locations are still saved often enough to be considered accurate.

The position history page allows a user to view all their own position histories in a list. When a position history is selected, specific details are shown including the time and date they visited each position. This page automatically updates each time a new position history is saved (if beacon scanning is enabled) allowing the user to see in real time which location they are in. The account settings page allows users to manage their account details such as name and password. The user account id and username are also shown (to use if admin support is required) and validation errors result in icons like those on the login and registration pages. The URL that the application uses can be changed here. The reset button allows a user to reset any changes they have made.

The manage user accounts page displays all information relating to existing user accounts such as usernames, user account types and full names. From this part, an administrative user can create, update or delete user accounts.

To assign beacons to positions, there must be locations that the positions are associated with. The manage locations page allows administrative users to manage locations including the ability to create, update and delete location data. Deleting a location also deletes any positions which are assigned to that location, with the user made aware of this while confirming deletion. There is no limit on the number of locations however some manufacturers do have limits on the number of beacons assigned to a network (Fig. 3).

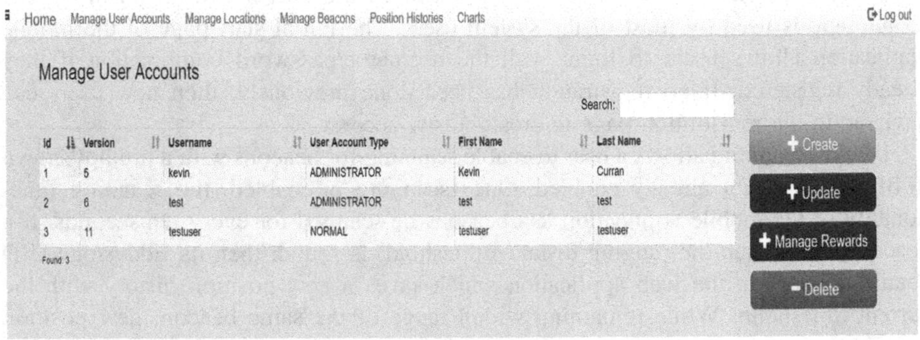

Fig. 3. Manage user accounts page

Managing beacons is one of the key features of the web portal and the manage beacon page allows administrators perform this. Beacons can be created, modified and deleted as shown in Fig. 4. During creation or modification, beacons can be optionally assigned to a location via a dropdown which is populated with locations saved through the manage locations page. If a beacon is assigned a location, then a position name must be entered.

The view position histories page can be used by admins to view position histories for all users within the system (see Fig. 5). The admin can also elect to auto refresh this page meaning table data will automatically update every few seconds, displaying any new position histories which may have been saved.

Viewing charts allows an admin user to see various statistics about the system including the most popular locations by total linger time, most visited positions for a location by location linger time, average linger time for a location, and the overall top system users by reward points. Linger time and session times are calculated based on the amount of continuous time each user spends at a position, referred to as a position history session. All position history sessions are first determined to produce the chart data displayed on this page. This section also has auto refresh functionality which can be enabled or disabled to view a snapshot or real time statistics for the system whereby all chart data automatically updates as changes happen on the network when beacons send back updates. An administrative user can hover over each chart which displays

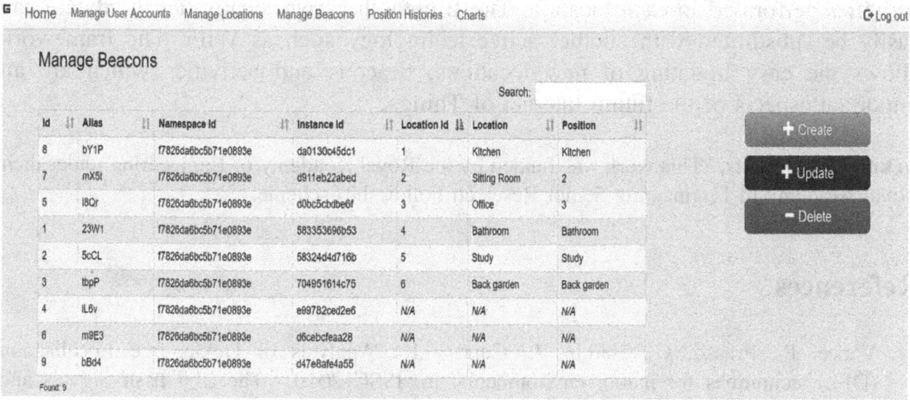

Fig. 4. Manage beacons page

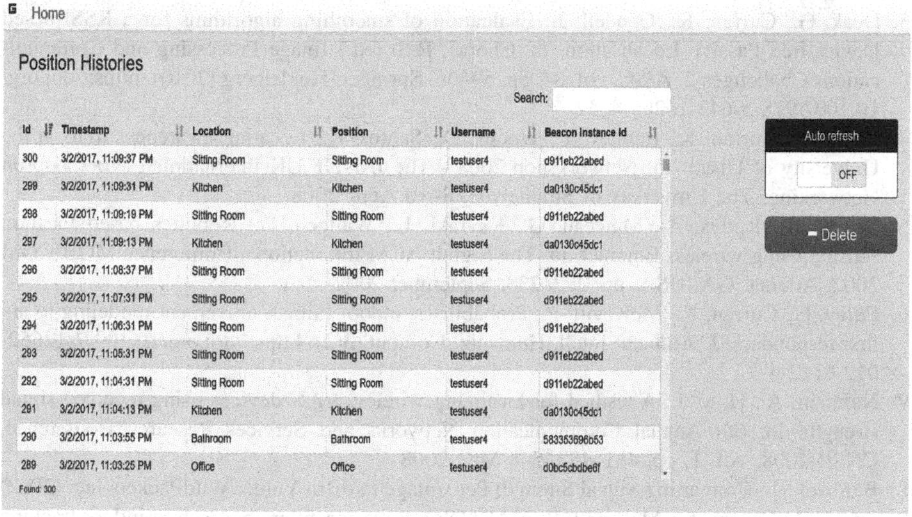
Home
Position Histories
Search:
Id Timestamp Location Position Username Beacon Instance Id
300 3/2/2017, 11:09:37 PM Sitting Room Sitting Room testuser4 d911eb22abed
299 3/2/2017, 11:09:31 PM Kitchen Kitchen testuser4 da0130c45dc1
298 3/2/2017, 11:09:19 PM Sitting Room Sitting Room testuser4 d911eb22abed
297 3/2/2017, 11:09:13 PM Kitchen Kitchen testuser4 da0130c45dc1
296 3/2/2017, 11:08:37 PM Sitting Room Sitting Room testuser4 d911eb22abed
295 3/2/2017, 11:07:31 PM Sitting Room Sitting Room testuser4 d911eb22abed
294 3/2/2017, 11:06:31 PM Sitting Room Sitting Room testuser4 d911eb22abed
293 3/2/2017, 11:05:31 PM Sitting Room Sitting Room testuser4 d911eb22abed
292 3/2/2017, 11:04:31 PM Sitting Room Sitting Room testuser4 d911eb22abed
291 3/2/2017, 11:04:13 PM Kitchen Kitchen testuser4 da0130c45dc1
290 3/2/2017, 11:03:55 PM Bathroom Bathroom testuser4 583353696b53
289 3/2/2017, 11:03:25 PM Office Office testuser4 d0bc5cbdbe6f
Found: 300
Auto refresh
OFF
Delete

Fig. 5. View position histories page

additional detail about the targeted data in a popup. A user can select locations on this chart which updates the other charts with data specifically for the chosen location.

4 Conclusion

We presented an overview of an extensible indoor location determination framework. It utilizes Bluetooth beacons for active positioning which can determine location of individuals. These beacons can use both ibeacon or Eddystone standards. The framework also facilitates device free passive localisation techniques for determination of activities performed in each location. The Bluetooth active localisation hardware can easily be substituted with another active technology such as WiFi. The framework allows the easy updating of new locations, beacons and activities which are all important aspects of the future Internet of Things.

Acknowledgements. This work was funded by the Royal Academy of Engineering under their Royal Academy of Engineering Senior Research Fellowship scheme.

References

1. Vance, P., Prasad, G., Harkin, J., Curran, K.: Analysis of Device-free Localisation (DFL) techniques for indoor environments. In: ISSC 2010 - The 21st Irish Signals and Systems Conference, University College Cork, 23–24 June 2010
2. Vance, P., Prasad, G., Harkin, J., Curran, K.: A wireless approach to Device-Free Localisation (DFL) for indoor environments. In: Assisted Living 2011 - IET Assisted Living Conference 2011, IET London: Savoy Place, UK, 6 April 2011
3. Deak, G., Curran, K., Condell, J.: Evaluation of smoothing algorithms for a RSSI-based Device-free Passive Localisation. In: Choraś, R.S. (ed.) Image Processing and Communications Challenges 2. AISC, vol. 84, pp. 59–66. Springer, Heidelberg (2010). https://doi.org/10.1007/978-3-642-16295-4_52
4. Furey, E., Curran, K., Lunney, T., Woods, D., Santos, K.: Location awareness trials at the University of Ulster. In: Networkshop 2008 - The JANET UK International Workshop on Networking, The University of Strathclyde, 8–10 April 2008
5. Bekris, K., Rudys, A., Marceau, G., Kavraki, L., Wallach, D.: Robotics based location sensing using wireless Ethernet. In: The Eighth ACM International Conference MOBICOM 2002, Atlanta, GA, USA, pp. 227–238, September 2002
6. Furey, E., Curran, K., McKevitt, P.: Probabilistic indoor human movement modeling to aid first responders. J. Ambient Intell. Humaniz. Comput. **3**(2). https://doi.org/10.1007/s12652-012-0112-4
7. Nafarieh, A., How, J.: A testbed for localizing wireless LAN devices using received signal strength. In: 6th Annual Communication Networks and Services Research Conference, CNSR 2008, vol. 1, pp. 481–487, 5–8 May 2008
8. Bardwell, J.: Converting Signal Strength Percentage to dBm Value. WildPackets Inc. (2002)
9. Furey, E., Curran, K., McKevitt, P.: HABITS: a Bayesian filter approach to indoor tracking and location. Int. J. Bio-Inspired Comput. (IJBIC) **4**(1), 79–88 (2011). ISSN (Print): 1758-0366
10. Carlin, S., Curran, K.: An active low cost mesh networking indoor tracking system. Int. J. Ambient Comput. Intell. **6**(1), 45–79 (2014). https://doi.org/10.4018/ijaci.2014010104
11. Curran, K., Furey, E.: Pinpointing users with location estimation techniques and Wi-Fi hotspot technology. Int. J. Netw. Manag. **18**(5), 395–408 (2008). https://doi.org/10.1002/nem.683
12. Deak, G., Curran, K., Condell, J., Deak, D.: Detection of multi-occupancy using Device-free Passive Localisation (DfPL). IET Wirel. Sens. Syst. **4**(2), 1–8 (2014). https://doi.org/10.1049/iet-wss.2013.0031
13. Mansell, G., Curran, K.: Location aware tracking with Beacons. In: IPIN 2016 - The 7th International Conference on Indoor Positioning and Indoor Navigation, Madrid, Spain, 4–7 October 2016

A Hybrid Architecture for Cooperative UAV and USV Swarm Vehicles

Salima Bella[1(✉)], Assia Belbachir[2], and Ghalem Belalem[1]

[1] Department of Computer Science, Faculty of Exact and Applied Sciences, Université Oran1 Ahmed Ben Bella, Oran, Algeria
bella.salyma@gmail.com, ghalemldz@gmail.com

[2] Mechatronics Department, Polytechnic Institute of Advanced Sciences, IPSA, Ivry-Sur-Seine, France
belbachir.assia@yahoo.fr

Abstract. This paper is interested in the problem of monitoring and cleaning dirty zones of oceans, dealing with the notion of path planning for semi-autonomous unmanned vehicles. We present a hybrid cooperative architecture for unmanned aerial vehicle (UAV) to monitor ocean region and clean dirty zones with the help of swarm unmanned surface vehicles (USVs). In the path planning problem, unmanned vehicles must plan their path from the starting to the goal position. In this article, we propose a solution to handle the problem of trajectory planning for semi-autonomous cleaning vehicles. This solution is based on the proposed Genetic Algorithm (GA). In order to optimize this process, our proposed solution detects and reduces the pollution level of the ocean zones while taking into account the problem of fault tolerance related to these vehicles.

Keywords: UAV · USV · Swarm · GA

1 Introduction

Swarming is a collective behavior observed on animals of similar size which aggregate together the same behavior. In nature, we can find ant that are using this swarm behavior and sharing their pheromones to find the food place [10]. In the context of robotics, swarm robots are generally used to coordinate multi-robot systems [12]. The interest of the swarm is that the system shares at any time simple information related to the environment with other member of the system. Thus, it is used for cooperative behavior to resolve complex problems such as monitoring, search and rescue. In order to perform these tasks, the robot swarm system should have the ability to dynamically train itself whenever communication is done between its members and plan their paths.

Most studies on cooperation with dynamic unmanned vehicles such as unmanned aerial, surface and ground focused on formation control. This formation control refers to the task of controlling a group of autonomous mobile unmanned vehicles to follow a predefined path or trajectory while maintaining the desired formation pattern. One important aspect of swarm is reflected in the ability to interact with global path planning. Path planning refers to search an optimal or a sub-optimal path from the

É. Renault et al. (Eds.): MLN 2018, LNCS 11407, pp. 341–363, 2019.
https://doi.org/10.1007/978-3-030-19945-6_25

starting position to the target position. Path planning has two key technologies: environmental modeling and planning algorithms. Environmental modeling is an important aspect. It characterizes all the actions to be used for the path planning algorithm. Several work on USV environmental modeling methods used the grid representation, the geometry of space law, topological method and the electronic chart and so on [8].

The main consideration in environmental modeling is the environmental data storage, query and update. Features of grid method are simple, easy to implement, can be applied to different algorithms. USV path planning algorithms commonly includes two ways: conventional planning methods and intelligent planning. Traditional methods include: artificial potential field method [15], dynamic planning [20], A* method [9], and so on. Intelligent planning methods include: genetic algorithm [2, 10], colony algorithm [9], particle swarm algorithm [14]. According to [10], heuristic algorithms that implement search in the solution space can be classified as instance-based or model-based. The instance-based algorithms generate new candidate solutions using the current solution or the current population of solutions, such as genetic algorithms. Genetic algorithms are iterative stochastic optimization algorithms based on the mechanisms of natural selection and genetics. They provide the solutions to problems that do not have computational solutions in reasonable time analytically or algorithmically.

The interest of this work is to present a hybrid approach which allows easier management between different unmanned vehicles. Centralized management has a central node with deterministic decision-making capability and easy coordination to implement. This coordinator has a global view of the unmanned vehicle activities of its appropriate system. These unmanned vehicles: the monitoring vehicles and the cleaning swarm vehicles cooperate with each other to execute a mission of monitoring of the oceanic regions and the cleaning of their dirty zones, and to follow the requests (tasks) of its coordinator. Distributed management begins when the surveillance vehicle is assigned to a region. The monitoring vehicle has a lower decision than its coordinator. It has the ability to coordinate its own cleaning swarms to accomplish the cleansing mission. These swarms plan their movement to get to the dirty zones from their storage bases. In this context, we proposed a genetic algorithm as a solution to solve the problem of path planning. Since this cleaning vehicle is a mechanical, electronic and computer system which may at some point fail at the hardware/software level in the performance of its tasks. For this our solution allows to select and replace the failed cleaning vehicles by the competent cleaning vehicles during the realization of the cleaning mission.

The paper proposes a hybrid cooperative approach for heterogeneous unmanned vehicles. This approach includes a solution for managing the trajectory planning problem for the USV swarm. This solution is based on a genetic algorithm. While the proposed approach allows to cooperate an unmanned aerial vehicle (UAV) to monitor an oceanic region and an unmanned surface vehicle (USV) swarm to clean dirty oceanic zones. We suppose that the UAV has on-board sensors that allow it to locate dirty zones. The UAV discretizes its environment map and updates this map by the collected information related to dirty zones. The UAV sends its environment map to its general coordinator (represented by a laptop, guided by a human operator). After an analysis of this collected data, the general coordinator allocates the explored map (way-points) to the USV swarm to clean each dirty zones. This swarm navigates to the assigned dirty zone and cleans it based on the proposed solution. The novelty in this work is that our approach is extended by a fault tolerance service for these cleaning vehicles (USVs).

The paper is organized as follows: in Sect. 2, we present some related works; we describe, in Sect. 3, the proposed approach for the different unmanned vehicles with the proposed solution, and model them by logical formalization; we illustrate, in Sect. 4, an example to simulate the operation of our approach. We finally conclude our work in Sect. 5 and show some future directions.

2 Related Works

Many heuristic and meta-heuristic algorithms have been applied to the path planning problem, such as genetic algorithms. In addition, there are many methods for coordinating multiple unmanned vehicle in a swarm or formation for the path tracking. These methods can be roughly categorized into three basic approaches: behavioral, virtual structure, and leader follower [2]. In this section, we discuss several studies on this topic.

CADDY [3] is a cooperative guidance system to monitor the behavior of human divers and assist them during the execution of demanding missions in complex 3D environments at sea. The CADDY functionalities rely on the coordination of two robotic platforms (an extended description in [5]) and a guide; one operating on the surface "Charlie ASV (autonomous surface vehicle)", called dive buddy "observer". It allows geo-referenced positioning of the underwater segment and acting as a communication relay. Another operating underwater "AUV(Autonomous Underwater Vehicles)", called dive buddy "slave" at close distance from the diver, to meet the specific objectives for the diver operation support. As a dive buddy "guide", the system is in charge of guiding (upon request) the human diver from one spot to another safely towards its goal. The cooperative path-following method is based on the virtual target approach (A description of this method is defined in the work of [4, 5]).

An improved swarm-based path-following guidance system for an autonomous multivehicle marine system has been introduced in the work of [6]. In the seminal idea a team of Unmanned Surface Vehicles (four USV charlie vehicles) is required to join into a formation by means of a potential-based swarm aggregation methodology, while a virtual-target based guidance module drives the whole formation towards a desired reference path. This latter goal is fulfilled by means of a virtual target based path following approach developed in [7]. The initial idea of combining the swarm behavior with the path-following guidance system was initially proposed in [4] where the feasibility of the approach was proven. The evaluation of the proposed approach is carried out simulating the vehicles, relying on the kinematical and dynamical model of the Charlie USV, described in [7].

A method of planning the exploratory pathway was applied in the article of [10]. The method is based on the genetic algorithm (GA) for autonomous mobile robots, where GA is used to find a sequence of actions that robots must perform to reach the goal as fast as possible. In the proposed approach, the robot did not know in advance the disposition of the environment and has only a rough estimate of starting positions and objective. At first, a set of actions are generated randomly. The robots execute these actions, then their physical condition is evaluated. The fitness function is based on the distance traveled and the euclidean distance from the goal. Individuals are selected by tournament to breed. Then, a new sequence of actions is generated by applying cross-over and mutation operators. However, the evolution continues only for the sequence of actions related to the robots that did not reach the goal. The proposed GA has been compared to A* and a proposed algorithm C*. GA has a better average performance than A* and C* with the better smallest distances traveled by the solutions.

A potential-based genetic algorithm for the motion planning of heterogeneous holonomic robot swarms has been proposed in the paper of [2]. The proposed algorithm consists of a global path planner (GPP) and a motion planner (MP). GPP algorithm looks for a path, that the robot swarm center should move, in a Voronoi scheme of free space (2D workspace). MP is a genetic algorithm based on an artificial potential field The repulsion keeps robots away from obstacles and the "spring" function maintains the robot swarm within a certain distance from each other. Since the GA searches for an optimal configuration which has lower potential, the obtained paths are safe. The robot swarm moves toward the goal by sequentially traversing a sequence of positions along the Voronoi diagram. The results of the simulation that are presented by a PC. They demonstrated that the proposed algorithm can plan collision less paths for swarm robots.

Table 1. Comparison between the related works.

Work/criteria	[3]	[6]	[10]	[2]	Proposed approach
System type	Cooperative orientation system	USV Swarm path following guidance system (improved)	Robot exploratory path planning System	Movement planning system for robot swarms	Monitoring-cleaning and USV swarms path planning system
Main objective	Monitor human divers at seamission execution	Combine swarm behavior with following guidance system	Find a path to goal	Plan collision-free paths	UAV-monitor region, clean a dirty zone and plan path-swarm
Work type/decision	Coordination of two robots with the guide	USV team to join a formation	Not cited	Cooperation between the global path planning and motion planning algorithm	Cooperate UAV-USV swarm, coordinate by a coordinator and planning path
Used type of method	Virtual target (VT)	Aggregation potential, guidance module-VT	Genetic algorithm (GA)	Potential-based GA	GA System
Gear type environment	1 ASV, 1 AUV 3D marine complexes	4 charlie USV Disruptmarine	1 robot Not cited	8 robots 3D workspaces	1 UAV-Five USV Maritime-Atmosphere space
Algorithm/functions	Function Lyapunov	Function Lyapunov	A*, GA	Repulsion-spring function, Voronoi and GA algorithm	GA, calculate the needed number of USV, USVs placement to clean an zone and nodirty function
Selection	Not cited	Not cited	Tournament	Highest %: 10%	Tournament
Fitness	Not cited	Not cited	Not cited	Configuration formula (defined)	Total distance-minipath & direction-cost
Metrics to measure	Horizontal motion-speed of vehicles	Robots movements-speed during the swarm aggregation-following path	Distance traveled by robots in each algorithm	Behavior of swarm trajectories in optimizing the configuration	The average of TEC of each USV_{cz} swarm, average of TEC of USV selected
Simulator	HIL [11] ROS environment	Kinematic-dynamic simulation-model	PS Tools	PC (Core2Duo1.7 GHZCPU-WXP)	PC(Ci55200U-C2.20 GHz-W7Pr)-java

In this article, the genetic algorithm is used to find an optimal or near optimal path between the starting position of a unmanned cleaning vehicles swarm and the goal. This path planning between the base of life and the dirty zone is carried out based on the metric map of an oceanic region. This map is built by an unmanned surveillance vehicle. To this end, we try to contribute to a maritime pollution problem by proposing a hybrid approach based on the proposed solutions "GA". A co-operation between heterogeneous air-sea vehicles is presented to monitor ocean regions and clean dirty zones. Each study above its characteristics/parameters that differentiate it from others. For this, we introduced in our approach a fault tolerances service for unmanned cleaning vehicles. Table 1 shows a comparison study of the previous cited work and our proposed approach.

3 Proposed Approach

In this part, we will describe the architecture and constituent entities of our proposed approach, as well as its functioning.

3.1 Architecture of the System

The hybrid architecture of our system, shown in Fig. 1 consists of a central unit, surveillance vehicle to monitor maritime region, swarms of cleaning vehicles to clean dirty zones, and recovery vehicles are presented to recover defective vehicles. The central unit is composed of a general coordinator, a base of life and a database. Coordinator General stores and consults the data in the database. It interacts with the base of life and the surveillance vehicle via a communication link of the IEEE.802.11a standard (5000 m outside range). While the Wi-Fi network of the IEEE.802.11b standard (35 m–140 m indoor range) allows messages to be connected between the surveillance vehicle and the swarm of cleaning vehicles.

3.2 Hierarchical Role of Each Vehicle

Figure 2 shows a sequence diagram of decision-making hierarchy in our proposed approach where the general coordinator has the highest decision, namely the launch of the tasks to initiate and control unmanned vehicles (monitoring, cleaning and recovery) that are located in the base of life. The monitoring drone is loaded of a lower decision, because it is introduced by the coordinator. It has the role to oversee its cleaning vehicle swarm. These latter are composed by leaders vehicles with their cleaning following vehicles. Their purpose is to carry out the cleaning operation of the dirty zone according to the energy availability of each member. Each leader of the swarm has two necessary roles. It is responsible for the tasks/characteristics of its followers, and also, it shares (cooperates) with them the cleaning action. The coordinator launches a recovery vehicle to recover the failing vehicles.

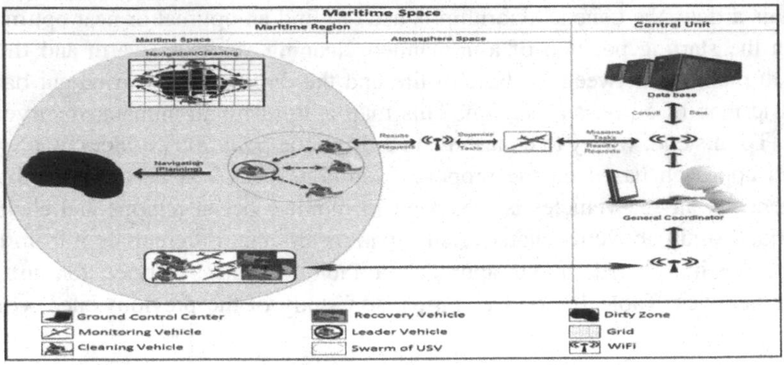

Fig. 1. Hybrid architecture of our system.

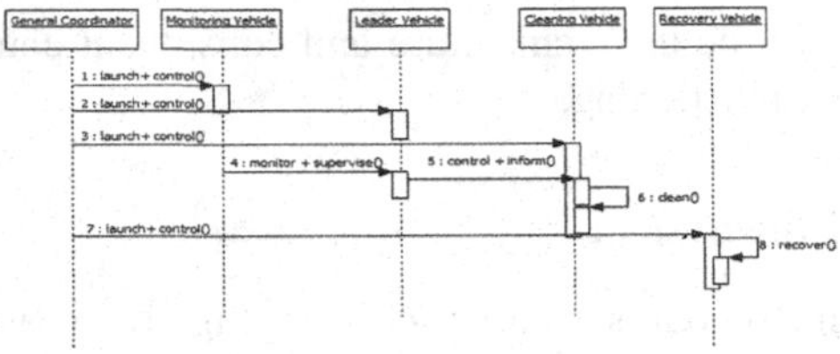

Fig. 2. Sequence diagram for the decision-making hierarchy of the approach.

3.3 Environment Modeling

This section describes the model of our environment.

1. *Set of tasks.* We define the five high level tasks that are used by the general coordinator, the monitoring and cleaning vehicles.
   - *Monitoring task*: t_{mr} represents a task of monitoring the region r.
   - *Cleaning task*: t_{cz} represents the task of cleaning a dirty zone z.
   - *Supervising cleaning task*: t_{sz} represents the task of supervising the cleaning of a dirty zone z.
   - *Allocating task*: t_a represents thc task of allocating unmanned vehicle to different regions and dirty zones.
   - *Launching task*: t_l represents the launching spot of the different previous tasks.
2. *Set of vehicles.* We represent different used vehicles with their related roles.
   - UAV_{mr}: Monitoring, homogeneous and semi-autonomous aerial vehicle. This vehicle incorporates a camera, an ultrasonic sensor, a GPS (Global Positioning System) and an autopilot software. It features a speed and an energy capacity to allow it to: monitoring the regions, the dirty zones, the cleaning vehicles, supervising the cleaning vehicles swarm(USV_{cz}), asking/informing the leader of each swarm by tasks and launching and return the data and the results to the coordinator.

- USV_{cz}: Cleaning, homogeneous and semi-autonomous surface vehicle. It has the same hardware and features of *UAVmr* except it does not embed a camera. USV_{cz} is answerable of: cleaning the dirty zones, following/request the tasks of its leader ($Leader_{cz}$), informing and returning the data and the results to its leader and coming-back to the base of life.
- $Leader_{cz}$: Cleaning, homogeneous and semi-autonomous surface vehicle. This vehicle is similar to the USV_{cz} in hardware and software components. It allows to: sending the requests for its followers (USV_{cz}), sending the requests for its followers, receiving and saving the characteristics of each follower (USV_{cz}) (and save its features) and informing and returning the data and the results to its supervisor.
- $Vehicle_{rec}$: it is a special vehicle named recovery vehicle ($Vehicle_{rec}$). It is similar as the USV_{cz} in hardware and software components. It allows bringing back the faulty vehicles (out of order) towards the base of life (The detailed description of this vehicle is presented in our article [19]).

3. *Set of agents.* We show here a description for each agent used:
   - $General_{crd}$: General coordinator. $General_{crd}$ is represented by a laptop that contains a coordination software, guided by a human operator. It is responsible for: the base of life, data of the regions and the dirty zones, use (treatment) and the storage of data in the database, launch of the tasks/missions and allocate tasks to vehicles.
   - $leader_{cz}$: Leader vehicle. It is an unmanned surface vehicle that has two roles: it is an intermediary between the supervisor Sup_{mr} and USV_{cz} of the swarm, and also cooperates in the cleanup operation.
   - USV_{cz}*(discharge)*: Cleaning vehicle in discharge state. It is a cleaning vehicle, that has not yet completed its task and energy capacity are low during cleaning.
   - USV_{cz}*(free)*: Cleaning vehicle in free state. It is a cleaning vehicle that has completed its task.
   - USV_{cz}*(prepared)*: Cleaning vehicle in prepared state. It is a cleaning vehicle prepared in the base of life by the $General_{crd}$ which will replace USV_{cz}*(discharge)*.
   - Sup_{mr}*(*USV_{cz}*_discharge)*: Supervisor of USV_{cz}*(discharge)*. It is the surveillance vehicle (USV_{cz}) of the cleaning vehicle in discharge state.
   - Sup_{mr}*(*USV_{cz}*_free)*: Supervisor of USV_{cz}*(free)*. It is the surveillance vehicle (USV_{cz}) of the cleaning vehicle in free state.
   - Sup_{mr}*(*USV_{cz}*_prepared)*: Supervisor of USV_{cz}*(prepared)*. It is the surveillance vehicle (USV_{cz}) of the cleaning vehicle in prepared state.
   - $Leader_{cz}$*(*USV_{cz}*_discharge)*: Leader of USV_{cz}*(discharge)*. It is the leader of the cleaning vehicle in discharge state.
   - $Leader_{cz}$*(*USV_{cz}*_free)*: Leader of USV_{cz}*(free)*. It is the leader of the cleaning vehicle in free state.
   - $Leader_{cz}$*(*USV_{cz}*_prepared)*: Leader of USV_{cz}*(prepared)*. It is the leader of the cleaning vehicle in prepared state.

4. *Set of regions.* The monitored maritime space is divided into maritime regions. The region is composed of two sub-spaces; an atmosphere sub-space where we can find the UAV_{mr} and a maritime sub-space where we can see the swarm of USV_{cz}, $Vehicle_{rec}$, base of life and dirty zones.
5. *Base of life.* It is a zone (can be a boat, ship and an island, …) to store a fixed number of UAV_{mr}, USV_{cz} and $Vehicle_{rec}$.
6. *Database.* It is a basis for storing and saving all the data and features of the maritime space as well as the different unmanned vehicles used.
7. *Dirty zones.* They represent dirty part where we find the water pollution, for example oil slicks. The proposed metric in this work is the degrees of dirt for each zone where they are broken down into four types, namely, strong dirt, average-strong dirt, average dirt and low dirt. Each zone is characterized by a list "$List_{zone}$" which delimits its borders by the coordinates. These last are composed by the list of degrees of dirt "$List\_Degree_{cell}$", list of cell positions of these degrees "$List\_Position_{cell}$" and they are attached to a zone by "$Position_{zone}$".

Used Parameters for Our Proposal. We used a set of parameters in the realization of our approach.

- *Threshold*: It is a fixed threshold that allows classifying the zone coordinates according to the degrees of the cells "$List\_Degree_{cell}$".
- $Threshold_{average\_energycap}$: The threshold of the average energy capacity that has a fixed value and it allows to determine the average energy capacity.
- $Threshold_{low\_energycap}$: The threshold of low energy capacity that has a fixed value and it allows to determine the low energy capacity.
- $List_{ID}(Id_{USV\ cz})$: The list of USV identifiers composing the swarm.
- $List_{energy\_usv(t)}(Id_{USV\ cz}, Cap_{energy}CZ)$: The list of the USV energy capacity in instant t during cleaning as well as its identifier.
- $List_{zone}(List\_Degree_{cell}, List\_Position_{cell}, Position_{zone})$: The list of coordinates of the dirty zone.
- $List_{zoneS}$: The $List_{zone}$ sorted by the degree of dirt.
- $List_{threshold\_degree}Z$: The coordinates list of the dirty zone that is compared with the threshold of dirt compared to the dirty degrees ($List\_Degree_{cell}$).
- Nbr_{CZ}: A function that gives the number of selected USV_{cz} by $General_{crd}$.
- $Parameters_{start\text{-}up(M)}(Id_{region}, Position_{region}, Path_{region})$: A triplet of start parameter for each UAV_{mr} which represents the identifier, position, and path of a region.
- $Parameters_{start\text{-}up(C)}(Id_{region}, Id_{zone}, Position_{zone}, Path_{zone}, Id_{UAVMR})$: A triplet of start parameters for each USV_{cz} which represents the identifier of a USV_{cz}, identifier of a region, identifier of its supervisor Sup_{mr}, identifier of an zone thus its position and path.
- $Parameters_{cleaning}(PosUSV\ g, Cell_{cz}, Id_{region}, Id_{zone}, Position_{zone}, Path_{zone}, Id_{SUPMR})$: The cleanup parameters for the USV_{cz} that replaces the USV_{cz} in discharge state. This list contains: the position of USV_{cz} in the grid, the cell to be cleaned, the identifier, the position and the path of the zone, the identifier of the region, the Sup_{mr} identifier.

- $List_{characteristics}(Id_{zone}, Id_{USV\ CZ}, Dur_{EC}, Cons_{EC}, Dur_{ED}, Cons_{ED})$: The list of the USV_{cz} characteristics: identifier of an zone (Id_{zone}), identifier of a $USV(Id_{USV\ CZ})$, the duration and the energy consumption of cleaning (Dur_{EC}, $Cons_{EC}$) and the duration and the energy consumption of displacement (Dur_{ED}, $Cons_{ED}$).

3.4 Main Phases of the Proposed Approach

In this section, we will describe the key steps of our hybrid approach. For this, we have decomposed the course of our approach in two essential steps: monitoring and cleaning. We proposed a genetic algorithm as a solution to solve the problem of path planning. This solution is applied in the cleaning step.

First Step: Monitoring. In this step, we presented the actions of the monitoring step of each UAV_{mr}, namely, triggers the monitoring of the region and analyzing the captured data from each dirty zone. This step is done, for example, twice a week in the maritime space to be monitored. The general coordinator $General_{crd}$ prepares the monitoring drones according to the regions numbers, allocating a UAV_{mr} for each region. In addition, $General_{crd}$ checks periodically the maritime space statistics every day, which are saved in the database. So, if it finds a region statistic that shows that this last contains a high percentage of dirty zones in certain period. So, it launches a UAV_{mr} for a full day to watch this dirty zone. $General_{crd}$ triggers the monitoring step according to the following actions:

1. Prepare each UAV_{mr} with a $Parameters_{start\text{-}up(M)}(Id_{region}, Position_{region}, Path_{region})$, energy (battery charged) and associated speed.
2. Each UAV_{mr} is launched from the base of life, and it follows the path with a rectilinear movement (See Fig. 3) to reach its region.
3. One $UAV_{mr}(Sup_{mr})$ arrives at its region, it captures the data of its environment from its initial position with a sweep movement (See Fig. 3). Also, it discretizes its region, then plans its movement through the model representation for that it can fly in the right conditions and perform its monitoring task.

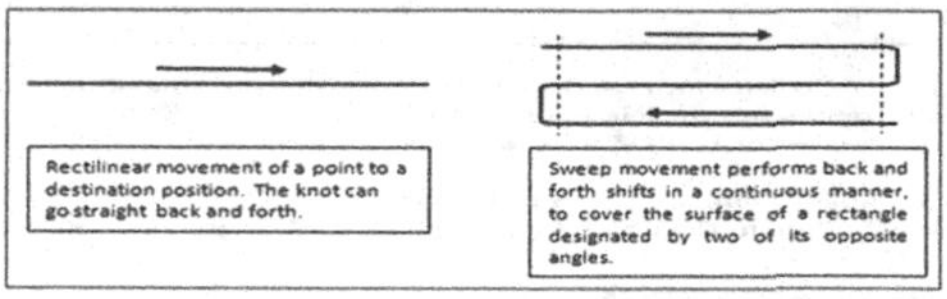

Fig. 3. UAV movement to achieve.

When the UAV_{mr} arrives at its region, it captures the data from its region and stores it in the $List_{zone}$. The collection of these data is done by the environment discretization by using two grids (a discrete space of dimension 2) of square form. This map is constructed from the obtained sensory data from UAV different sensors by a camera and an ultrasonic sensor. A square grid to discretise the atmosphere sub-space. This grid is used to plan the UAV_{mr} move in this environment. The maritime sub-space grid is used so that the UAV_{mr} can identify dirty zones. Once the sea level data collection is complete, the UAV_{mr} sorts the collected list and saves it in a list $List_{zonesS}$. Then, it compares the *List_Degree*$_{cell}$ of this Lis_{zonesS} with a fixed threshold *Threshold*. After the comparison, it saves the result in *List_threshold*$_{degreeZ}$ and sends it to the general coordinator.

Second Step: Cleaning. We will quote the solution proposed in this step. This solution is applied exactly in the trigger phase of the cleaning process. In addition, it guides the USV_{cz} swarm to execute the move phase to dirty zones. Before starting the solution, select the number of USV_{cz} containing the swarm. For this, $General_{crd}$ analyzes the received data (the *List_threshold*$_{degreeZ}$) from each UAV_{mr}. Then, $General_{crd}$ determines the USV_{cz} number containing the swarm for each dirty zone. This action is described by the Algorithm 1 (See Fig. 4).

After the execution of Algorithm 1, we presents a description of our proposed. "Genetic Algorithm" to trigger the cleaning process. It consists of three phases:

1. *Phase 1: Trigger Phase of the Cleaning Process.*
   A. *Solution: Using a genetic algorithm.* In this article, we have inspired an algorithm presented in Yakoubi et al. [18] to propose our own genetic algorithm (GA). In this proposed GA, the swarm of USV_{cz} must search for an optimal path between the starting position (the base of life) and the goal (the dirty zone). GA was used to select a leader for the swarm, and to plan the leader movement. Generally, the GA has the disadvantage of taking a lot of computing time, and it is worthwhile to reduce the execution time while keeping a good quality and a good result, so it consumes more energy [1]. Then the AG stops when the swarm arrives at its dirty zone. At this moment the swarm executes a proposed algorithm (Algorithm 2) to move and clean its dirty zone. In addition, we made a comparative table (Table 2) of our approach to the work cited in article [18] to see the difference in use of the proposed GA. The proposed GA develops the positions sequences that direct a group of USV_{cz} from the source (the base of life) to the goal (the dirty zone). USV_{cz} must reach the goal as soon as possible.

```
Algorithm 1
Inputs
• List_threshold_degreeZ[C]: list which contains the degree of dirty cells of an zone compared with
Threshold, where C: represents the length of List_threshold_z[],
• M: represents the number of USVCZ in base of life,
Outputs
• y: the solution variable to find the number of USVCZ.
• Nbr_CZ[U]: list which contains the USVCZ numbers, where U: represents the length of Nbr_CZ[U].
begin
for all List_threshold_degreeZ received do
   Calculate the sum(List_threshold_degreeZ);
   Calculate the average(List_threshold_degreeZ;
   Calculate the min of the set List_threshold_degreeZ;
   % calculate the solution variable to find the number of USV_cz
   calculate x ← average(List_threshold_degreeZ)/min;
   calculate the solution of equation (y) m ← x * y;
   for all u ∈ U do
      Nbr_CZ[u] ← y * x;
   end for
end for
```

Fig. 4. Calculate the needed number of USV_{cz}.

B. *Presentation of the used algorithm.*

- *Environment modeling*. The environment modeling depends on the map of the discretized region (metric map) by the supervisor UAV_{mr}. Workspace W is discretized by a square grid G. G is composed of square cells. A node is located inside each cell to facilitate the movement of USV. These nodes build a dynamic graph where the arcs are presented by R_{ij} connection links. The R_{ij} between two neighboring cell-nodes represents the distance between two positions as illustrated in Fig. 5. This distance is represented by the energy value E_{ij} that the USV_{cz} can consume in the displacement. Each cell-node ith represents a free/occupied position by a position or an obstacle in the environment. These positions are identified by cartesian coordinates (X, Y) in a 2D plane. The R_{ij} between these positions have a maximum of eight links with the neighboring positions j_{th}.
- *The path planning method. General$_{crd}$* assigns a start position Ps (x, y) and the list of goal positions Pg (x, y) to each USV_{cz}. This list sets the positions to arrive at the dirty zone. Then, the path traveled by USV_{cz} from a Ps (x, y) to a Pg (x, y) passing through all accessible positions is called the global path planning for the region. To get this global path, it should be divided into a mini-path, the latter must not exceed the range of the sensor of USV_{cz} (an ultrasonic sensor). This mini-path contains the start position Ps and the end position Pe (x, y). If the end position Pe (x, y) is equal to one of the positions of the list Pg (x, y), then the USV_{cz} has arrived in the dirty zone. The purpose of our work is to achieve an effective global path, for this the mini-path should be as efficient automatically and it will have a short distance and a minimum of energy consumed.
- *The evolutionary approach.* To minimize energy consumption and trajectory distance, we proposed an evolutionary approach based on genetic algorithms (GA). GA helps the USV find an efficient path from the Ps (x, y) to Pe (x, y). GA was invented by Holland [13]. This consists of several steps in which the first step is the solution initialization (genome or chromosome). After that, the evaluation of these populations is launched using the fitness functions. From this evaluation, we can move on to the next generations who have the lowest fitness values. Genetic Operators are applied by replacing existing chromosomes with new ones to preserve a constant number of chromosomes. Those operators that are popular are selection, crossover and mutation.
- *Encoding and representation.* After the environment modeling is finished. Each node has its own number. The gene is made of one part which represents this number nodes. In the path planning of the region, the gene corresponds to a position of USV_{cz}. The chromosome is consists of genes which represent a mini-path. In this study, the chromosome contains a number of genes, the latter should belong to the rayon of the sensors. To more clarification, an example is shown in Fig. 6 in which we notice that any gene which is located between the position of USV_{cz} and the circular zone is taken into consideration to form a chromosome. The order of genes in chromosomes should be obligatory because it represents the different positions that take the USV_{cz}.

Table 2. Comparison of our approach with the work of the article [18].

Concept/work	The path planning of cleaner robot for coverage region using Genetic Algorithms [18]	Our approach
Robot type	Vacuum cleaner robot	Swarm of USV
Mission	Move and clean all accessible areas in the environment avoiding obstacles	Move in the environment to get to the dirty zone
Task	Path planning of coverage region (PPCR)	Path planning to get to the dirty zone
Result	Obtain an effective global path with a short distance, a minimum of turns and the repeated clean positions	Obtain an effective global path with a short distance and a minimum of consumed energy
Environment	Unknown	Known
Tool-discretization	Use sensory sensors to discretise the workspace	Use the explored map of the UAV
Discretization of the work space	Discrete by disks as a form to facilitate cleaning	Discretize by a square grid G of square-shaped cells. A node of a dynamic graph is built inside each cell
Connection type	Connection R_{ij} represents the Euclidean distance between two positions between two neighboring disks	R_{ij} represents the distance between two positions between two neighboring cell-nodes. This distance is represented by the energy value E_{ij} that the USV can consume in the displacement
Position type used	Start position Ps(x, y), exit position Pe(x, y)	Start position Ps(x, y), goal position Pg(x, y) and list of end positions Pe(x, y)
Fitness parameters/evaluation	Total distance of the mini-path, the number of consecutive unclean cells and the total distance of each position cell relative to the current robot position. The highest fitness value is considered	Total distance of the mini-path, the direction cost for each USV_{cz}. Takes into account the lowest fitness value

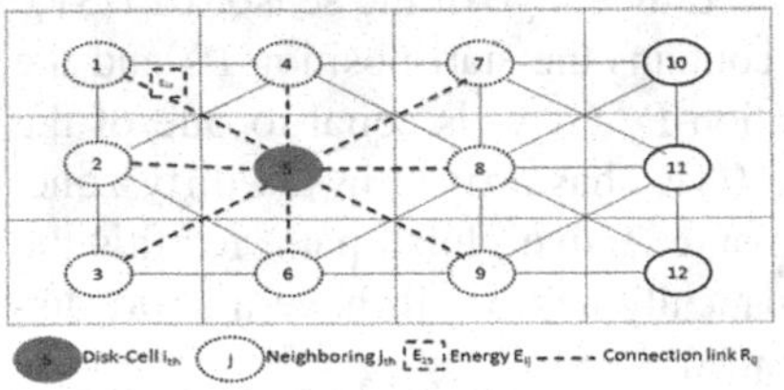

Fig. 5. Environment modeling.

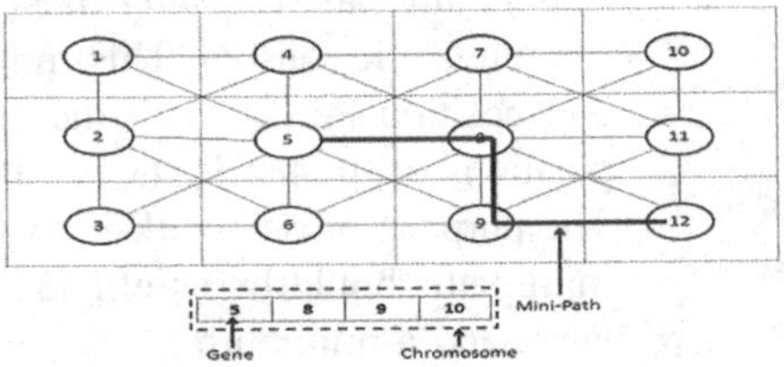

Fig. 6. Gene presentation.

– *Generation of initial population.* The population has a fixed length of p individuals and each individual has a sequence of initial positions of length l. The positions represent the cell-nodes that the USV_{cz} can explore in the known environment. In this phase, the mini-path planning is random by choosing neighboring positions. As a result, individuals (chromosomes) have their sequences of positions filled randomly, and each sequence is executed by the USV_{cz} associated with that individual. We can get different proposed paths with a standard zigzag movement [18] for the mini-path. Among these mini paths, we choose a mini path for USV_{cz} in order to execute it. Once the USV_{cz} has the sequence of positions, a fitness value will be updated based on the path traveled in environment.

– *Fitness*. Function and Selection. Fitness function is necessary to know the details description and solution of the problem. It is directly related to the constraints of navigation coverage, ie the energy reduction and the path size. Two parameters are taken into consideration to evaluate the fitness of the mini-paths: the total distance of mini-path, the direction cost for each USV_{cz}. An appropriate fitness function of mini-path (1) is constructed as:

$$F_i = A * Dist(i) + B * \sum Cost(i) \tag{1}$$

$$Cost(i) = \sum left_{cost} + \sum right_{cost} \tag{2}$$

where A, B, C are the constant number, *Dist(i)* et *Cost(i)* are the distance of mini-path and the direction cost for each USV respectively.

- Total distance of the mini-path: It is calculated by the sum of each distance between two cell-nodes. This distance is represented by an energy value.
- Direction cost for each USV_{cz}: It is calculated according to the direction of the USV_{cz}. The displacement of USV_{cz}: It is composed in two directions:
  - Direction to the right ($right_{cost}$): USV_{cz} turns to the right of its position to get to another position.
  - Direction to the left ($left_{cost}$): USV_{cz} turns to the left of its position to reach another position.

The management costs are fixed in advance and the ideal form is the solution that does not contain redundant cells visited.

– *Genetic operators*. New position sequences (new individuals) are created by applying crossover and mutation operators. Crossover is performed positions sequence provided by two parents. These parents (two mini-paths) are selected by tournament, where t individuals are chosen at random and who with a low fitness value becomes a parent. We opted for a one-point crossover techniques and the two children are then added to the population. The mutation randomly defines a quantity of genes that is randomly chosen to be mutated. In this case, each gene is replaced by one of the previous neighbor genes that are also randomly selected. At each generation t, p new individuals of length l can be created to replace the previous population. Each new position sequence is executed by an associated USV_{cz}, so that the fitness value of each new individual is calculated.

C. *GA-based trigger process*. The steps of this cleaning trigger process are defined as follows:

(a) When the $General_{crd}$ chooses the swarm of USV_{cz}, it sends the list of their identifiers $List_{idusv}(Id_{USV\ CZ})$ to their supervisor (Sup_{mr}).

(b) Upon receipt of this message, Sup_{mr} sends its discrete maritime space exploration map to $General_{crd}$.

(c) $General_{crd}$ broadcasts a set of parameters to the selected swarm, namely:

- $Parameters_{start\text{-}up(C)}$, Explored Map (EM), Energy Capacity (EC) and Speed (S).
- *List_tabou(IdUSV, Cell(x; y))*: This list is set to store cells that are already cleaned by the USV. It has the identifier of the $USV(Id_{USV})$ that has already cleaned the cell where the coordinate of the latter is the pair (x, y) in the grid.

(d) $General_{crd}$ launches the execution of AG and each USV_{cz} of the swarm plans to move from the above mentioned steps of GA to the dirty zone.

2. *Phase 2: Cleaning*

This phase allows the swarm of USV to move into the dirty zone and clean it. The maritime space is discretized in square grid (G). Each box of the G can contain an object (a dirty cell/a clean cell). Objects are points in a numerical space with M dimensions. These objects to be partitioned are positioned. USV_{cz} move on G and perceive a detection region Rs of s * s boxes in their neighborhood. These USV_{cz} can clean the dirty cell type objects. This phase is started when the swarm arrives at its dirty zone. $General_{crd}$ stops running the GA. $Leader_{cz}$ of the swarm launches Algorithm 2 (Figs. 7 and 8) to generate a graph (cycle) on the grid using the final list of positions/cities. Also, it starts the recording the characteristics of himself and its followers. The followers of the swarm follow the algorithm so that they can position well around the dirty zone, and they share their positions with each other. Each USV_{cz} determines its detection region and it selects the cell to clean afterwards.

```
Algorithm 2
List_USV_z[U]: table which contains the ID of each used USV
to clean an zone. List_threshold_positionDZ[X, Y]: table which
contains the coordinate of dirty cells. Direction (r, l):
boolean variable, where r, l are right and left direction
respectively,
Nodirty(List_threshold_positionDZ[X,Y], Direction) : procedure
to know the next position of USV,
Outputs:
Position_USV[] : list which contains the new position of
USV (u),
Position_CellU[]: list which contains the dirty cell to be
cleaned by the USV(u).
begin
for all u ∈ U do
Calculate identify the non-dirty cells by Nodirty
(List_threshold_positionDZ[x, y], Direction);
calculate the new Position_USV [List_USV_z[u]] ←
Direction;
calculate Position_CellU[List_USV_z[u]] ←
List_threshold_positionDZ[x, y];
end for
```

Fig. 7. USVs Placement to clean an zone.

```
Nodirty(GR, List_threshold_positionDZ[x, y], Direction)
for all k = 1toK do
  % determine if the cell List_threshold_positionDZ[x-k, y] is not
  dirty and if it is not the position of a USV
  if List_threshold_positionDZ[x-k, y]=0 ∧ List_threshold_positionDZ[x-k, y]
  ≠ Position_USV_u then
    if x − k > 0 ∨ y > 0 then
      % place this position in right direction
      r ← List_threshold_positionDZ[x-k, y];
    end if
  end if
  % determine if the cell List_threshold_positionDZ[x, y-k] is not dirty
  and if it is not the position of a USV
  if List_threshold_positionDZ[x, y-k]=0 ∧ List_threshold_positionDZ[x, y-k]
  ≠ Position_USV_u then
    if y > 0 ∨ y − k > 0 then
      % place this position in left direction
      l ← List_threshold_positionDZ[x, y-k] ;
    end if
  end if
end for
```

Fig. 8. Nodirty function.

3. *Phase 3: Finish Cleaning*

This phase consists of identifying the end of cleaning and collecting the current characteristics of each USV_{cz}. It consists of three following cases.

Case 1: The cleaning task is completed.

a. When the USV_{cz} finishes its cleaning task, it informs $Leader_{cz}$ by its mission completeness by sending a message.

b. Upon reception of the message, $Leader_{cz}$ informs the USV_{cz} to return to the base of life.
c. When the USV_{cz} arrives at the base of life, it informs $Leader_{cz}$ by sending a message. When $Leader_{cz}$ receives this message, it stops saving the characteristics of the USV_{cz}.
d. $Leader_{cz}$ sends $List_{characteristics}$ of USV_{cz} to its Sup_{mr} for recording.
e. Sup_{mr} sends this $List_{characteristics}$ to the $General_{crd}$.

Case 2: The cleaning task is not complete and $Energy_{capCZ}$ is average. In this case, we illustrated two situations.

– *First situation is illustrated as follows*:

a. At some point, the energy capacity ($Energy_{capCZ}$) of USV_{cz} is greater than a $Threshold_{average\_energycap}$, ie its energy capacity is average.
b. USV_{cz} sends its $List_{energy\_usv}(Id_{USV\ CZ}, Energy_{capCZ})$ to its $Leader_{cz}$ to inform it that its energy is in a average level. After saving this list, $Leader_{cz}(USV_{cz}\_discharge)$ sends this $List_{energyusv}$ and $Parameters_{cleaning}(PosUSV\ g, Cell_{cz}, Id_{region}, Id_{zone}, Position_{zone}, Path_{zone}, Id_{UAVMR})$ to its Sup_{mr}.
c. After a deadline of wait, if Sup_{mr} does not receive another message (a detection message of $USV_{cz}(free)$ in the same swarm by $Leader_{cz}(USV_{cz}\_discharge)$) before exceeding this deadline, then it passes to the second situation. Nb: $Leader_{cz}$ plays two roles at the same time in this situation; once $Leader_{cz}(USV_{cz}\_discharge)$ and once $Leader_{cz}(USV_{cz}\_free)$.
d. Otherwise, $Leader_{cz}(USV_{cz}\_discharge)$ sends their $List_{characteristics\_usvs}$ with their $List_{energy\_usv}$ to its Sup_{mr}. This latter processes the received message and parses these lists. Then, it begins to select the competent USV that has a higher energy capacity. If it does not find any USV that can complete the task of $USV_{cz}(discharge)$ then it moves to the second situation.
e. Sup_{mr} informs its $Leader_{cz}(USV_{cz}\_free)$ by the competent USV with the complete cleanup parameters (Sup_{mr} changes the new position of USV_{cz}(free), the new cell that $USV_{cz}(free)$ will clean it with the same parameters $Id_{region}, Id_{zone}, Position_{zone}, Path_{zone}, Id_{UAV\ mr}$).
f. $Leader_{cz}(USV_{cz}\_free)$ returns the received message to the competent USV who presents USV_{cz}(free). $Leader_{cz}(USV_{cz}\_discharge)$ or $Leader_{cz}(USV_{cz}\_free)$ waits for the energy capacity message of $USV_{cz}(discharge)$) becomes low to start $USV_{cz}(free)$.

– *Second situation is illustrated as follows*: If there is not a USV_{cz}(free) in the same field, then $Sup_{mr}(USV_{cz}\_discharge)$ sends a message to $General_{crd}$ to inform it of need for a USV with cleanup settings. $General_{crd}$ processes the message and then fills the path box of the zone. At the same time, it chooses a USV_{cz} according to the cleaning parameters in the base of life. Also, it is waiting for the signal to launch the USV_{cz}(prepared). $Leader_{cz}(USV_{cz}\_discharge)$ waits for the message of USV_{cz}(discharge) capacity becomes low to launch $USV_{cz}(free)$.

Case 3: The cleaning task is not complete and $Energy_{capCZ}$ *is low*. This case is realized when the energy of USV_{cz}*(discharge)* becomes low, that is, the energetic capacity ($Energy_{capCZ}$) of USV_{cz} is greater than a $Threshold_{low\_energycap}$ (following the situations mentioned above). USV_{cz}*(discharge)* sends this information to its supervisor through its leader. Upon receipt of this information, Sup_{mr}*(*USV_{cz} *_discharge)* sends a message to $Leader_{cz}$*(*USV_{cz}*_free)* to launch USV_{cz}*(free)* and to prepare it by the $Parameters_{cleaning}$ and change $List_{characteristics\_usvs}$ at the same time (the case of the first situation). Sup_{mr}*(*USV_{cz}*_discharge)* sends a message to $General_{crd}$ to launch USV_{cz}*(prepared)* and to prepare it by the $Parameters_{cleaning}$ and change the $List_{characteristics\_usvs}$ at the same time (the case of the seconde situation). In addition, Sup_{mr}*(*USV_{cz}*_discharge)* informs the USV_{cz}*(discharge)* through its $Leader_{cz}$*(*USV_{cz}*_discharge)* to return to the base of life.

3.5 Logical Formalization of the Proposal

Conceptual Model for Planning. A conceptual model is a simple theoretical device for describing the main elements of a problem [16]. Most of the planning approaches described in [17] rely on a general model, which is common to other areas of computer science, the model of state-transition systems (also called discrete-event systems) [16, 17].

Example 1. This example shows the state transition systems defined for three domains: UAV Monitoring (UAV-M), Swarm(USV)-Cleaning (SUSV-C) and USV (substitute)-Cleaning (USVS-C). These domains are presented by the following:

- *"UAV-Monitoring" domain:* Fig. 9 shows a state transition system for a region involving two locations, a dirty zone, a base of life (a boat for example) and a UAV_{mr}. The set of states is s0, s1, s2, s3 and the set of actions is stayinbase, flaputbase, move1^startmonitor, move2^end-monitor, discover, undiscover. The arc (s2, s3) is marked by the action "discover", the arc (s1, s0) with the action "stay-inbase", and so on.
- *"Swarm(USV)-Cleaning" domain:* The system of region (Fig. 10) involves three locations, two dirty zones, a base of life, object of the crane type for picking up, putting down and releasing unnamed vehicles. Here, the set of states s0, s1, s2, s3 and the set of actions is take, put, start-clean, end-clean, move1, move2. The arc (s0, s1) with "put", (s3, s2) with the action "end-clean", and so on.
- *"USV(substitute)-Cleaning" domain*: The system of Fig. 11 is similar to the previous system but the difference is the use of a replacement cleaning vehicle (in free/prepared state), which will replace the USV in discharge state. The set of states is s0, s1, s2, s3, s4, s5 with takeD^move2, putD^putD, moveD1^start-clean, moveD2^endclean, moveD1^move2, moveD2^move1, move1, move2. The arc (s1, s2) is marked by "moveD2^end-clean", (s3, s4) with "move1", and so.

A Running Example "UAV-M", "SUSV-C" & "USVS-C". The planning procedures and techniques are illustrated on three scenarios, namely, the scenario UAV-Monitoring, Swarm(USV)-Cleaning and USV(Substitute)-Cleaning. A version of the used domains are defined using ten finite sets of constant symbols:

- A set of regions *(R_r)* where a region r (r = 1,...R, R > 0) is a space of two sub-spaces (atmosphere and maritime) which contains a base of life, one or several dirty zones, one or several vehicles and locations.
- A set of locations for each region *(L_{lr})* where a location l (l = 1,...L, L > 0) is part of r.
- A set of base of life *(B_{br})* where a base of life b (b = 1,...B, B > = 0) can be found in each r.
- A set of dirty zones *(Z_{zl})* where a z (z = 1,...Z, Z > = 0) is located in one or more l.
- A set of monitoring vehicle (UAV_{mr}) where each monitoring vehicle m (m = 1,...M, M > 0) may be in a certain location of a r. It may move to another adjacent l of same r.
- A set of cleaning vehicle swarm (USV_{cz}) where each cleaning vehicle swarm c (c = 1,...C, C > 0) may be in a certain z or l, it can move to another adjacent z or adjacent l either in the same r or in another adjacent r.
- A set of cleaning vehicle in discharge state ($USVD_{cz}$) where each $USVD_{cz}$ can be transformed into an discharge state when performing its task.
- A set of cleaning vehicle in free state ($USVF_{cz}$) where each $USVF_{cz}$ can replace the *USV_{cz}(discharge)*.
- A set of cleaning vehicle in prepared state ($USVP_{cz}$) where each $USVP_{cz}$ can replace the *USV_{cz}(discharge)*.
- A set of cranes *(C_{nb})* where a crane-type object n (n = 1,...N, N > = 0) is found in b.

The topology of the UAV-M, SUSV-C and USVS-C domains is noted using the instances of predicates:

- adjacent(E, E′): Localization E = $\{L_{lr}, R_r\}$ is adjacent to the localization E = $\{L'_{lr}, R'_r\}$.
- belong (C, L): Crane C = $\{C_{nb}\}$ belongs to location L.
- belong (B, R): Base of life B = $\{B_{br}\}$ belongs to region R = $\{R_r\}$.

The current configuration of the domains is denoted using instances of the following predicates, which represents the relationships that changes over time:

- at(NAME, E): Vehicle NAME = NAME1, NAME2, namelly, NAME1 = $\{UAV_{mr}, USV_{cz}\}$, NAME2 = $\{USVF_{cz}, USVP_{cz}, USVD_{cz}, SUSV_{cz}\}$ i currently at E.
- occupied (L): Location L is already occupied by NAME2 vehicles, Z = $\{Z_{zl}\}$ and B.
- start-clean(NAME, Z): Vehicles NAME1: USV_{cz}, NAME2: $USVF_{cz}$ and NAME2: $USVP_{cz}$ is ready for cleaning in the dirty zone z.

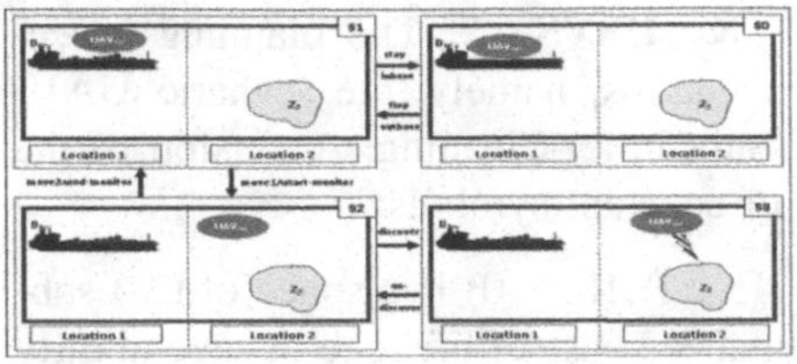

Fig. 9. A state transition system "UAVM domain".

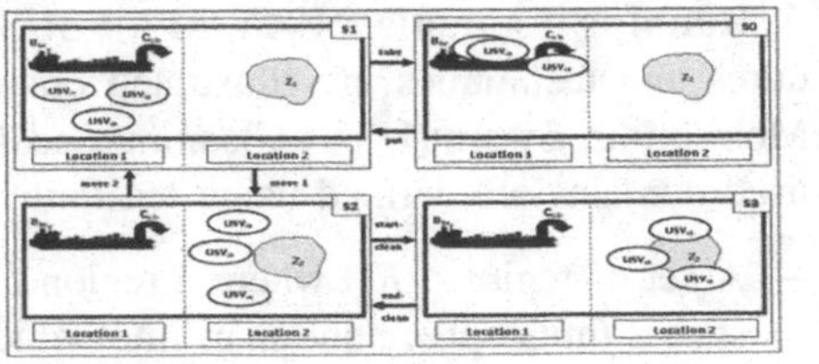

Fig. 10. A state transition system "SUSVC domain".

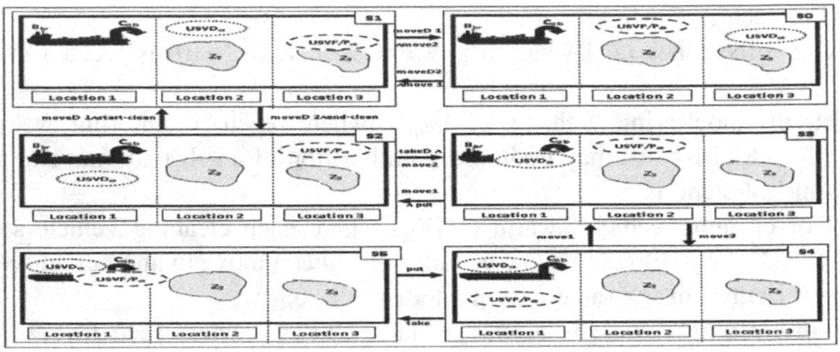

Fig. 11. A state transition system for USVS-C domain.

- end-clean(NAME, Z): NAME1: USV_{cz}, NAME2: $USVF_{cz}$ is finished the cleaningZ.
- start-monitor(NAME, R): NAME1: UAV_{mr} is ready to start monitoring in R.
- end-monitor(NAME, R): NAME1: UAV_{mr} has now completed monitoring of its R.
- replace(Name, Name): NAME2: $USVF_{cz}$ replaces the NAME2: $USVD_{cz}$.
- notreplace(Name, Name): NAME2:$USVF_{cz}$ does not replace NAME2: $USVD_{cz}$.
- discover(NAME, Z): Z is discovered by the drone NAME1: UAV_{mr}.
- undiscover(NAME, Z): R does not contain a Z or the Z is not discovered by the NAME1: UAV_{mr}.
- supervise(NAME, NAME): NAME1: UAV_{mr} supervises the swarm NAME2: $SUSV_{cz}$.
- notsupervise(NAME, NAME): NAME1: UAV_{mr} does not supervise the swarm NAME2: $SUSV_{cz}$.
- holding(C, NAME): C is currently holding vehicle NAME1: USV_{cz} and NOM2.
- stayinbase(NAME, B): NAME1: UAV_{mr} is currently in the base of life B.
- flapoutbase(NAME, B): NAME1: UAV_{mr} is outside its base B. It flies in the air.

We can enumerate the possible actions in the three domains UAV-M, SUSV-C and USVS-C:

- start-clean(NAME, Z), end-clean(NAME, Z), start-monitor(NAME, R), end-monitor(NAME, R), discover(NAME, R, Z), undiscover(NAME, R, Z), stayinbase(NAME, B) and flapoutbase(NAME, B) (description cited below).

- move(NAME, E1, E2): Vehicle NAME moves from a location E1 = $\{L_{lr}, R_r\}$ to some adjacent and unoccupied location E2 = $\{L'_{lr}, R'_{r}\}$.
- moveD(NAME, E1, E2): NAME2: $USVD_{cz}$ moves from a location E1 = $\{L_{lr}, R_r\}$ to some adjacent and unoccupied location E2 = $\{L'_{lr}, R'_{r}\}$.
- take(NAME, C, R): Empty C takes the NAME1: USV_{cz}, NAME2: $USVF_{cz}$ and NAME2: $USVP_{cz}$ in the same R.
- takeD(NAME, C, R): Empty C takes a NAME2: $USVD_{cz}$ in the same R.

NB: The action "moveD" is executed in parallel with the action "start-clean", "end-clean" and "move". The actions "takeD" and "putD" are executed in parallel with the action "move".

Representations for Classical Planning. There are three different ways to represent classical planning problems [16], namely: (i) Set-theoretic Representation, (ii) Classical Representation, (iii) State-Variable Representation. In this work, we focus on the second representation (classical) to apply its planning on our proposition.

Example 2. This example illustrates a classical representation of the scenario "SUSV-C domain combined with USVS-C domain" which are described in *Example 1*. We suppose that:

<u>*For the SUSV-C Domain Combined with USVS-C Domain:*</u> We want to illustrate an SUSVC domain with USVS-C domain which there is a region (R_1), three locations (L_{11}, L_{21}, L_{31}), eleven cleaning vehicles ($USVR_{34}$, USV_{22}, USV_{52}, $USVR_{62}$, USV_{74}, USV_{44}, USV_{25}, USV_{55}, USV_{85}, USV_{66}, USV_{76}), a crane (C_{11}) and five dirty zones (Z_{12}, Z_{42}, Z_{63}, Z_{22}, Z_{53}). The set of constant symbols is (R_1, L_{11}, L_{21}, L_{31}, $USVR_{34}$, USV_{22}, USV_{52}, $USVR_{62}$, USV_{74}, USV_{44}, USV_{25}, USV_{55}, USV_{85}, USV_{66}, USV_{76}). A state of this domain is illustrated in Fig. 12.

S3 = belong(C_{11}, L_{11}), holding(C_{11}, $USVR_{34}$), end-clean(USV_{52}, USV_{62}, USV_{22}, Z_{42}), start-clean(USV_{74}, USV_{44}, Z_{42}), start-clean(USV_{55}, USV_{25}, USV_{85}, Z_{53}), start-clean(USV_{66}, USV_{67}, Z_{63}), at(USV_{22}, USV_{52}, USV_{62}, L_{21}), replace($USVR_{34}$, $USVD_{44}$), replace($USVR_{34}$, $USVD_{44}$), at(USV_{34}, L_{11}), adjacent(L_{11}, L_{21}), adjacent(L_{21}, L_{11}), adjacent(L_{21}, L_{31}), adjacent(L_{31}, L_{21}), occupied(L_{11}), occupied(L_{21}), occupied(L_{31}).

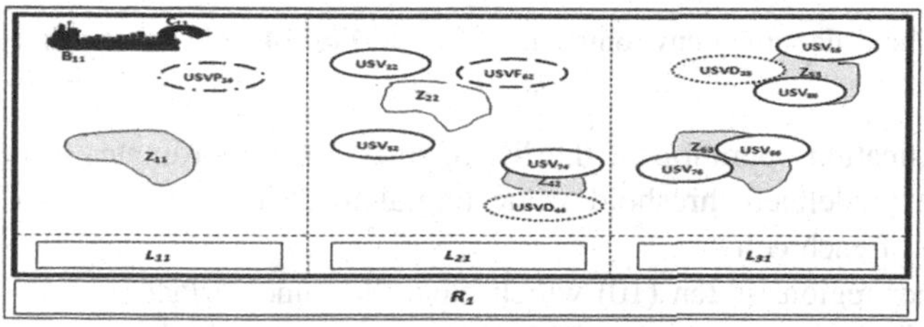

Fig. 12. SUSV-C planning domain combined with USVSC planning domain.

4 Simulation Example

We propose an illustrative example to simulate the functioning of our proposal. To highlight the contributions of our approach, we focus on the following metrics: The average of total energy consumption of USV_{cz} swarm, average of total energy consumption of USV_{cz} selected. We have developed two approaches "Direct Displacement to Another Zone (DDZ)" and "Indirect Displacement to Another Zone (IDZ)". DDZ is when a free USV_{cz} moves directly from its dirty zone to another dirty zone. On the other hand, the IDZ approach is when the free USV_{cz} moves to the base of life first and then to another dirty zone. In order to study the behavior of our approach "DDZ" and to analyze its obtained results on simulation, we compare them to the approach "IDZ" (Passage through the base of life). A series of simulations were realized according to different parameters. Before starting the experiments, we describe our virtual environment and its discretization.

4.1 Discretization of the Virtual Environment

We took a simplistic example of our virtual environment, shown in Fig. 13. We assume that the maritime space of a region "Region1" includes two dirty zones "Zone1 and Zone2", and the central unit consists of a base of life. After defining randomly the degrees of dirt in the example, and launching the discretization step on this environment (Region1), we obtain the following matrix (see Fig. 14). The black cells represent Zone1 and Zone2. For our environment, we make some assumptions:

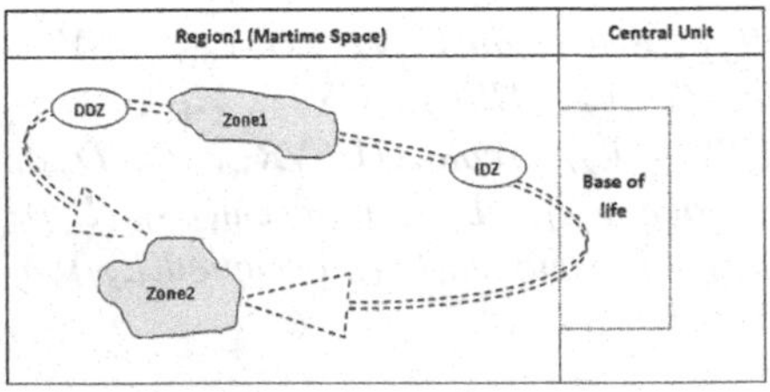

Fig. 13. A simplistic virtual environment.

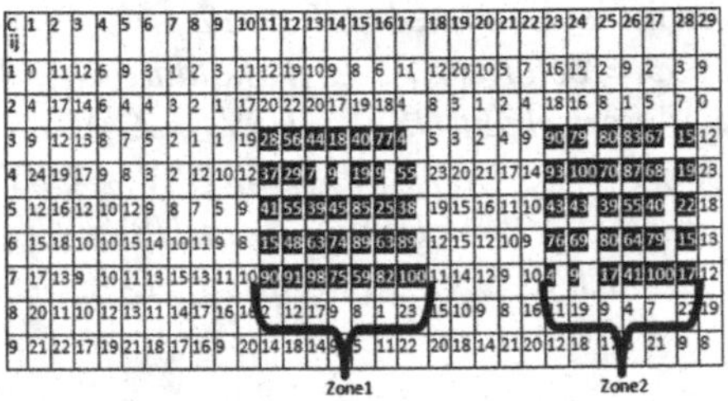

Fig. 14. Matrix of discretization.

- The classification operation of the list of dirty zone coordinates is done by comparing the predefined threshold value (equal to 25% of degree of dirt) with the $Degree_{cell}$ of each cell.
- The number regions is ten (10) which have the same surface.
- Solid (hard) obstacles are unavailable that can prevent vehicles from navigating and plan its trajectory in this environment.
- Total Energy Consumption (TEC) = Cleaning Energy Consumption (CEC) + Displacement Energy Consumption (DEC) to the zone or base of life + Displacement Energy Consumption between Cells (DECC).

– It is assumed that the needed energy for the USV to turn left/right from its position to another position is 0.03%.
– It is assumed that the needed energy to clean a black cell (strong dirt) is 0.9%, for a average-strong dirt cell is 0.5% and for an small dirt cell is 0.2%.

Result of the Average of Total Energy Consumption of USV_{cz} Swarm. Figure 15 presents the simulation to evaluate the USV_{cz} swarm behavior with respect to the average of the total energy consumption of USV_{cz}. We note that the average TEC curve that contains the displacement energy (DEC and DECC) and cleaning energy (CEC) of zone 1 is below the curve of average TEC which includes DEC and DECC and CEC of zone 2. Also, the average TEC decreases when the number of USV_{cz} containing the swarm increases. We conclude that each swarm consumes less energy in carrying out its tasks from the base of life to zone 1 compared to its second operation from the base of life to zone 2. Therefore our proposal could reduce the total energy consumption for each swarm which is estimated by an average gain of +3.76%.

Result of the Average of Total Energy Consumption of USV_{cz} Selected. Figure 16 shows the simulation to evaluate the behavior of USV_{cz} swarm with respect to the average of total energy consumption (TEC) for both approaches (DDZ and IDZ). The selected USV_{cz} is the vehicle that has consumed less energy compared to its neighbors in the swarm in zone 1. So, it is selected to complete the cleaning of zone 2 (the case of a free USV_{cz}). We notice that our approach (the gray curve) is below the curve of the IDZ approach (black curve). So, our proposal is better compared than the IDZ approach and it allows a significant reduction in the TEC of selected USV_{cz}) with an average gain of +2.87%.

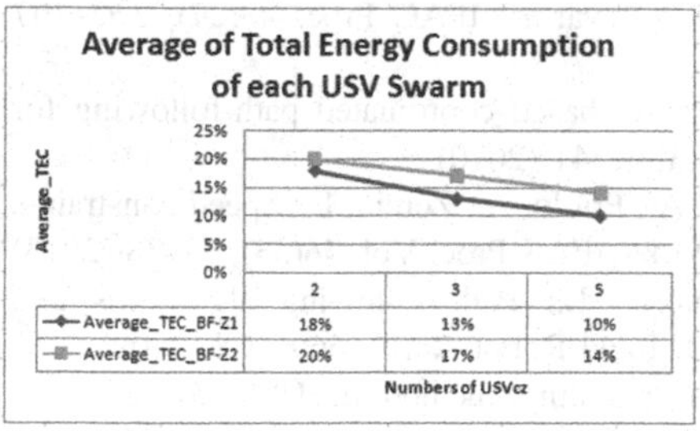

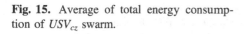

Fig. 15. Average of total energy consumption of USV_{cz} swarm.

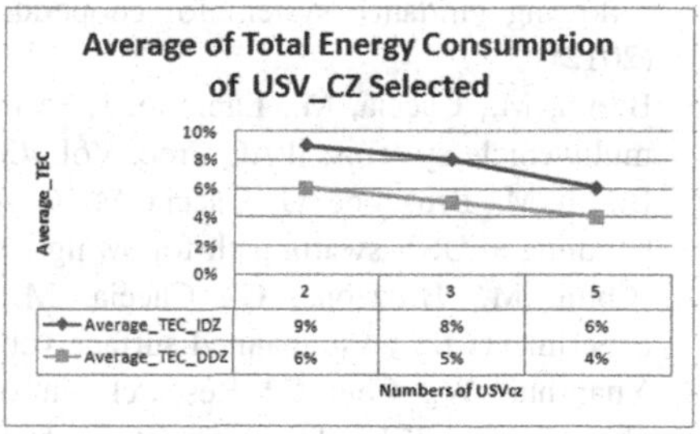

Fig. 16. Average of total energy consumption of USV_{cz} selected.

5 Conclusion

In this article, we presented a hierarchical decision-making system for the coopertive air-sea systems. This hybrid approach represents a cooperation of several semi-autonomous unmanned vehicles (UAV_{mr} and swarm of USV_{cz}) and their coordination by a general coordinator. Moreover, it allows to manage the path-planning and the fault tolerance problem for the USV swarms while performing their tasks. This proposed approach uses an UAV_{mr} for each region and a USV_{cz} swarm to clean dirty zones. In the monitoring step, we choose the color as the degree of dirt so that the UAV_{mr} can detect the dirty zones and initiate the cleaning. In the cleaning step, the swarm of USV_{cz} plans its movement through a proposed genetic algorithm to reach its dirty zone. Each swarm of USV_{cz} measures it's the energy amount by an energy threshold and then sends its information to its leader. The latter sends back information to its supervisor (Sup_{mr}) to make decisions to find a free USV. Sup_{mr} launches this USV_{cz}*(free)* which will replace the USV_{cz}*(discharge)* when the energy amount of the latter is low. We formalize our proposal by means of a classical representation. Measured metrics are the average of total energy consumption of USV_{cz} swarm and average of total energy consumption of selected USV_{cz}. The realized approach shows that the proposed method provides encouraging results. In this work, we did not deal with the problem of the central unit failure. Thus, as a future work, we plan to develop our hybrid approach to take into account this problem.

References

1. Seo, J.H., Ko, E.S., Kim, Y.H.: Performance comparison of GPUs with a genetic algorithm based on CUDA. Adv. Sci. Technol. Lett. **65**, 36–40 (2014)
2. Lin, C.C., Chen, K.C., Hsiao, P.Y., Chuang, W.J.: Motion planning of swarm robots using potential-based genetic algorithm. Int. J. Innov. Comput. Inf. Control. ICIC **9**, 305–318 (2013)
3. Bibuli, M., Bruzzone, G., Caccia, M., Ranieri, A., Zereik, E.: Multi-vehicle cooperative path-following guidance system for diver operation support. IFAC **48**(16), 75–80 (2015)
4. Bibuli, M., Gasparri, A., Priolo, A., Bruzzone, G., Caccia, M.: Virtual target based path-following guidance system for cooperative USV swarms. IFAC Proc. **45**(27), 362–367 (2012)
5. Bibuli, M., Caccia, M., Lapierre, L.: Virtual target based coordinated path-following for multivehicle systems. IFAC Proc. Vol. **43**(20), 336–341 (2010)
6. Bibuli, M., Bruzzone, G., Caccia, M., Gasparri, A., Priolo, A., Zereik, E.: Speed constraints handling in USV swarm path-following frameworks. IFAC Proc. Vol. **46**(33), 73–78 (2013)
7. Bibuli, M., Bruzzone, G., Caccia, M., Lapierre, L.: Path following algorithms and experiments for an unmanned surface vehicle. J. Field Robot. **26**(8), 669–688 (2009)
8. Yuan-hui, W., Cen, C.: Research on optimal planning method of USV for complex obstacles. In: IEEE International Conference on Mechatronics and Automation (ICMA), pp. 2507–2511 (2016)
9. de Carvalho Santos, V., Osório, F.S., Toledo, C.F., Otero, F.E., Johnson, C.G.: Exploratory path planning using the Max-min ant system algorithm. In: 2016 IEEE Congress on Evolutionary Computation (CEC), pp. 4229–4235 (2016)
10. de Carvalho Santos, V., Toledo, C.F.M., Osório, F.S.: An exploratory path planning method based on genetic algorithm for autonomous mobile robots. In: 2015 IEEE Congress on Evolutionary Computation (CEC), pp. 62–69 (2015)
11. Controllab Products B.V.: What is HIL Simulation? University of Twente by young and enthusiastic control engineers (1995). http://www.hil-simulation.com/home/hil-simulation.html. Accessed 20 Dec 2017

12. Kloetzer, M., Belta, C.: Temporal logic planning and control of robotic swarms by hierarchical abstractions. IEEE Trans. Robot. **23**(2), 320–330 (2007)
13. Goldberg, D.E., Holland, J.H.: Genetic algorithms and machine learning. Mach. Learn. **3**(2), 95–99 (1988)
14. Guéret, C., Monmarché, N., Slimane, M.: Ants can play music. In: Dorigo, M., Birattari, M., Blum, C., Gambardella, L.M., Mondada, F., Stützle, T. (eds.) ANTS 2004. LNCS, vol. 3172, pp. 310–317. Springer, Heidelberg (2004). https://doi.org/10.1007/978-3-540-28646-2_29
15. Labroche, N., Monmarché, N., Venturini, G.: Visual clustering with artificial ants colonies. In: Palade, V., Howlett, R.J., Jain, L. (eds.) KES 2003. LNCS (LNAI), vol. 2773, pp. 332–338. Springer, Heidelberg (2003). https://doi.org/10.1007/978-3-540-45224-9_47
16. Ghallab, M., Nau, D., Traverso, P.: Automated Planning: Theory and Practice. Elsevier, Amsterdam (2004)
17. Ghallab, M., Nau, D., Traverso, P.: Automated Planning and Acting. Cambridge University Press, Cambridge (2016). PDF of manuscript posted by permission of Cambridge University Press
18. Yakoubi, M.A., Laskri, M.T.: The path planning of cleaner robot for coverage region using genetic algorithms. J. Innov. Digit. Ecosyst. **3**(1), 37–43 (2016)
19. Bella, S., Belbachir, A., Belalem, G.: A centralized autonomous system of cooperation for UAVs-monitoring and USVs-cleaning. Int. J. Softw. Innov. (IJSI) **6**(2), 50–76 (2018)
20. Connell, D., La, H.M.: Dynamic path planning and replanning for mobile robots using RRT. In: 2017 IEEE International Conference on Systems, Man, and Cybernetics (SMC), pp. 1429–1434. IEEE, October 2017

Detecting Suspicious Transactions in IoT Blockchains for Smart Living Spaces

Mayra Samaniego and Ralph Deters(✉)

Department of Computer Science, University of Saskatchewan, Saskatoon, Canada
mayra.samaniego@usask.ca, deters@cs.usask.ca

Abstract. The idea of connecting physical things and cyber components to enable new and richer interactions is a key component in any smart space concept. One of the central challenges in these new smart spaces is the access control of data, services and things. In recent years, Distributed Ledger technology (DLT) like Blockchain Technology (BCT), emerged as the most promising solution for decentralized access management. Using capability-based access control, access to data/services/things is achieved by transferring tokens between the accounts of a distributed ledger. Managing how the access tokens are transferred is, of course, a major challenge. Within the IoT space, smart contracts are at the center of most of the proposals for DLT/BCT networks targeting access control. The main problem in using smart contracts as a means for checking if and what access token can be transferred from one account to another is their immutability and accessibility. Smart contracts and chain code are by design meant to be immutable since they represent a binding *contract between parties*. In addition, they need to be accessible since they are to be executed on many nodes. This allows an attacker to study them and design the attack in a manner that passes the rules of the smart contract/chain code. This paper focuses on the use of metadata as a more effective means to prevent attackers from gaining access to data/services/things in a smart living space.

Keywords: Smart space · Smart living · Distributed ledger · Blockchain · Malicious transactions · Fraud · Awareness · Metadata

1 Smart Living Spaces

The idea of connecting physical things and cyber components to enable new and richer interactions is a key component in any smart living space environment. Connecting things, services and data in the context of smart living offers a variety of advances for inhabitants, communities and vendors.

- Inhabitants
  Smart living spaces e.g. smart homes, allow the inhabitants of the space to better customize and control their environment. This, in turn, allows personalization of living spaces to offer richer, more engaging experiences and to ensure that the needs of the inhabitants are met.

É. Renault et al. (Eds.): MLN 2018, LNCS 11407, pp. 364–377, 2019.
https://doi.org/10.1007/978-3-030-19945-6_26

- Communities
  Particularly in urban settings, the negative impacts of uncontrolled resource consumption are increasingly obvious. Using the concept of smart living, in which the physical environment is linked to cyber-components, it becomes possible to enforce policies on the way inhabitants interact with their space. A city might enforce policies regarding the use of electricity e.g. incentivize the use of air conditioning by only preventing excessive use with the exception of people with legitimate reasons e.g. elderly.
- Vendors
  Producers of equipment targeting living spaces are able to continuously optimize their products by updating settings. A fridge may be re-configured to meet new environmental norms with respect to energy efficiency. In addition, a connected living space will open new marketing strategies vendors may choose to enable functionality upgrades as a fee for service.

Smart Living Spaces is a particular type of the Internet of Things (IoT). IoT aims to integrate physical devices (aka "things") via Internet protocols. Enabling (physical) devices to form loosely-coupled connections with each other and Internet services/resources allows for new and richer interactions between devices, internet enabled services/resources and users.

The idea of enhancing the capabilities of physical devices by connecting them to remotely hosted software components is also at the center of the cyber-physical system (CPS) paradigm. However, unlike IoT that supports loose-coupling between physical and cyber components, CPS favours "... the tight conjoining of and coordination between computational and physical resources. ..." [1]. Gubbi et al. [2] differentiate between a thing-centric and cloud-centric view in the IoT space. The cloud-centric-view [1, 2] moves the focus away from the interactions with things towards the services and applications that process large data streams. The cloud-centric view is primarily concerned with the requirement to scale e,g, handle/manage large numbers of connected devices and is thus the underlying concept of Smart City approaches. The thing-centric view on the other side, emphasizes the enhancement of things and richer user experiences when engaging them. Smart objects [3] or enchanted devices [4] are the most prominent examples in this category. The thing-centric-view seems particularly well suited for the Smart Living scenario in which various IoT devices communicate with each other and backend-services to provide the best possible experiences for inhabitants of a given space. This, in turn, leads to the challenge of how to control access to the IoT devices and maybe more importantly, how to manage access to the user data.

The remainder of the paper is structured as follows. Section 2 focuses on the access control issues in IoT. Section 3 introduces a smart living example. This is followed by a section on stolen keys and how to detect transactions that have been signed by an attacker. In Sect. 5 the results of a performance evaluation are presented. The paper concludes with a summary and discussion of future work.

2 Beyond Devices -Access Control in IoT

IBM's ADEPT system [5] that is built on Bluemix is one of the earliest examples of using blockchain within IoT. However, it is focused primarily the tracking and updating of appliances. More recent examples of using BCT/DLT to handle access control in IoT scenarios are [9–14]. In the context of Smart Living, it is important to note that the IoT devices connect to backend services and user data. This, in turn, leads to the question of how to manage access to the very heterogeneous set of devices, services and data. By exposing devices, services and data as RESTful web services it becomes possible to establish a uniform interface and thus simplify the access management. Interaction in REST follows the HTTP request/response pattern in which each side assumes that all information is contained in the request and response. Robinson [7] identifies within his 4-level web maturity model two patterns that have become popular within REST, namely CRUD and Hypermedia. The most widely used REST pattern is CRUD [6–8] (Create Read Update Delete) approach that follows a basic data-centric style. The CRUD commands are mapped to the HTTP verbs POST, GET, PUT/PATCH and DELETE. Unlike the data-centric CRUD pattern, the hypermedia pattern focuses on embedding links into the responses of requests. By offering links, the server provides the client with possible state-transitions and ways to obtain further information. Therefore the client drives and maintains the application state by selecting from the past and current choices presented in the server responses.

To avoid centralization and thus performance and security bottlenecks, a decentralized access control is generally seen as the best approach for REST web services. Blockchain Technology (BCT) (aka Blockchain) is a subclass of Distributed Ledger Technology networks, that deploy consensus algorithms that create blocks of verified transactions. The term Distributed Ledger Technology (DLT) refers to a class of P2P networks that are fully decentralized and physically distributed transaction-processing systems. Due to their trustless design, these P2P transaction-processing networks are designed to be highly resilient towards malicious/faulty nodes and messages. As indicated in the name, the ledger (record of all transactions) is distributed across the autonomous nodes that form the DLT network. While the specifics of the distributed ledger structure varies within DLT systems (e.g. use of sharding), the distributed ledger is expected to contain a record of the accounts and the approved transactions. Re-Executing the approved transactions allows a node to calculate the current state of each account. Compared to DLTs, BCTs networks tend to face higher latency and lower throughput when processing transactions. However, they remain popular due to the larger number of deployments and simpler consensus algorithms (typically PoW).

Access control in DLT/BCT networks can be achieved by implementing a capability-based access control [41]. In such a system, tokens are used to represent certain access privileges on a service/device/data. By transferring a token the access privileges associated with this token can be transferred between participants. For example, the owner of a device can issue a token granting read access to the device and then transferring it to a software agent. The software agent can use this token to access the device or maybe transfer the token to another agent. Typically smart contracts are used to govern if and how tokens can be transferred and used (Fig. 2).

Fig. 1. Samsung Smart Home vision [42]

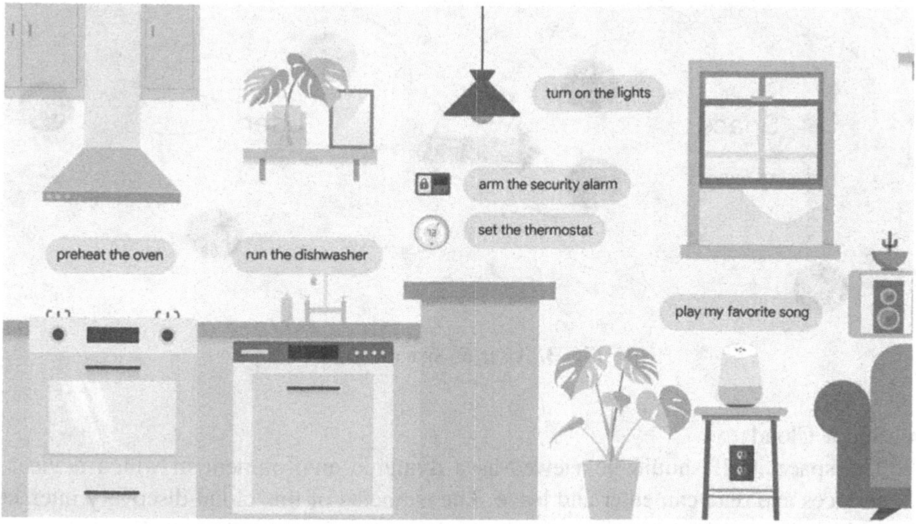

Fig. 2. Google Smart Home [43]

3 Smart Living Scenario

Typical smart living space scenarios like Samsung's and Google's "Smart Home" centers on the presence of

- Stationary Sensors in the living space (e.g. apartment/house)
  - e.g. light, power, temperature, moisture, proximity, occupancy touch sensors etc.
- Connected appliances
  - fridge, washing machine, dishwasher,
- Entertainment devices
  - Video, Audio
- Connected climate controls
  - Heating, AC
- Handheld devices
  - Smartphone, Tablets

While focussing on a fixed set of "things" they tend to marginalize the importance of software components e.g. agents that act on behalf of the user, and tend to downplay the importance of backend services as sources and sinks for user/device data.

Unlike Fig. 1, which displays the users interacting via a smartphone, Fig. 3 views the smart living space as the interaction of highly dynamic user and space clouds.

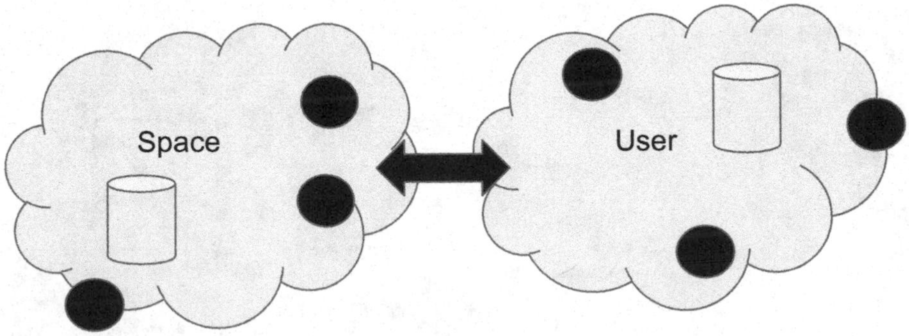

Fig. 3. User & Space Cloud

- Space Cloud
  The space itself should be viewed as a dynamic environment in which devices, services and data can enter and leave. The elements of this cloud discovery interact with each other in dynamic ways (e.g. perpetual updates, the establishment of new functional dependencies).
- User Cloud
  The user itself is also surrounded by a myriad of devices, services and data and thus forms a cloud.

We view the smart living space as a highly dynamic digital ecosystem consisting of devices, services/agents and data of the space itself and the inhabitants. Consequently, the functional dependencies of all the elements in this digital ecosystem are subject to constant change. It is also important to note that the elements of the digital ecosystem are not all governed by a single authority. This, in turn, leads to serious challenges with respect to access control since a large number of heterogeneous elements in the ecosystem tend to be very difficult to monitor. Understanding if the behaviour of an element in this digital ecosystem is a result of a malware attack or emerging behaviour is very difficult to determine.

4 Stolen Keys and DLT/BCT

Security [32, 33, 35] like stealing signing keys is a major issue in the deployment of DLT/BCT systems. Once attackers obtain the signing keys, they obtain full control over the account. In a capability-based access control system, this could allow attackers to issue their own access tokens and thus gain control over devices/services/data. CHAINALYIS [15] & BlockSci [16] are examples of a fast-growing set of tools to monitor account activity. Unfortunately, these tools are of limited use in scenarios where software components exchange access tokens.

The most common approach for dealing with irregularities within DLT/BCT systems is the use of Smart Contracts. Ethereum [17] is generally credited with introducing the concept of smart contracts in the area of BCT/DLT. While it was possible to define in Forth [18] code snippets that are linked to the transaction in Bitcoin [19], Ethereum generalized and extended the notion of using code in BCT networks. Enabling code to be executed within a distributed ledger increases its use significant. The rapid development of countless decentralized applications (DAPPs) demonstrates the use of this concept. The main problem in using smart contracts as a means for checking if and what access token can be transferred is their immutability and accessibility.

Smart contracts and chain code are by design meant to be immutable since they represent a binding *contract between parties*. This, in turn, means that once they are established they can't be changed and thus an attacker can analyze them to identify vulnerabilities and to plan attacks accordingly. Even if a flaw is later discovered by the parties using the contract there is no accepted way of patching it. Trying to hide smart contracts is impossible since they need to be visible to all nodes of the DLT/BCT network. Consequently, we don't consider smart contracts as effective tools for access-management in IoT.

5 Detecting Suspicious Transactions with Metadata

Once attackers obtain the signing keys they are able to produce formally correct transactions. Similar to credit card transactions it is possible to detect false/suspicious transactions if a model of standard behaviour is used. Suspicious transactions within DLT/BCT systems are transactions that meet all formal criteria of a valid transaction but don't conform to expected behaviour. In other words, a suspicious transaction is an anomaly. Detection of anomalies [15], especially in the field of financial transactions [20], is a well-established field. To detect anomalies and thus patterns that indicate suspicious transactions a context-dependent analysis of transactions is needed. The key challenge in the detection of suspicious/abnormal transactions is in the development of a classifier with sufficient accuracy. This, in turn, leads to the question of how to obtain the "knowledge" that can be used for the classification. Generally speaking "a posteriori" (past experiences) or "a priori" (existing knowledge) can be used. In other words, machine learning [21] or existing theories/model [22] can be used. The key challenge in this context-dependent classification is the availability of detailed data e.g. user behaviour. This, in turn, leads to the dilemma of trading privacy for security [34].

Instead of revealing data and establishing a context to evaluate transactions, it is also possible to define policies and associating them with accounts as *metadata*. The policies that govern the classification of transactions can be presented as workflows. While this appears to be similar to the concept of chain code or smart contracts [23–25] there is a major difference. Smart Contracts represent agreements between parties that don't change. Policies that define when to consider a transaction suspicious and when to accept it are subject to change. By changing policies frequently it becomes harder for an attacker to successfully inject fake transactions. Generally speaking, the use of metadata accounts allows us to introduce to pre-processing step to check the transaction should be sent to an analysis component for further investigation. Please note that a more detailed description of the algorithms and data structures is available in [26]. In our implementation metadata accounts consist of

- version number
- verification workflow
- event processor

The version number is a counter is like an ETag that defines the version of the metadata account. The verification workflow defines the process of verification/validation of transactions. Upon receiving a transaction, each node will perform a basic check if the transaction is well-formed and properly signed. The next step is to identify the accounts involved in the transaction. Upon identifying the accounts the related metadata is retrieved and if necessary the new verification workflow is downloaded and executed If the workflow creates an {approved, transaction} event, the transaction is considered non-suspicious and can be processed by the node. However, if the workflow creates an {suspicious, transaction, explanation} event, the transaction is not processed and the transaction flagged and will not be processed any further by the node. The messages produced by the workflow are always sent to the specified event processors [37–39]. While approval events are used to track the processing of the transaction in the network, suspicious transaction events indicate a potential problem. Upon receiving a suspicious transaction event, the event processor can decide to still approve the transaction by changing the verification workflow.

6 Evaluation

Given that a DLT/BCT network needs to exceed several hundred nodes for any realistic evaluation, simulation emerges as the only viable approach. However, evaluation of DLT/BCT networks is a still-evolving area of research [27–30]. The absence of workloads, benchmarks, and simulation approaches makes the evaluation of DLT/BCT networks challenging.

The following evaluation results are obtained by use of the Google Compute Engine [31] VMs. The experiments use 5 VMs with 30 GB Ram and 8 cores each. Each VM is located in a different zone (us-east (2), us-west (2) us-central (1)) to ensure "realistic" network traffic. In each experiment, one VM is used as a load generator, one as the host of the event processors and the rest as host for the accounts and smart contracts. Each experimental run consists of an initial warm-up period, and a measurement phase. Once an experiment has been concluded all VMs are restarted. Please note that the Erlang language is used. Erlang is a highly efficient concurrency-oriented programming language. Communication between the Erlang machines was ensured by the use of a message daemon that uses TCP connections and in different runs by simply running on each machine a standard UDP server. The following results are based on using the TCP message daemon. The event workflows are as Erlang processes that receive messages from the simulated nodes. Using a load generator 1000 messages are sent into the network to test the ability to respond to suspicious transactions. The delays between the messages were fluctuating between 0 (no added delay) and 100 ms.

The network connections between the DLT/BCT nodes were configured as

A. Fully Meshed
   Each node sees every other node. Nodes send messages in one step to all other nodes.
   or
B. Kademlia
   The kademlia algorithm [40] is used. Nodes see only a subset of nodes and messages are routed.

Experiment 1 - Evaluation of Overhead on Nodes
As expected the costs for evaluating a transaction is very low and thus presents negligible costs in both network configurations. The main difference between the network configurations is the time it takes to inform all nodes. Kademlia is slower since messages are routed and require log(number of nodes) hops.

Experiment 2 - Performance of Event Processor
The second set of experiments focussed on the performance of the event processor under various loads. In these experiments, the event processors were exposed to traffic from the verification workflows (Figs. 4, 5, 6, 7 and 8).

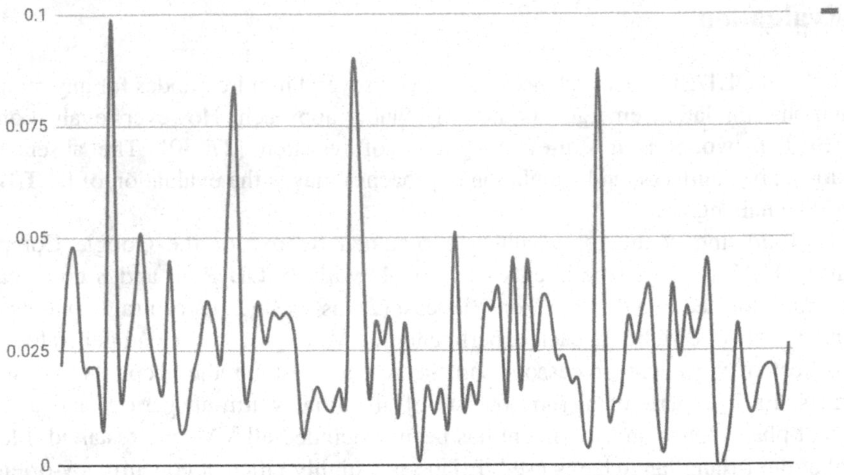

Fig. 4. Single node exposed to Suspicious Traffic contacting Event Processor

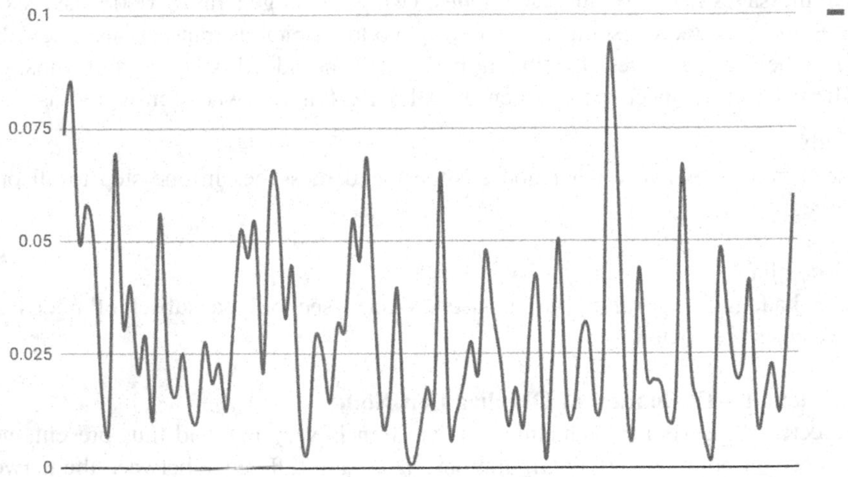

Fig. 5. 10 Nodes exposed to Suspicious Traffic contacting Event Processor

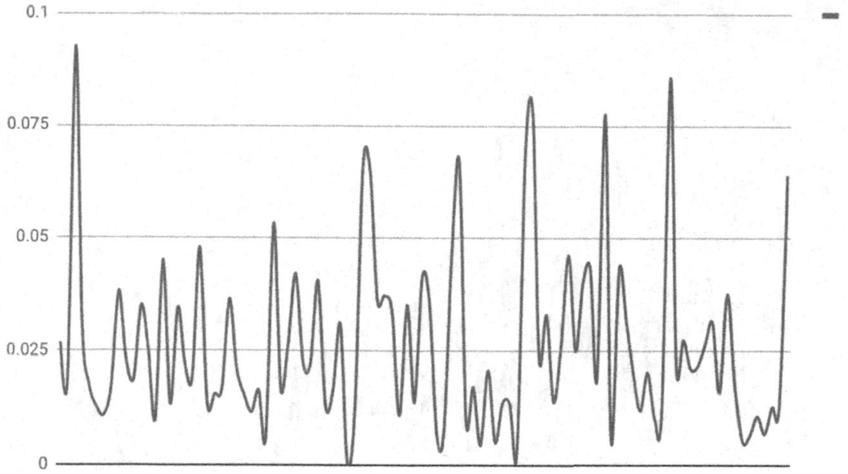

Fig. 6. 100 Nodes exposed to Suspicious Traffic contacting Event Processor

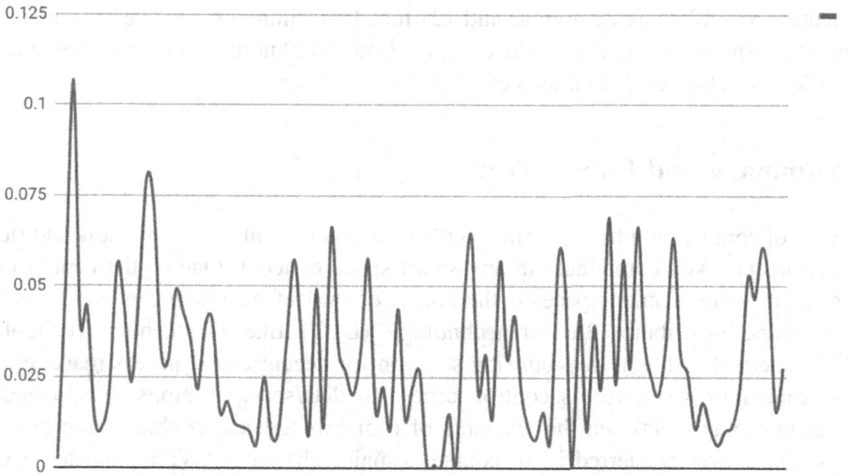

Fig. 7. 1000 Nodes exposed to Suspicious Traffic contacting Event Processor

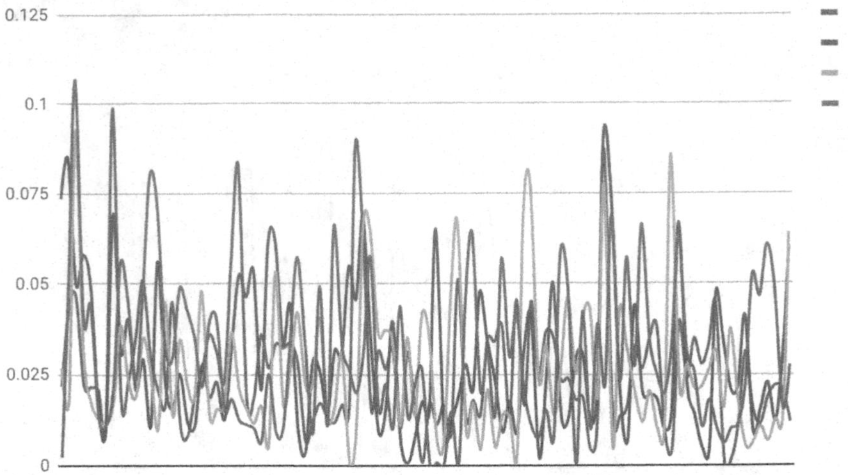

Fig. 8. 1–1000 Node(s) exposed to Suspicious Traffic contacting Event Processor

The results of the two experiments show that (1) the overhead for executing a verification workflow is negligible and (2) that large numbers of events sent to the event processor don't seem to produce high delays (1000 nodes each reporting at a rate of 1–100 ms delay result in a latency of 0.1 s.

7 Summary and Future Work

The idea of connecting physical things and cyber components to enable new and richer interactions is a key component in any smart space concept. One of the central challenges in these new smart spaces is the access control of data, services and things. In recent years, Distributed Ledger technology (DLT) like Blockchain Technology (BCT), emerged as the most promising solution for decentralized access management. Using capability-based access control, access to data/services/things is achieved by transferring tokens between the accounts of a distributed ledger. Managing how the access tokens are transferred is, of course, a major challenge. Within the IoT space, smart contracts are at the center of most of the proposals for DLT/BCT networks targeting access control. The main problem in using smart contracts as a means for checking if and what access token can be transferred from one account to another is their immutability and accessibility. Smart contracts and chain code are by design meant to be immutable since they represent a binding *contract between parties*. In addition, they need to be accessible since they are to be executed on many nodes. This allows an attacker to study them and design the attack in a manner that passes the rules of the smart contract/chain code. This paper focuses on the use of metadata as a more effective means to prevent attackers from gaining access to data/services/things in a smart living space. The simulation showed that there is a minimal overhead indicating that this is a viable option. Compared with the use of smart contracts and chain code, the use of metadata and event processors offers better protection since the metadata can be changed easily at runtime.

One of the most interesting aspects of using workflows for verification and event processors for dealing with potentially harmful transactions is the possibility to enable different evaluation and processing workflows based on account on perceived threat status. This, in turn, leads to the possibility of heterogeneous DLT in which different accounts can enforce different workflows for evaluating and processing transactions. We also consider the use of hierarchical control loops very interesting and plan to investigate this further. Using the extensive research on autonomic computing [36] and linking it to accounts within DLT networks promises interesting future research.

References

1. NSF: Cyber-physical systems (CPS) (2010). https://www.nsf.gov/pubs/2010/nsf10515/nsf10515.htm
2. Gubbi, J., Buyya, R., Marusic, S., Palaniswami, M.: Internet of Things (IoT): a vision, architectural elements, and future directions. Future Gener. Comput. Syst. **29**(7), 1645–1660 (2013)
3. Sanchez, T., Ranasinghe, D.C., Harrison, M., McFarlane, D.: Adding sense to the Internet of Things—an architecture framework for smart object systems. Pers. Ubiquit. Comput. **16**(3), 291–308 (2012)
4. Rose, D.: Enchanted Objects: Design, Human Desire, and the Internet of Things. Simon and Schuster, New York (2014)
5. Panikkar, S., Nair, S., Brody, P., Pureswaran, V.: ADEPT: An IoT Practitioner Perspective (2015). http://static1.squarespace.com/static/55f73743e4b051cfcc0b02cf/55f73e5ee4b09b2bff5b2eca/55f73e72e4b09b2bff5b3267/1442266738638/IBM-ADEPT-Practictioner-Perspective-Pre-Publication-Draft-7-Jan-2015.pdf?format=original
6. Fielding, R.: Architectural Styles and the Design of Network-based Software Architectures. Dissertation University of Irvine, vol. 7 (2000)
7. Robinson, L.: Richardson Maturity Model. https://martinfowler.com/articles/richardsonMaturityModel.html
8. CRUD: “Create Read, Update and Delete”. http://en.wikipedia.org/wiki/Create,_read,_update_and_delete
9. Samaniego, M., Deters, R.: Blockchain as a Service for IoT. In: 2016 IEEE International Conference on Internet of Things (iThings) and IEEE Green Computing and Communications (GreenCom) and IEEE Cyber, Physical and Social Computing (CPSCom) and IEEE Smart Data (SmartData), pp. 433–436. IEEE (2016)
10. Samaniego, M., Deters, R.: Using blockchain to push software-defined IoT components onto edge hosts. In: Proceedings of the International Conference on Big Data and Advanced Wireless Technologies, p. 58. ACM (2016)
11. Samaniego, M., Deters, R.: Management and Internet of Things. Procedia Comput. Sci. **94**, 137–143 (2016)
12. Samaniego, M., Deters, R.: Internet of Smart Things-IoST: using Blockchain and CLIPS to make things autonomous. In: 2017 IEEE International Conference on Cognitive Computing (ICCC), pp. 9–16. IEEE (2017)
13. Samaniego, M., Deters, R.: Virtual resources & blockchain for configuration management in IoT. J. Ubiquit. Syst. Pervasive Netw. **9**(2), 01–13 (2017)

14. Samaniego, M., Deters, R.: Zero-trust hierarchical management in IoT. In: 2018 IEEE International Congress on Internet of Things (ICIOT), pp. 88–95. IEEE (2018)
15. Chandola, V., Banerjee, A., Kumar, V.: Anomaly detection: a survey. ACM Comput. Surv. (CSUR) **41**(3), 15 (2009)
16. Chainalysis. https://www.chainalysis.com/
17. A Next-Generation Smart Contract and Decentralized Application Platform. https://github.com/ethereum/wiki/wiki/White-Paper
18. Forth. https://www.forth.com/forth/
19. Nakamoto, S.: Bitcoin: A Peer-to-Peer Electronic Cash System. https://bitcoin.org/bitcoin.pdf
20. Debreceny, R.S., Gray, G.L.: Data mining journal entries for fraud detection: an exploratory study. Int. J. Account. Inf. Syst. **11**(3), 157–181 (2010)
21. Lane, T., Brodley, C.E.: An application of machine learning to anomaly detection. In: Proceedings of the 20th National Information Systems Security Conference, Baltimore, USA, vol. 377, pp. 366–380 (1997)
22. Valdes, A., Skinner, K.: Adaptive, model-based monitoring for cyber attack detection. In: Debar, H., Mé, L., Wu, S.F. (eds.) RAID 2000. LNCS, vol. 1907, pp. 80–93. Springer, Heidelberg (2000). https://doi.org/10.1007/3-540-39945-3_6
23. Christidis, K., Devetsikiotis, M.: Blockchains and smart contracts for the Internet of Things. IEEE Access **4**, 2292–2303 (2016)
24. Ouaddah, A., Abou Elkalam, A., Ait Ouahman, A.: FairAccess: a new Blockchain-based access control framework for the Internet of Things. Secur. Commun. Netw. **9**(18), 5943–5964 (2016)
25. Ouaddah, A., Abou Elkalam, A., Ait Ouahman, A.: Towards a novel privacy-preserving access control model based on blockchain technology in IoT. In: Rocha, Á., Serrhini, M., Felgueiras, M.C. (eds.) Europe and MENA Cooperation Advances in Information and Communication Technologies, pp. 523–533. Springer, Cham (2017). https://doi.org/10.1007/978-3-319-46568-5_53
26. Deters, R.: How to detect and contain suspicious transactions in distributed ledgers. In: Qiu, M. (ed.) SmartBlock 2018. LNCS, vol. 11373, pp. 149–158. Springer, Cham (2018). https://doi.org/10.1007/978-3-030-05764-0_16
27. Rouhani, S., Deters, R.: Performance analysis of Ethereum transactions in private blockchain. In: 2017 8th IEEE International Conference on Software Engineering and Service Science (ICSESS), pp. 70–74 (2017)
28. Rouhani, S., Butterworth, L., Dimmond, A.D., Humphery, D.G., Deters, R.: MediChainTM: a secure decentralized medical data asset management system. In: 2018 IEEE Conference on Internet of Things, Green Computing and Communications, Cyber, Physical and Social Computing, Smart Data, Blockchain, Computer and Information Technology, Congress on Cybermatics, pp. 1533–1538 (2018)
29. Rouhani, S., Pourheidari, V., Deters, R.: Physical access control management system based on permissioned blockchain. In: 2018 IEEE Conference on Internet of Things, Green Computing and Communications, Cyber, Physical and Social Computing, Smart Data, Blockchain, Computer and Information Technology, Congress on Cybermatics, pp. 1078–1083 (2018)
30. Pourheidari, V., Rouhani, S., Deters, R.: A case study of execution of untrusted business process on permissioned blockchain. In: 2018 IEEE Conference on Internet of Things, Green Computing and Communications, Cyber, Physical and Social Computing, Smart Data, Blockchain, Computer and Information Technology, Congress on Cybermatics, pp. 1588–1594 (2018)
31. Compute Engine. https://cloud.google.com/compute/

32. Eyal, I., Sirer, E.G.: Majority is not enough: Bitcoin mining is vulnerable. Commun. ACM **61**(7), 95–102 (2018)
33. Heilman, E., Kendler, A., Zohar, A., Goldberg, S.: Eclipse attacks on Bitcoin's peer-to-peer network. In: USENIX Security Symposium, pp. 129–144 (2015)
34. Dorri, A., Kanhere, S.S., Jurdak, R., Gauravaram, P.: Blockchain for IoT security and privacy: the case study of a smart home. In: 2017 IEEE International Conference on Pervasive Computing and Communications Workshops (PerCom Workshops), pp. 618–623. IEEE (2017)
35. Man-in-the-middle attacks on wallets. http://news.bitcoin.com/ledger-addresses-man-in-the-middle-attack-that-threatens-millions-of-hardware-wallets/
36. Huebscher, M.C., McCann, J.A.: A survey of autonomic computing—degrees, models, and applications. ACM Comput. Surv. (CSUR) **40**(3), 7 (2008)
37. Nygate, Y.A.: Event correlation using rule and object based techniques. In: Sethi, A.S., Raynaud, Y., Faure-Vincent, F. (eds.) International Symposium on Integrated Network Management IV. ITIFIP, pp. 278–289. Springer, Boston, MA (1995). https://doi.org/10.1007/978-0-387-34890-2_25
38. Buchmann, A., Koldehofe, B.: Complex event processing. IT-Information Technology Methoden und innovative Anwendungen der Informatik und Informationstechnik **51**(5), 241–242 (2009)
39. Deters, R.: Case-based diagnosis of multiple faults. In: Veloso, M., Aamodt, A. (eds.) ICCBR 1995. LNCS, vol. 1010, pp. 411–420. Springer, Heidelberg (1995). https://doi.org/10.1007/3-540-60598-3_37
40. Maymounkov, P., Mazières, D.: Kademlia: a peer-to-peer information system based on the XOR metric. In: Druschel, P., Kaashoek, F., Rowstron, A. (eds.) IPTPS 2002. LNCS, vol. 2429, pp. 53–65. Springer, Heidelberg (2002). https://doi.org/10.1007/3-540-45748-8_5
41. Mahalle, P.N., Anggorojati, B., Prasad, N.R., Prasad, R.: Identity authentication and capability-based access control (IACAC) for the Internet of Things. J. Cyber Secur. Mobility **1**(4), 309–348 (2013)
42. https://www.samsung.com/sg/smarthome/
43. https://cdn.mos.cms.futurecdn.net/exdTX6QGDyg8hDausRwzhJ-970-80.jpg

Intelligent ERP Based Multi Agent Systems and Cloud Computing

Nardjes Bouchemal[1,2(✉)] and Naila Bouchemal[3]

[1] University Center of Mila, Mila, Algeria
n.bouchemal.dz@ieee.org
[2] LIRE Laboratory of Constantine2, Constantine, Algeria
[3] Ecole Centrale d'Electronique Paris, Paris, France
naila.bouchemal@ece.com

Abstract. The use of agents makes ERP systems more intelligent. They provide capabilities to independently interact with their surrounding environment and performing autonomous actions while cooperating with other systems Regarding the cloud computing, its use is almost obvious. It provides access to full-function applications at a rational cost without a substantial upfront for hardware and software spending. In this paper, we propose an Intelligent ERP system based on new technologies: agents and cloud computing.

This combination makes ERP users (managers, customers, employees, etc.) expect data to be made available to them via the most widely used communication mobile device.

Keywords: Multi agent system · Intelligent ERP · Cloud computing · JADE-Leap

1 Introduction

ERP (Enterprise Resource Planning) as a colossal recording system is disappearing and giving way to intelligent ERP or iERP, where actions and information are important. Intelligent ERP combines artificial intelligence and machine learning, blockchain technology, big data, user experience, cloud and Internet of Things.

Against this background of increasing connectivity, these new intelligent systems particularly include three major characteristics: intelligence, autonomy and real-time behavior (Fig. 1).

Furthermore, iERP solution generates results from the data it collects and augments them through technologies such as machine learning and advanced analytics.

Multi-Agent Systems (MAS) are one of the major technological paradigms that are used to implement such systems.

According to the definition of these intelligent entities, software agents provide capabilities to independently interact with their surrounding environment and performing autonomous actions while cooperating with other systems [5].

Moreover, in the context of ubiquitous environment; agents are designed to be embedded in mobile devices having limited capabilities of battery lifetime, data and applications storage.

É. Renault et al. (Eds.): MLN 2018, LNCS 11407, pp. 378–386, 2019.
https://doi.org/10.1007/978-3-030-19945-6_27

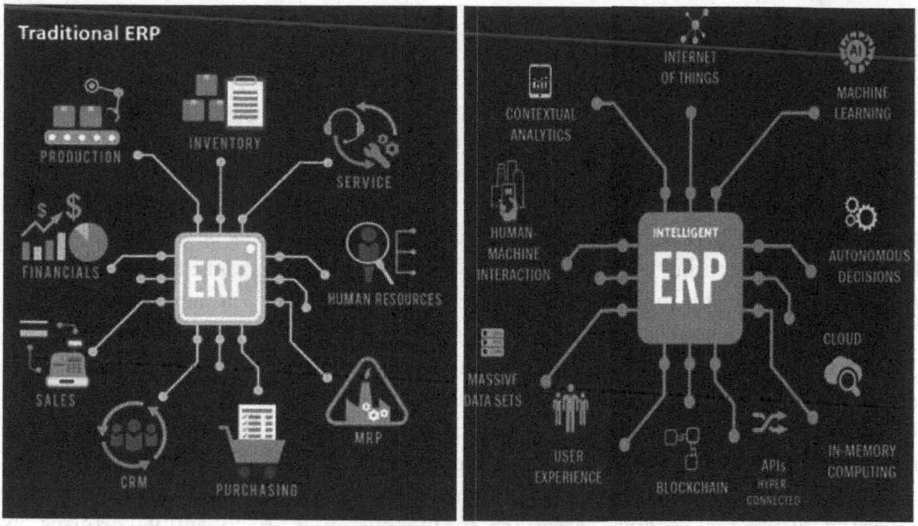

Fig. 1. Traditional ERP vs Intelligent ERP

In the other hand, companies rely on ERP systems to help aggregate and organize data that is spread across each of its independent departments. Traditional ERP solutions are often housed within a company's own server infrastructure and require updating and servicing to stay current. Cloud ERP, as the name suggests, is based in the cloud much like SaaS (Software as a Service). Unlike traditional ERP software, Cloud-based ERP relies on the cloud rather than proprietary server infrastructure to help companies share information across departments. Cloud ERP software integrates some or all of the essential functions to running a business, e.g. accounting, inventory and order management, human resources, customer relationship management (CRM), etc. – into one complete ERP system.

Essential to Cloud ERP systems (and on-premises ERP systems) is a shared database that supports multiple functions used by different business units. That allows employees and managers in different divisions and also customers to access and rely on the same information for their specific needs. And with Cloud ERP this is accomplished without requiring an extensive on-premises server presence [18].

The rest of paper is organized as follow: Section two gives an overview on agents and cloud in iERP systems. Section three presents the contribution based on the combination of two concepts: Agent Technology and Cloud Computing. In section four, we give implementation results using JAD-LEAP.

2 Related Work

Over the time, ERP adapts and introduces more and more new technologies, as well as other areas such as health; education. In this work, we are interested to agents and cloud computing technologies.

Agents in ERP

Intelligent agents have the ability to make decisions defined by their inherent properties, which are: reactivity, pro-activity and sociability. These properties are intended to enable the agent to meet the objectives for which they were designed, following rules of behavior that enable them to communicate with their environment [13]. These principles have tempted many researchers in the field of ERP.

Dice the beginning of 90, Pan and Tenenbaum [12] propose an Intelligent Agent based framework to integrate people and computer systems in large geographically dispersed manufacturing enterprise. In the framework, each agent supports a clearly discernible task or job function, and interacts with each other via messages through a shared, distributed knowledge base.

The framework is built by dividing complex enterprise operations into a collection of elementary tasks, and each task is modeled in cognitive terms and entrusted to an IA for execution. Their preliminary experimental results indicate that agent-based systems are a practical way to integrate a complex enterprise.

Lin et al. [9] define an enterprise to be a collection of business entities working toward delivering a product or service to the customer. Each entity performs its independent processes organized in a hierarchy. A high-level process consists of a set of low-level processes. The authors propose a multiagent information system (MAIS) for the supply chain network for capturing both the structure and the processes of an enterprise.

Jennings et al. [5] suggest a community of negotiating agents, and propose the Advanced Decision Environment for Process Tasks (ADEPT) for conceptualizing, designing, and implementing business process management systems.

In the framework, various functions of the business process are delegated to a number of autonomous problem-solving agents, and these agents interact and negotiate with each other in order to coordinate their actions and to buy in the services they require.

Another example of multi-agent system related to ERP is presented by Kehagias [8]. He mentioned the connection with the ERP database designing a new functional layer. This layer corresponds to a MAS architecture which control sales orders and give a recommendation on how could be fulfilled. This system consists of five agents: the client orders agent, the recommender agent, the customer profile identifier agent, the inventory profile identifier agent and the supplier profile identifier agent.

Jingrong [6] refers to the use of a MAS in order to identify the standards used by the flows of processes regarding an already implemented ERP system, since the use of agents allows that each functional unit are defined as an agent than could interacts with other agents in order to identify the exchange of knowledge and how they should handle coordination of functional units.

Authors in [14] research how exactly does ERP systems interact using Application Programing Interface or "API" specified by a credit card clearing house. They present an implementation of the AIM API for merchant credit card processing is seamless, transparent, and unobtrusive to users. They use agent technology to have maximum control over the merchants credit card processing.

This solution is innovative, scalable, and compatible with today's modern trends for optimization and effectiveness. However, authors don't introduce the cloud computing concept to have a permanent connection between customers and the enterprise.

The goal of the authors in [15] is to enable a autonomous, reactive production by means of intelligent communication in autonomous systems. They propose an incorporation of multi-agent systems and OPC Unified Architecture that bears the potential to use adaptive behavior embedded into small devices on the field of a production more effectively by combining the knowledge and capabilities of agents with information from the enterprise layers such as ERP or other high level systems.

The proposed system in [16] is based on ubiquitous agents to assist ERP costumer surrounded by many mobile devices, any-time, any-where and on any-device. The idea is to connect the ERP database with the mobile client, any-where and any-time, through an ubiquitous agent that can migrate from one device to another and can work offline.

They design an ubiquitous agent embedded in one of the customer devices (ex. smart phone) and knows all other devices surrounding him (ex. Tablet, laptop, smart TV, etc.). The proposed agent is mobile, can migrate on all customer devices and takes advantages from all of them especially in case of energy insufficiency, limited storage space or failure calculating capabilities.

Cloud ERP

Cloud-based computing (also called Software as a Service, or SaaS) allows users access to software applications that run on shared computing resources through Internet.

Cloud ERP is Software as a Service that allows users to access Enterprise Resource Planning (ERP) software over the Internet. Cloud ERP generally has much lower upfront costs and offer an access to business-critical applications at any time from any location.

Comprise the work of [11] on the use of cloud by the review of development of Low cost ERP solution to Indian industries on Mobile, using latest technologies such as Mobile computing, SaaS, Cloud Computing.

Authors in [10] define: Cloud ERP is an approach to enterprise resource planning (ERP) that makes use of cloud computing platforms and services to provide a business with more flexible business process transformation.

Cacciagrano [2] focus their interest on the use of ontology, web semantic and ubiquitous computing. They propose architecture based on an expandable 'Business Intelligence 2.0' Enterprise Resource Planning (ERP) prototype, with the aim to lead Public Administration toward Business Intelligence and information maturity. Distributed and heterogeneous knowledge through semantic-driven GUI (Graphical User Interface)-based components is integrated on a common semantic knowledge model and embedded in a Cloud-based middleware.

Authors in [14], investigate the phenomenon of cloud computing and its importance for the operation of ERP systems. They argue that the phenomenon of cloud computing could lead to a decisive change in the way business software is deployed in companies. Their reference framework contains three levels (IaaS, PaaS, SaaS) and clarifies the meaning of public, private and hybrid clouds. The three levels of cloud computing and their impact on ERP systems operation are discussed. From the literature they identify areas for future research and propose a research agenda.

Discussion
Proposed solutions based on agents and multi agents systems introduce ubiquitous devices and agents concept, but the problem is when these devices are offline (ex. problem of connection or insufficient battery), not available or does not have enough resources. Furthermore, ERP systems demand a very large storage space considering the huge amount of data.

In the other hand, presented solutions based cloud computing, offer very talented research in the cloud computing applications as it is the result of the collective profits of mobile and cloud technologies. But these solutions present several disadvantages such as the introduction of AI techniques and the consideration of limited devices resources. There are no complete systems which offer complete solutions for ERP systems.

3 Intelligent ERP Based Multi Agent Systems and Cloud Computing

Today users around the world are increasingly connected to world of digital information through smartphones and other mobile devices. That means that ERP employees and managers access to vital information when it is still fresh and opportunity to customers to observe events as they occur. Our idea is to facilitate access and management of information to iERP by combining agents technology and cloud computing, such as:

- Cloud technology: for storage of a large amount of information.
- Agent technology: for intelligent management and connectivity between ERP users.

The system is composed essentially of three levels: The ERP modules, ERP users (employees, managers and customers) and the cloud.

At the ERP level, we create Intelligent **ERP Manager Agent** to assure the communication between ERP modules the users through the cloud.

In user's level, we design **User Agent** for assistance, information collection and communication through the cloud.

We use cloud computing concept to assure data storage and mobile communications. For that, we create **Cloud Manager Agent** to manage messages, information and to guarantee the communication between **ERP Manager Agent** and **User Agent** (Fig. 2).

ERP Manager Agent provides all services of the underlying ERP system and enables direct interaction with different modules. When new orders are placed in the ERP system, the agent is automatically notified about the order by an event. It immediately proceeds to do the updates in the cloud by contacting **Cloud Manager Agent.**

User Agent, on its part, notifies its owner on one of his surrounding devices. The proposed agent is mobile, can migrate on all users devices and takes advantages from all of them especially in case of energy insufficiency, limited storage space or failure calculating capabilities. So, even if a device is off or has no connection, **User Agent** can complete its tasks.

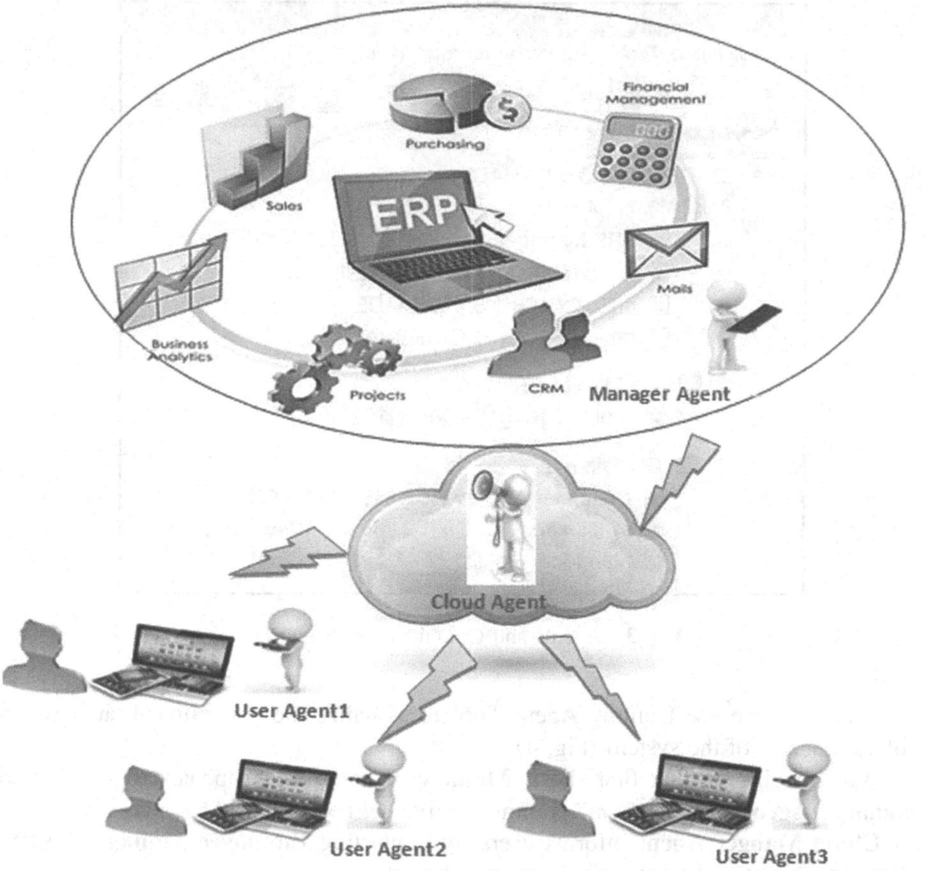

Fig. 2. ERP based Cloud and Multi Agent Systems

4 Implementation Using JADE-LEAP

JADE is a set of Java classes that allow a developer to build a FIPA-compliant multi-agent system quite easily, [3]. An add-on to JADE, called LEAP, was released. It replaces some parts of the JADE kernel, creating a modified environment called JADE-LEAP [4], which allows the implementation of agents in mobile devices with limited resources.

Note that in JADE-LEAP each agent is represented by a container. The rest of programming (behaviors, messages, etc.) is identical to JADE.

To implement our proposal, we create three containers: Main Container hosts the ERP Manger Agent, Container1 simulates users and hosts three agents (Customer Agent, Employee Agent and Manager Agent). Container 2 simulates the cloud and hosts Cloud Manager Agent (Fig. 3).

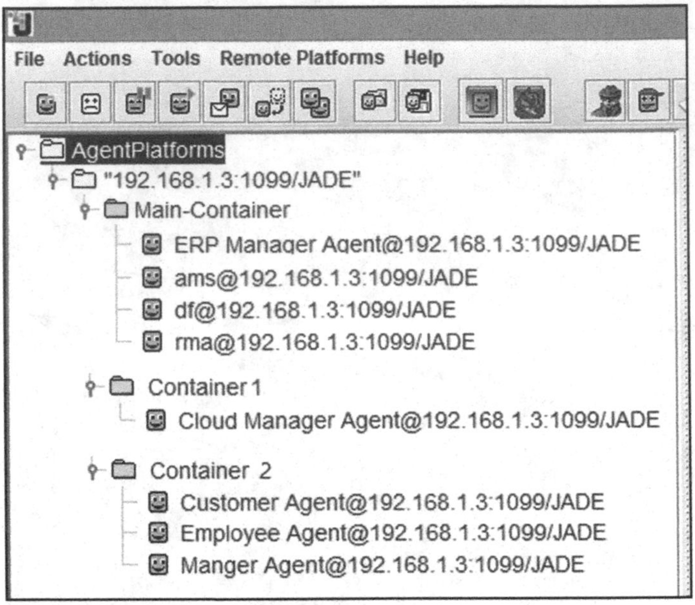

Fig. 3. Agents and Containers of the system

After that, we use Dummy Agent Tool to visualize the communication between different agents of the system (Fig. 4).

We can observe that first, **ERP Manager Agent** sends updates to the **Cloud Manager Agent** using "*inform*" communicative act from FIPA ACL massages.

Cloud Manger Agent informs users by contacting **Employee Manager Agent**, **Manager Agent** and **Customer Manager Agent**.

Then, the **Employee Manager Agent** requests the **Cloud Manager Agent** by sending "*request*" communicative act.

Cloud Manager Agent transmits the request to the **ERP Manager Agent** which performs treatments and then responds.

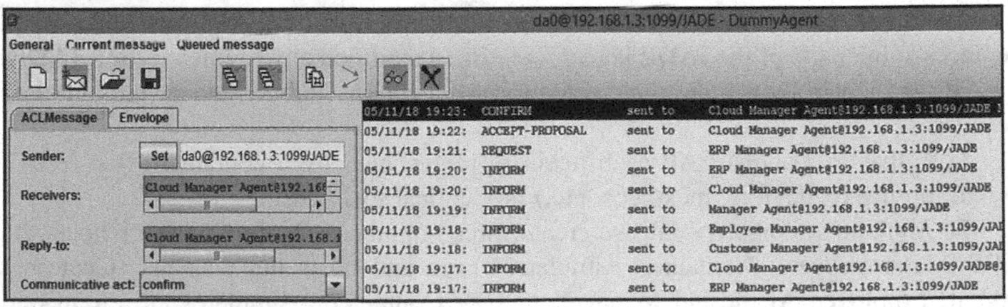

Fig. 4. Agents communication by Dummy Agent from JADE Leap

5 Conclusion

We proposed in this paper an approach based on multi agent systems and cloud computing concept in order to assist users (employees, managers and customers) to be permanently connected to ERP system. Being mobile, the user hands several mobile devices. We design User Manager Agent at this level to assist the user.

At enterprise level, ERP Manager Agent is responsible for communicating and informing User Manager Agent of any update, or to respond to his queries. We propose a Cloud Manager Agent at the cloud level to play the role of mediator between User Manager Agent and the ERP Manager Agent.

We have done a preliminary implementation using JADE-LEAP framework. In future works, we would like to test our proposal in real enterprises.

References

1. Alur, S.J.: Enterprise Integration and Distributed Computing: A Ubiquitous Phenomenon. Microsoft Corporation (2008). http://msdn.microsoft.com/en-us/library/cc949110.aspx
2. Cacciagrano, D.: Semantics on the cloud: toward an ubiquitous business intelligence 2.0 ERP desktop. In: SEMAPRO 2012: The Sixth International Conference on Advances in Semantic Processing (2012)
3. Caire, G.: JADE tutorial-JADE programming for beginners. TILAB (2001)
4. JADE-LEAP. http://sharon.cslet.it/project/jade
5. Jennings, N.R., Norman, T.J., Faratin, P., O'Brien, P., Odgers, B.: Autonomous agents for business process management. J. Appl. Artif. Intell. **14**(2), 145–189 (2000)
6. Jingrong, Y.: Research on Reengineering of ERP System Based on Data Mining and MAS, pp. 180–184 (2008)
7. Kay, R.: Surrounded By Devices, We Inhabit a World of Increasing User-Centricity. Forbes, 4 June 2013. http://www.forbes.com/sites/rogerkay/2013/01/04/surrounded-by-devices-we-inhabit-a-world-of-increasing-user-centricity/
8. Kehagias, D.: Information agents cooperating with heterogenous data sources for customer-order management. In: Proceedings of the 2004 ACM Symposium (2004)
9. Lin, F.R., Tan, G.W., Shaw, M.J.: Multi agent enterprise modeling. J. Organ. Comput. Electron. Commer. **9**(1), 7–32 (1999)
10. McKendrick, J.: When ERP goes to the cloud, You Know Things Are Getting Serious. Forbes, 18 June 2014. http://www.forbes.com/sites/joemckendrick/2014/06/18/when-erp-goes-to-the-cloud-you-know-things-are-getting-serious/
11. Saini, S.L.: Cloud and ERP. In: Proceedings of the World Congress on Engineering (2011)
12. Pan, J.Y.C., Tenenbaum, J.M.: An intelligent agent framework for enterprise integration. IEEE Trans. Syst. Man Cybern. **21**(6), 1391–1408 (1991)
13. Wooldridge, M.: An Introduction to Multi Agents Systems. John Wiley & Sons Ltd., Sussez (2002)
14. Galante, A.T.: Intelligent agent technologies: the work horse of ERP e-commerce. Int. J. Intell. Sci. **5**, 173–176 (2015). https://doi.org/10.4236/ijis.2015.54015
15. Hoffmann, M., Meisen, T., Jeschke, S.: OPC UA based ERP Agents: enabling scalable communication solutions in heterogeneous automation environments. In: Demazeau, Y., Davidsson, P., Bajo, J., Vale, Z. (eds.) PAAMS 2017. LNCS (LNAI), vol. 10349, pp. 120–131. Springer, Cham (2017). https://doi.org/10.1007/978-3-319-59930-4_10

16. Bouchemal, N., Maamri, R., Sahnoun, Z.: ERP costumer assistance using ubiquitous agents. In: Boumerdassi, S., Bouzefrane, S., Renault, É. (eds.) MSPN 2015. LNCS, vol. 9395, pp. 55–62. Springer, Cham (2015). https://doi.org/10.1007/978-3-319-25744-0_5
17. Schubert, P., et al.: Cloud Computing for Standard ERP Systems: Reference Framework and Research Agenda, Fachbereich Informatik Nr. 16/2011 (2011)
18. https://www.financialforce.com/resources/what-is-cloud-erp/

Author Index

Printed in the United States
By Bookmasters